NONLINEAR FUNCTIONAL ANALYSIS
AND APPLICATIONS

Publication No. 26
of the Mathematics Research Center
The University of Wisconsin

Nonlinear Functional Analysis and Applications

Edited by Louis B. Rall

Proceedings of an Advanced Seminar
Conducted by the Mathematics Research Center,
The University of Wisconsin, Madison
October 12-14, 1970

Academic Press
New York · London   1971

COPYRIGHT © 1971, BY ACADEMIC PRESS, INC.
ALL RIGHTS RESERVED
NO PART OF THIS BOOK MAY BE REPRODUCED IN ANY FORM,
BY PHOTOSTAT, MICROFILM, RETRIEVAL SYSTEM, OR ANY
OTHER MEANS, WITHOUT WRITTEN PERMISSION FROM
THE PUBLISHERS.

ACADEMIC PRESS, INC.
111 Fifth Avenue, New York, New York 10003

United Kingdom Edition published by
ACADEMIC PRESS, INC. (LONDON) LTD.
Berkeley Square House, London W1X 6BA

LIBRARY OF CONGRESS CATALOG CARD NUMBER: 73-152749

AMS (MOS) 1970 Subject Classifications: 46-02, 47-02, 49-02, 65-02

PRINTED IN THE UNITED STATES OF AMERICA

Contents

PREFACE vii

Some Applications of Functional Analysis to Analysis,
Particularly to Nonlinear Integral Equations 1
   L. Collatz
      University of Hamburg, Hamburg, Germany

The Differentiation and Integration of
Nonlinear Operators 45
   R. A. Tapia
      Rice University, Houston, Texas

Differentiability and Related Properties of
Nonlinear Operators: Some Aspects of the
Role of Differentials in Nonlinear Functional Analysis 103
   M. Z. Nashed
      Mathematics Research Center, The University of Wisconsin,
      Madison, Wisconsin, and Georgia Institute of Technology,
      Atlanta, Georgia

Generalized Inverses, Normal Solvability, and
Iteration for Singular Operator Equations 311
   M. Z. Nashed
      Mathematics Research Center, The University of Wisconsin,
      Madison, Wisconsin, and Georgia Institute of Technology,
      Atlanta, Georgia

On Polynomial Operators and Equations 361
   Patricia M. Prenter
      Colorado State University, Fort Collins, Colorado

Applications and Methods for the Minimization of Functionals . . . 399
   James W. Daniel
      University of Texas, Austin, Texas

CONTENTS

Toward a Unified Convergence Theory for
Newton-Like Methods 425
   J. E. Dennis, Jr.
      Cornell University, Ithaca, New York

Operator Solutions of Nonlinear Equations in
Optimal Control Problems 473
   David L. Russell
      University of California at Los Angeles,
      Los Angeles, California

Complementary Variational Principles 507
   Peter D. Robinson
      Bradford University, Yorkshire, England

INDEX 577

Preface

The subject of nonlinear functional analysis and its applications is naturally, after more than four decades of development, too large to be encompassed in a single volume. The intent of the present collection of papers is to give a reasonably self-contained introduction to the basic concepts and techniques of this field, highlighted by a few significant applications. To this end, the first two articles give an introduction which is accessible to persons equipped with an acquaintance with the fundamental ideas of Hilbert and Banach spaces, and linear operators in them. The remaining papers, which are essentially independent of each other, give extensions of the theory, or deal with an important application selected from the many possible. Each author has included ample bibliographical material to assist those wishing to extend their knowledge of the topic considered. The papers are published in the order in which they were presented at the Advanced Seminar.

Many persons, including a number who did not attend, are to be thanked for the pleasant atmosphere and smooth operation of this conference. Mrs. Gladys Moran served again with distinction as secretary of the program committee. The staffs of the Wisconsin Center and the University Club were particularly helpful. The preparation of what is now one of the larger books in the Mathematics Research Center series was in the capable hands of Mrs. Doris Whitmore.

Some Applications of Functional Analysis to Analysis, Particularly to Nonlinear Integral Equations

L. COLLATZ

In this survey, a few possibilities for applying ideas from functional analysis to problems in analysis are explored. Paragraphs 1 and 2 deal with the two important principles of measuring and ordering and paragraphs 4 - 12 are devoted to illustrating and composing different principles in the case of the special topic of nonlinear integral equations. Of course, here only a few examples can be mentioned.

The bibliography is not meant to be comprehensive, and I apologize for any inadvertent omissions.

1. Some Distances

Metric spaces occur in so many fields of analysis, that we refer to the many books of functional analysis which exist, and we give here only some additional examples.

a) Intervals

On the real x-axis we consider two intervals $A = [a,b]$, $B = [c,d]$ with $a < b$, $c < d$ and define <u>a distance</u> $\rho(A,B)$ to be the length of the union of the two intervals minus the length of the intersection. See Figure 1. Let x_1, x_2, x_3, x_4 denote the numbers a, b, c, d where these numbers are then ordered according to the classical ordering of real numbers, $x_1 \leq x_2 \leq x_3 \leq x_4$, then the distance is given by

(1.1) $$\rho(A,B) = x_2 + x_4 - x_1 - x_3$$

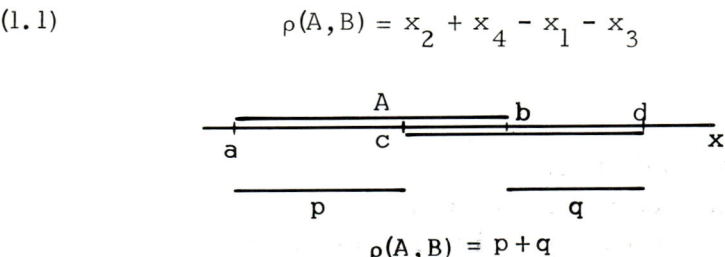

$$\rho(A,B) = p+q$$

Figure 1

The same idea can be used in higher dimensions (Figure 2).

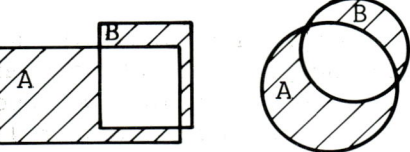

Figure 2

b) Polynomials

Let n be a fixed natural number. Consider the class, C, of all polynomials $A(x) = \sum_{\nu=0}^{n} a_\nu x^\nu$ with $a_n = 1$ and complex a_ν, $\nu = 0, \ldots, n-1$. Let $B(x) = \sum_{\nu=0}^{n} b_\nu x^\nu \in C$. Denote the zeros of $A(x)$ by x_j and the zeros of $B(x)$ by y_j, $j = 1, \ldots, n$. Let P be a permutation of the numbers $1, \ldots, n$ into numbers $p_1, \ldots, p_n$ and introduce

(1.2) $$\mu_P = \max_j |x_j - y_{P_j}|.$$

Then we can define a distance by taking the minimum of μ_P for all permutations P (Figure 3).

(1.3) $$\rho(A,B) = \min_P \mu_P.$$

APPLICATIONS OF FUNCTIONAL ANALYSIS

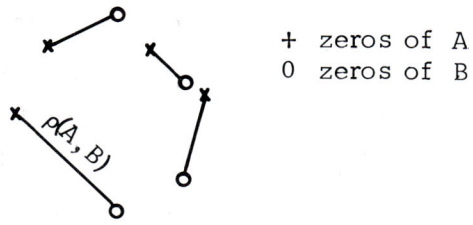

+ zeros of A
0 zeros of B

Figure 3

c) Parabolic distance

Let $x = \{x_1,\ldots,x_n\}$, $y = \{y_1,\ldots,y_n\}$ be points in the real n-dimensional space R^n and introduce as distance

$$(1.4) \quad \rho(x,y) = \left[\sum_{j=1}^{k}(x_j - y_j)^2 + \sum_{j=k+1}^{n}|x_j - y_j|\right]^{1/2}$$

for a fixed k with $1 \leq k \leq n-1$. This distance was used for parabolic equations. See Friedman [64] and Mustaza [69].

d) Part metric, Bauer-Bear [69]

Let C be a convex set in a real linear space, for instance in the point space R^n, and suppose that C doesn't contain a whole straight line, Figure 4. Let x,y be any fixed pair ϵ C and call r a "possible" number for x,y if and only if $x + r(x-y) \epsilon$ C and $y + r(y-x) \epsilon$ C. We define as "part-distance":

$$(1.5) \quad \rho(x,y) = \inf\{\ln(1+\frac{1}{r}),\ r \text{ possible for } x,y\}.$$

If y tends to the boundary Γ of C and x is fixed, then the distance goes to ∞. There are connections with the Cayley-Klein-distance (W. Blaschke, Proj. Geom. Basel 1954).

The distance is not isotropic. This distance was applied to potential theory and the theory of measures. Figure 4 gives an example: C as square, x fixed, the rectangle R contains points y with $\rho(x,y) \leq \ln 2$.

3

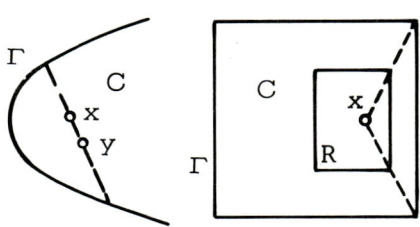

Figure 4

e) Distance of functions, Sendov [70]

First let $f(x)$, $g(x)$ be given continuous functions of x on a real interval Δ and let $k > 0$ be a given constant. Let us introduce

(1.6) $$[f(x), g(y)]_k = \text{Max}\{\tfrac{1}{k}|x-y|, |f(x)-g(y)|\}$$

$$|f(x) \stackrel{k}{-} g(x)| = \text{Max}\{\underset{y \in \Delta}{\text{Min}}[f(x), g(y)]_k, \underset{y \in \Delta}{\text{Min}}[f(y), g(x)]_k\}$$

(1.7) $$r(\Delta, k, f, g) = \underset{x \in \Delta}{\text{Max}} |f(x) \stackrel{k}{-} g(x)|$$

$k \to 0$ gives the classical values

(1.8) $$\lim_{k \to 0} |f(x) \stackrel{k}{-} g(x)| = |f(x) - g(x)|,$$

$$\lim_{k \to 0} r(\Delta, k, f, g) = \rho(f, g) = \underset{x \in \Delta}{\text{Max}} |f(x) - g(x)|.$$

With this distance $r(\Delta, k, f, g)$, the topology defined is Hausdorff and one can define in the usual way, a derivative f^* for a given function $f(x)$; this derivative f^* depends on k. It is not longer necessary that $f(x)$ be continuous in Δ. For every bounded function $f(x)$, f^* exists; if $f(x)$ has a derivative f' in the classical sense, then

$$f^{\bullet} = \frac{f'}{1+k|f'|}.$$

One can develop rules for differentiation, for instance

$$(f+g)^{\bullet} = \frac{f^{\bullet} + g^{\bullet} - 2kf^{\bullet}g^{\bullet}}{1-k^2 f^{\bullet} g^{\bullet}}.$$

The "polynomial" $P(x,k)$ of degree 2 with $P^{\bullet} = x$ is not $\frac{1}{2}x^2$, but

$$P(x,k) = -\frac{x}{k} - \frac{1}{k^2} \ln|1-kx|.$$

It is intended to apply this idea to physical problems.

f) Many other examples could be added. (See for instance Collatz [66], p. 21-45)

2. Some Orderings

As in Section 1, books on functional analysis give so many examples of orderings, that we mention here only a few additional examples.

a) Ordering the knots of an asymmetric graph. Duyvestijn (Enschede, Netherlands 1970) determines weights $a_{j\nu}$ ($j = 1,\ldots,k$; $\nu = 0,1,2,\ldots$) to the knots $P_1,\ldots,P_k$ of a graph by the following matrix-iteration:

$$(2.1) \quad a_{j,\nu+1} = \sum_{\rho=1}^{k} \varepsilon_{j\rho} a_{\rho,\nu}, \quad j=1,\ldots,k, \quad a_{j0}=1 \text{ for all } j,$$

here $\varepsilon_{j\rho}$ is 1, if P_j and P_ρ are connected by a line, otherwise $\varepsilon_{j\rho} = 0$, and also $\varepsilon_{jj} = 0$ for all j.

If for a certain ν all $a_{j\nu}$ are different, then one has a total ordering of the knots. In case of symmetry, of course, some knots have the same weight for every ν. Figure 5 gives an example with knots $P_1, \ldots, P_8$ and the table shows that the natural orderings of the knots would be $\varphi_1, \ldots, \varphi_8$. In this case already for $\nu = 2$ all $a_{j\nu}$ are different.

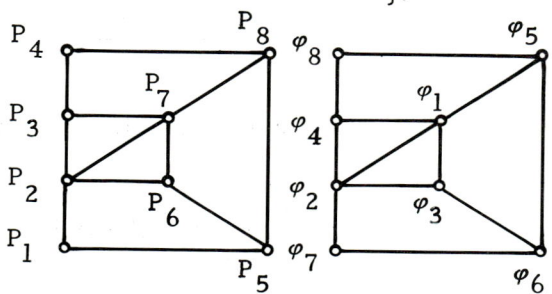

Figure 5

Take the graph as the map of a city and the lines as streets. Then this ordering finds the "most important" point φ_1 of the city in a certain sense.

j	1	2	3	4	5	6	7	8	Control sum
a_{j0}	1	1	1	1	1	1	1	1	8
a_{j1}	2	4	3	2	3	3	4	3	24
a_{j2}	7	12	10	6	8	11	13	9	76

b) Finite connected sets of squares

In the infinite net of squares defined in the usual way by the straight lines $x_1, x_2 = 0, \pm 1, \pm 2, \ldots$ of an x_1, x_2-plane we select connected sets of a finite number of these squares. Sets which can be produced from another by moving or reflecting are considered as equal. By comparing two sets f,g we introduce the ordering. We write f < g, if f = g or if f can be produced by taking off one or more squares of g.

In Figure 6 we have $f_1 \prec f_2 \prec f_3$; the considered sets of squares are not a lattice, for instance f_1, f_4 in Figure 6 have different upper bounds, f_2, f_5, but no supremum.

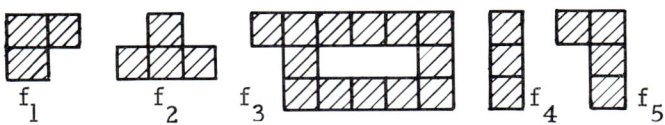

Figure 6

c) Partitions

For this classical example for partitions of a natural number n, one can introduce different orderings. If one takes n individual things (Case I) one has a lattice, if one looks only on the numbers of things (Case II), one has no lattice. It seems to be unnecessary to describe this in detail and it may be sufficient to illustrate these facts by the example n = 5 in Figure 7.

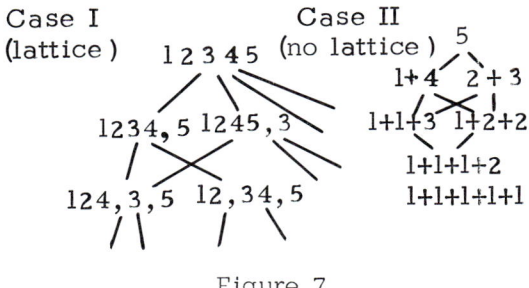

Figure 7

d) Real vectors R^n

We use exactly the same notation x,y as in Section 1, Example c. We write $x \prec y$, if one can order the components x_ν, y_ν so that

1) $x_1 \geq x_2 \geq \ldots \geq x_n$, $y_1 \geq y_2 \geq \ldots \geq y_n$

2) $x_1 + \ldots + x_k \leq y_1 + \ldots + y_k$ for all k with $1 \leq k \leq n$

3) $x_1 + \ldots + x_n = y_1 + \ldots + y_n$.

This ordering has different applications, for instance in the theory of probability. Then one can prove that $x < y$, if and only if x can be produced from y by taking convex means (Hardy-Littlewood-Polya [52], p. 45), or if y is more "scattered" or "dispersed" than x .

3) Functions of several variables

Let $R = C^1[Q]$ be the set of real valued functions $f(x,y)$ with continuous partial derivatives of first order in a closed bounded domain Q, e.g. in the domain $-1 \leq y \leq 0 \leq x \leq 1$. For $f, g \in R$, one can introduce the ordering

$f < g$ iff $f(P) \leq g(P)$, $\frac{\partial f}{\partial x}(P) \leq \frac{\partial g}{\partial x}(P)$, $\frac{\partial f}{\partial y}(P) \leq \frac{\partial g}{\partial y}(P)$

for all points $P \in Q$.

The sign $\leq$ refers to the classical ordering of real numbers. The ordering $<$ gives no lattice in this case, since for instance the functions $f = 0$, $g = xy$ have no infimum. For functions of only one independent real variable x, the corresponding ordering gives a lattice, and these facts are, perhaps, a deeper reason that many theorems in real analysis are valid for one but not for two independent variables.

3. General Applications to Differential Equations

There are many applications of functional analysis to analysis, which cannot be listed here. We mention only

a) Applications of the different fixed point theorems to differential and integral equations, see here Nos. 5-8. Let us mention only the famous applications of the Schauder-Leray fixed point theorem to nonlinear elliptic equations.

b) Very famous too are the applications of fixed point theorems to nonlinear vibrations, see No. 4.

c) Browder [70] gave at the International Congress of Mathematicians at Nice, Sept. 1970, a survey of the applications to partial differential equations, mostly about existence and uniqueness of solutions and about iteration procedures, also in the case of non compact operators and of non-expansive mappings, so that it is not necessary to deal here with applications of the mentioned type.

d) Many applications of nonlinear functional analysis occur in the calculus of variations, optimal control theory, nonlinear optimization, nonlinear approximation (for instance spline approximation with variable knots, Burchard [70]) and many other fields.

Therefore, in the following we may select as a special topic the nonlinear integral equations, which, perhaps, will increase in importance in the future. Even here we have chosen special fields and do not deal with singular integral equations (for these see for instance Muschelischwili [65], Michlin [69]).

4. Classification of Nonlinear Integral Equations

The following schedule gives names and forms of some types of integral equations:

Name	Form of Equation $u(x) = T\,u(x)$: $Tu =$	Examples	
		typical forms	special examples
Linear equation	(4.1) $f(x) + \mu \int_B K(x,t)\,u(t)\,dt$	$Lu = r(x)\,u(x)$ $Su = 0$	$-u''(x) = x^2\,u(x)$ $u(0) = 0, u(1) = 1$
Hammerstein–Equation [30]	(4.2) $f(x) + \int_B K(x,t)\,\varphi(u(t))\,dt$	$Lu = r(x)\,\varphi(u(x))$ $Su = 0$	$-u''(x) = [u(x)]^2$ $u(0) = 0, u(1) = 1$
Urysohn–Equation	(4.3) $\int_B K(x,t,u(t))\,dt$	$Lu = \varphi(x,u(x))$ $Su = 0$	$-u''(x) = [x^2 + u^2]^{1/2}$ $u(0) = 0, u(1) = 1$
Biargument–Equation	(4.4) $\int_B K(x,t,u(x),u(t))\,dt$	Chandrasekhar-Equation of radiative transfer (Ortega-Rheinboldt [70], p. 18) (4.6) $u(t) = 1 + \int_0^1 \dfrac{t}{s+t}\,u(s)\,u(t)\,\varphi(s)\,ds$	
Integrofunctional–Equation	(4.5) $\int_B K(x,t,u(x),u(t),\ldots, u(\varphi(t)),\ldots)\,dt$	Population theory (4.7) $\dfrac{du}{dt} = c_1 u(t)[1 - c_2 \int_{t-\tau}^{t} u(s)\,ds]$	

Notation: Let B be a domain in the n-dimensional real space R^n of points $x = \{x_1,\ldots,x_n\}$. $u(x)$ is the unknown function (single valued, but of course vector-valued functions can be considered too). $f(x)$ and $r(x)$: $B \to R^1$, φ: $R^1 \to R^1$, $K(x,t)$: $B \times B \to R^1$ and all listed functions K which depend on more independent variables are given functions.

If we integrate over a domain which depends on t instead of over a fixed domain B, we get an integral equation of the corresponding Volterra-Type. In all cases one can consider regular or weakly singular or strongly singular kernel functions K. Sometimes an equation (4.3) with a kernel of the form $K(x,t) \cdot \varphi(t,u(t))$ is also called a Hammerstein-Equation.

In the examples, L and S mean linear (differential) operators, $Su = 0$ being boundary conditions; we suppose that there exists a Green's function $G(x,t)$ which solves the boundary value problem

(4.8) $\qquad Lu(x) = r(x)$ in B, $Su = 0$ in the form

(4.9) $\qquad u(x) = \int_B G(x,t)\, r(t)\, dt$.

An important application of integral equations are the non-linear vibrations, (Reissig [69], Werner [70]). Consider the vector-differential equation

(4.10) $\qquad\qquad x'(t) = A(t)\, x(t) + g(t,x)$

with a given matrix $A(t)$ and a given function g which have the period P in t. Let $Y(t)$ be the solution of $Y' = AY$, $Y(0) =$ unit matrix I and suppose $\det |Y(P) - I| \neq 0$; then $x' = Ax + r(t)$ with $r(t) = r(t+p)$ has a periodic solution $x(t) = Tr$ with $Tr(t) = \int_0^p G(t,s)\, r(s)\, ds$, where $G(t,s)$ is the Green's function.

From (4.10) we come to the integral equation

(4.11) $\qquad\qquad x(t) = \int_0^p G(t,s)\, g(s,x(s))\, ds$

Another example of a biargument equation occurs in viscoelasticity, Huang-Lee [66]:

(4.12) $$S = \int_{-\infty}^{t} K_1(t-\tau) e(\tau) d\tau + \int_{-\infty}^{t} \int_{-\infty}^{t} K_2(t-\tau_1, t-\tau_2) e(\tau_1) e(\tau_2) d\tau_1 d\tau_2$$

where on the right side can be added also a triple integral; here S, K_1, K_2 are given functions and $e(\tau)$ is to be determined. S, K_1, K_2, e may be single valued or may be matrices (S as stress-tensor, $e = \frac{dE}{dt}$, E as strain tensor). Approximately one can take

$$K_1(t) = \alpha_1 \exp(-t/t_0)$$
$$K_2(t,0) = -\alpha_2[1 + \beta_2(t/t_0) - \frac{1}{2} \beta_2^2(t/t_0)^2]$$

in the case $e \geq 0$ one has monotonically decomposible operators (No. 6).

Other examples of Integrofunctional-Equations are the PY-Equation (Broyles [61] for the radial distribution function for a classical fluid of particles interacting with the Lennard-Jones potential

(4.13) $$u(r) = 1 + n \int_0^{\infty} f(s) u(s) [u(s-r) f(s-r) + u(s-r) - 1] ds .$$

The theory of population (Caswell [70] gives as its simplest model an equation of the form (4.7), but in refinement the system

(4.14) $$\frac{du_j(t)}{dt} = c_{j,0}[1 - \int_{t-\tau_j}^{t} \sum_{k=1}^{2} c_{j,k} N_k(s) ds] N_j(t) \quad j = 1, 2 .$$

We observe that we have often quadratic equations (Rall [61] [69a], Patricia Prenter [70]).

A nonlinear integro-differential equation describing the movement of the interface between two fluids in a porous medium, one displacing the other, has been derived and treated by Fulks and Guenther [69].

5. Contraction Mapping Theorems

We will not deal here with the very often used methods of discretization (summation, methods of finite differences), for which one has useful error estimations only in the linear case (Wielandt, Brakhage, Hämmerlin, Kussmaul-Werner.., Aitkinson and others) nor with variational methods (compare Michlin [69]).

Instead let us consider possibilities for getting exact error bounds for approximate solutions in the nonlinear case. The fixed point theorems which were first applied, were the theorems on contraction mappings.

a) Metric spaces

The theorems were first stated for Hilbert and Banach spaces and then generalized to metric spaces.

Theorem (Weissinger [52].) Let T be a (linear or nonlinear) operator, which maps a complete set D of a metric space R into R. We suppose that T satisfies a Lipschitz condition with a constant $K < 1$:

(5.1) $\qquad \rho(Tv, Tw) \leq K \rho(v, w)$ for all $v, w \in D$,

ρ means the distance in the space R. For the iteration procedure

(5.2) $\qquad u_{n+1} = Tu_n \quad (u = 0, 1, \ldots)$

we suppose that $u_0 \in D$, $u_1 \in D$ and that the whole sphere S,

(5.3) $$S = \{h/\rho(h,u_1) \leq \frac{K}{1-K} \rho(u_0,u_1)\},$$

is contained in D. Then the iteration u_n with (5.2) converges to an element u which solves the equation

(5.4) $$u = Tu.$$

In D exists no other solution of (5.4) and u lies in S (error estimation).

b) Pseudometric spaces

We refer to the generalization given by J. Schröder [56], described for instance by Collatz [66]; in pseudometric spaces the distances are not real numbers but elements of a coordinated linear partially ordered space.

c) Weakening the contraction condition (5.1)

Boyd and Wong [69] replace the condition (5.1) by the weaker condition

(5.5) $$\rho(Tv, Tw) \leq \varphi(\rho(v,w))$$

with $\varphi(r) < r$ for $0 < r < \infty$, which enlarges slightly the field of applicability. Browder [70] considers non-expansive mappings and non-compact operators.

d) Numerical example

Consider the Urysohn-Equation

(5.6) $$u(x) = Tu = \mu \int_0^1 \frac{dt}{1+x+u(t)}.$$

For $\mu = 1$ the trial $u_0 = \text{const.} = c$ with $c = 0.5$ gives $u_1 = \frac{1}{1.5+x}$, Figure 8.

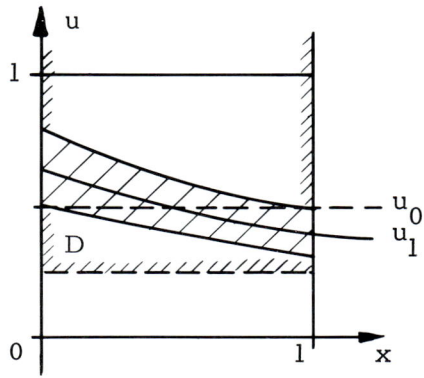

Figure 8

To calculate a Lipschitz constant K, we need bounds for a domain D, say $u \geq \alpha \geq 0$. As norm we may use $\|f\| = \underset{0 \leq x \leq 1}{\text{Max}} |f(x) \varphi(x)|$ with a continuous positive weight function $\varphi(x)$. We get

$$\|Tv - Tw\| = \underset{0 \leq x \leq 1}{\text{Max}} [\varphi(x)|Tv - Tw|] \leq K \|v - w\| \text{ with}$$

$$K = \mu \cdot \left[\underset{0 \leq x \leq 1}{\text{Max}} \frac{\varphi(x)}{(1+x+\alpha)^2} \right] \int_0^1 \frac{dt}{\varphi(t)} .$$

For $\mu = 1$ and for the simplest trial $\varphi(x) = 1$ the sphere-condition (5.3) is not satisfied for any α, $0 \leq \alpha \leq 0.4$. Therefore we have to be a little more careful.

The choice $\varphi(x) = 1+x$ with $\alpha = \frac{1}{4}$ and $K = \frac{16}{25} \ln 2$, $\frac{K}{1-K} < \frac{4}{5}$ is acceptable and gives the existence of a solution (5.6) for $\mu = 1$ and the error estimation

$$\left| u(x) - \frac{1}{1.5+x} \right| \leq \frac{0.16}{1+x} \text{ for } 0 \leq x \leq 1, \text{ Figure 8 .}$$

For large values of the parameter μ the equation (5.6) has a solution $u(x)$, as it will be shown in No. 6, d) Example II, but the operator T is then no longer contractive in the whole.

6. Schauder-Theorem and Monotonically Decomposible Operators

a) Topological fixed point theorems.

For applying the following theorems, no information is needed concerning the norms, distances, Lipschitz constants, etc.

<u>Theorem</u> (Brouwer [12].) Every continuous mapping T of a convex, bounded, closed set D of a finite dimensional space R^n into itself (TD $\subset$ D) has at least one fixed point u with u = Tu. One can substitute for D any set $\hat{D}$, which can be got by a one to one continuous mapping.

A generalization to infinitely many dimensions gives the following

<u>Theorem</u> (Schauder [30].) Let T be a continuous operator, which maps a closed convex bounded set M of a Banach space R into a relatively compact set, TM, which belongs to M. Then exists at least one fixed point u of T with u = Tu, u $\in$ M.

These theorems have been used very often; perhaps the Schauder theorem is one of the most important theorems for the numerical treatment of equations occurring in analysis.

Examples are given for finite systems of linear algebraic equations (Schröder [59]), for nonlinear vibrations (Reissig [69]), etc.

Let us give a simple example. In (4.10) let $g(t,x)$ satisfy $|g(t,x) - g(t,0)| \leq s(t) q(|x|)$, where $s(t)$ is a continuous function of period p and $q(x)$ is a continuous, monotone, non-decreasing, bounded function with $q(0) = 0$. Then the Brouwer theorem for a sufficiently large sphere gives the existence of a periodic solution of (4.10) (Reissig [69]).

b) Monotonically decomposible operators, for short MDO, (Schröder [59], see for instance Collatz [66], p. 282). The operator T mapping the domain D of an partially ordered space R_1 into a partially ordered space R_2 is called syntone, if $v \leq w$, v and $w \in D$ implies $Tv \leq Tw$ and is called antitone, if from $v \leq w$, it follows that $Tv \geq Tw$ for all $v, w \in D$.

Theorem: In the equation

(6.1) $$u = Tu + r = \hat{T}u$$

in a partially ordered Banach space R we suppose, that T has the form $T_1 + T_2$ where T_1 is syntone and T_2 is antitone, and that T_1, T_2 are continuous and defined in a convex domain D of R^2. Let the iteration procedure

(6.2) $$\left\{\begin{array}{l} v_{n+1} = T_1 v_n + T_2 w_n + r \\ w_{n+1} = T_1 w_n + T_2 v_n + r \end{array}\right\} \quad n = 0, 1, 2, \ldots$$

start with elements $v_0, w_0 \in D$ with

(6.3) $$v_0 \leq v_1 \leq w_1 \leq w_0 .$$

If $\hat{T}$ maps the interval $M_n = [v_n, w_n]$ for some $n \geq 0$ into a relatively compact set, then there exists at least one element $u \in M_n$ with $u = \hat{T}u$.

It is quite elementary that $\hat{T}$ maps M_n into itself; (M_n is the set of all elements z with $v_n \leq z \leq w_n$) one has

(6.4) $$M_n \supset M_{n+1} \supset \hat{T} M_n \quad n = 0, 1, 2, \ldots .$$

The existence of a fixpoint is given by Schauder's theorem. The condition of compactness is often satisfied for integral operators.

c) Basic examples

1) Let $A = (a_{jk})$ $(j,k = 1,\ldots,n)$ be a matrix with real elements a_{jk}. In the case $a_{jk} \geq 0$ is A syntone, for $a_{jk} \leq 0$ is A antitone; but every real matrix A is monotonically decomposible: $A = A_1 + A_2$ with A_1 syntone, A_2 antitone. Let $\hat{A} = (|a_{jk}|)$ be the matrix of the absolute modulus of the elements a_{jk}, then one can choose $2A_1 = A + \hat{A}$, $2A_2 = A - \hat{A}$.

2) In the same way, every real kernel $K(x,t)$ can be written as $K(x,t) = K_1(x,t) + K_2(x,t)$ with $K_1 \geq 0$, $K_2 \leq 0$. Then the operator

(6.5) $$T u = \int_B K_j(x,t) u(t) dt$$

is syntone for $j = 1$ and antitone for $j = 2$.

3) Every Hammerstein-Operator of the form (4.2) with real K and a function φ of bounded variation is monotonically decomposible. $\varphi(z)$ can be written as $\varphi(z) = \varphi_1(z) + \varphi_2(z)$ with monotone, non-decreasing φ_1 and monotone, non-increasing φ_2. Then the operator

(6.6) $$\hat{T} u = \int_B K(x,t) \varphi(u(t)) dt = T_1 u + T_2 u \quad \text{with}$$

$$T_1 u = \int_B [K_1 \varphi_1(u) + K_2 \varphi_2(u)] dt \quad \text{and}$$

$$T_2 u = \int_B [K_1 \varphi_2(u) + K_2 \varphi_1(u)] dt$$

is monotonically decomposible.

d) Numerical examples

I. This example and examples in Nos. 7, 9, 10 show that the theory of MDO mentioned in b) is very powerful.
Consider the problem of solving the Hammerstein-Equation

(6.7) $$u(x) = 1 + \int_0^1 |x-t| [u(t) - \frac{1}{2} u^2(t)] dt \; .$$

In the range $u(x) \geq 0$, $T_1 v = \int_0^1 |x-t| v(t) dt$ is syntone and $T_2 v = -\frac{1}{2} \int_0^1 |x-t| v^2(t) dt$ is antitone.

Here $v_0 = 0$, $w_0 = 2$ give with (6.2) $v_1 = 1 - \int_0^1 |x-t| \frac{1}{2} \cdot 4 dt = 2(x-x^2)$, $w_1 = 2(1-x+x^2)$ and (6.3) is satisfied, Figure 9, and a solution $u(x)$ exists with $v_1 \leq u \leq w_1$.

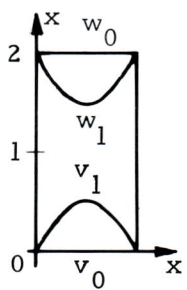

Figure 9

II. In the example of equation (5.6) the MDO-method is much more powerful than the contractive mapping theorem. The operator is antitone for $u \geq 0$, and we have $T = T_2$, $T_1 = 0$. We can take as w_0 a very large constant, formally $w_0 = +\infty$ and get $v_1 = Tw_0 = 0$. For $v_0 = 0$ we get $w_1 = Tv_0 = \frac{\mu}{1+x}$ and (6.3) is satisfied for all $\mu > 0$. We get the existence

19

of a solution $u(x)$ for <u>every</u> positive μ with $0 \leq u(x) \leq \frac{\mu}{1+x}$. Therefore we have $u(x) \leq \mu$ and with respect to $T\mu = \frac{\mu}{1+x+\mu}$ we have $\frac{\mu}{1+x+\mu} \leq u(x) \leq \frac{\mu}{1+x}$. Of course it is easy to improve these bounds.

e) Schauder-Leray-Theorem

A generalization of the Schauder-Theorem was stated by Leray and Schauder. It uses the topological degree of a mapping. In the application to nonlinear vibrations, it is used often in the form of the Ljapunov-Theorem (see Reissig [69]). For example, consider for $x \in R^m$ and $y \in R^n$, the system

(6.8)
$$\begin{cases} x' = Ax + f(t,y) + g(t) \\ y' = By + g(t,x) + h(t) \end{cases}$$

All coefficients have the period p; the matrices A, B may be stable and f may be bounded $|f(t,y)| \leq C_0$; then the Schauder-Leray-Theorem gives the existence of a periodic solution.

7. <u>The Fixed Point Theorem of Krasnoselskij</u>

a) A combination of the contraction mapping theorem and the Schauder theorem was given by Krasnoselskij [55]:

<u>Theorem</u>: Let S, T be (linear or nonlinear) operators which are defined in a bounded, closed, convex set D of a Banach space R, and suppose

1) $Sf + Tg \in D$ for every pair $f, g \in D$

2) S is contractive, i.e. there exists a constant $K < 1$ with $\|Sf - Sg\| \leq K \|f-g\|$ for all $f, g \in D$

3) T is completely continuous.

APPLICATIONS OF FUNCTIONAL ANALYSIS

Then there exists at least one fixed point $u \in D$ which satisfies

$$u = Su + Tu .$$

Numerical example for the biargument-equation

(7.1) $\quad u(x) = \frac{1}{3}[u^2(x) + x] + \frac{1}{3}\int_0^1 |u(t) - x|^{1/2} dt$

and the domain $D: 0 \leq u(x) \leq 1$ for $0 \leq x \leq 1$, the operator $Su = \frac{1}{3}[u^2(x) + x]$ is contractive but not completely continuous and the integral defining the operator T is completely continuous, but not Lipschitz-bounded. However, the condition 1) is obviously satisfied and the theorem of Krasnoselskij gives the existence of at least one solution $u(x)$ with $0 \leq u(x) \leq 1$.

b) Combination with monotonicity

In the example

(7.2) $\quad u(x) = \frac{1}{3}[u^2(x) + x] + \frac{1}{3}\int_0^1 |x-t| u(t) dt$

furthermore the operator $V = S+T$ is syntone for $u \geq 0$ and one gets two sequences v_n, w_n enclosing a solution. As in the case of (7.1) we also take for the set D, the functions $0 \leq u(x) \leq 1$, and condition 1) is satisfied. For $v_0 = 0$ and $w_0 = 1$ we get $v_1 = Vv_0 = \frac{1}{3}x$, $w_1 = Vw_0 = \frac{1}{2} - x + x^2$. We know the existence of a solution $u \geq 0$, and with respect to the monotonicity all v_n of the sequence $v_{n+1} = Vv_n$ $(n = 0, 1, 2, \ldots)$ are upper bounds to u and analogously for $w_{n+1} = Vw_n$, $\{n = 0, 1, 2, \ldots\}$ we have $u \leq w_n$ and so

(7.3) $\quad v_0 \leq v_1 \leq v_2 \leq \ldots \leq u \leq \ldots \leq w_2 \leq w_1 \leq w_0$.

For simplicity we use $w_1 \leq \frac{1}{2} = \hat{w}_0$ and calculate $w_{n+1} = V\hat{w}_n$, $\{n = 0,1,\ldots\}$. This gives the better inclusion $v_1 = \frac{1}{3}x \leq u(x) \leq \hat{w}_1 = \frac{1}{6}(1+x+x^2)$, Figure 10.

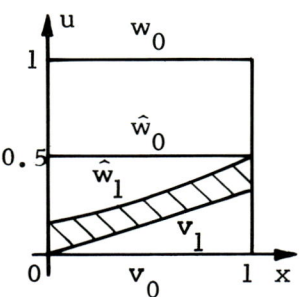

Figure 10

c) Weaker conditions for the Krasnoselskij-Theorem are given by Nashed and Wong [69]. If the space R under consideration is a Hilbert space, one can replace the strong condition 1) by a weaker condition

1') Sf + Tf ϵ D for every f ϵ D
(Krasnoselskij-Kachurovskij-Zabreiko [67]).

8. Cones

There is another combination of the principles of distances and of orderings: E. Bohl [69] [70] proved a fixed point theorem, which covers many of the applications, which were made from the fixed point theorem in pseudometric spaces and from the Schauder theorem;
The basic idea is the fact that one can introduce a norm in an ordered space under weak conditions:
Let R be a real linear space. A subset K of R is called a cone, if from f,g ϵ K and $\lambda > 0$ follows f+g ϵ K, λf ϵ K and if f ϵ K, -f ϵ K implies f = 0 = zero element of R.

Every cone K induces an ordering, namely $f < g$ (or $f \leq g$) is equivalent to $g - f \in K$.

An element $e \in K$ is called an ordering-unit, if to every element $f \in R$, one can find a nonnegative number $\rho = \rho(f)$ with $f + \rho e \in K$; then one can introduce a norm

$$(8.1) \qquad \|f\| = \|f\|_e = \inf\{\sigma, \pm f + \sigma e \in K\}.$$

A cone K is called normal, if there exists a positive constant τ such that $f, g \in R$ and $\theta < f < g$ implies $\|f\| \leq \tau \|g\|$.

Theorem (Bohl): Let K be a normal closed cone in the real linear space R and T an operator which maps a closed subset D of R into R. Let P be a linear, syntone operator with the property: For every ordering-unit e and all $f, g \in D$, $Tf - Tg < \|f-g\|_e \cdot Pe$ holds. Suppose there exist ordering-units $z, \hat{z}$ with $\hat{z} = (E-P)z$. Start an iteration $u_{n+1} = Tu_n$ $(n = 0, 1, 2, \ldots)$ with an element $u_0 \in D$. One of the conditions a)b)c) may be satisfied

a) T and P are completely continuous

b) T is completely continuous and $\lim_{n \to \infty} P^n \hat{z} = \theta$

c) R is a Banach space and $\lim_{n \to \infty} P^n = \theta$.

Then a fixed point $u \in D$ exists with $u = Tu$ and one has the error estimation:

$$(8.2) \qquad \|u - u_n\|_{P^n z} \leq \|u_0 - u_1\|_{\hat{z}} \text{ for } n \geq 0.$$

The condition c) corresponds to the classical contraction-mapping theorem, a) the Schauder-theorem and b) gives an intermediate theorem. Similar results and a numerical example are given in Schwetlick [69]; for applications to positive operators see Hadeler [66].

9. Monotonicity

a) Operators of monotonic type

An operator T, mapping a set D of an ordered space R_1 into an ordered space R_2 is said to be of monotonic type, if $Tv \leq Tw$ implies $v \leq w$ (Collatz [52]).
For instance let T be

$$(9.1) \qquad u(x) - \mu \int_B K(x,t) u(t) dt$$

with $\mu K(x,t) \geq 0$ for $x, t \in B$ and $\mu \int_B K(x,t) dt < 1$ for $x \in B$ or, more generally, let be $\mu \geq 0$, $K(x,t) \geq 0$, μ smaller than the first eigenvalue, then the resolvent $\Gamma(x,t,\mu) \geq 0$ and T is of monotonic type (for Volterra-equations the last condition is satisfied).

Numerical example: $Tu = u(x) - \int_0^1 |x-t| u(t) dt = 1$. Here
$\frac{1}{4} \leq \int_0^1 |x-t| dt = \frac{1}{2}(1 - 2x + 2x^2) \leq \frac{1}{2} \leq 1$.

For $w(x) = 29 - 24(x - x^2)$ we have $16.5 \leq Tw(x) \leq 16.5625$. This gives the error bound

$$\frac{w(x)}{16.5625} \leq u(x) \leq \frac{w(x)}{16.5}.$$

Of course operators of monotonic type may be nonlinear.

b) MDO

It was mentioned in Section 6, that the theory of monotonically decomposable operators can be applied very often. Here we give a very simple example for which no iteration procedure is necessary, but which shows, that one can cover the whole field of existence.

If

$$u(x) = 1 + \mu \int_0^1 (1 + [u(t)]^2)^{1/2} \, dt = T_1 u \;,$$

the solution is $u(x) = \text{const.} = a$, where a and μ are connected by, Figure 11,

$$\mu(1 + a^2)^{1/2} = a - 1 \;.$$

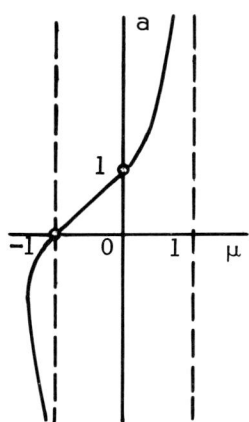

Figure 11

Let us consider $\mu > 0$. Then the operator T_1 is syntone in the domain B, the functions satisfying $u(x) \geq 0$. Starting with $v_0 = 0$, $w_0 = \text{"}+\infty\text{"}$ (as in Example No. 6d) II) one sees that (6.3) is satisfied for every $\mu > 0$ and the theory of MDO is applicable in all of B.

c) Different orderings

J. Werner [69] [70] had great success, especially in nonlinear vibrations by using several different orderings for the same problem, for instance one ordering in the classical ordering of real numbers (pointwise in t) and another ordering with respect to which the given operator is monotone. As an illustration, we give one of the many theorems of Werner.

In the non-autonomous Liénard-equation

(9.2) $\ddot{x} + f(x)\dot{x} + g(x) = e(t)$ with $x(0) = x(p), \dot{x}(0) = \dot{x}(p)$,

let $e(t)$ be a given, continuous function of period p . Let $\alpha(t), \beta(t) \in C^2[0,p]$ be functions with $\alpha(t) \leq \beta(t)$ for $t \in [0,p]$, which satisfy the same conditions of periodicity as $x(t)$, and which satisfy

(9.3) $\qquad -D\alpha(t) \leq 0 \leq -D\beta(t)$ for $t \in [0,p]$,

where the defect (residual) D is defined by

$$Dz(t) = -\ddot{z}(t) - f(z(t))\dot{z}(t) - g(z(t)) + e(t) .$$

Let I be the interval $[\underset{T}{\operatorname{Min}} \alpha(t), \underset{T}{\operatorname{Max}} \beta(t)]$ where T is the interval $[0,p]$. If $\gamma = \underset{I}{\operatorname{Max}} g'(x) > 0$ and $2\gamma^{1/2} \leq f(x)$ for $x \in I$, then exists a periodic solution $x(t)$ of (9.2) with

$$\alpha(t) \leq x(t) \leq \beta(t) \text{ in } T .$$

Numerical example (I thank Dr. J. Werner for this example)

The equation $\ddot{x} + 5(1-x^2)\dot{x} + x + 0.5x^3 = 0.5 \sin t$ has a 2π-periodic solution with

$$\alpha(t) = -0.1\cos t - 0.007 \leq x(t) \leq \beta(t) = -0.1\cos t + 0.007 .$$

Here $\gamma^{1/2} = 1.009 \leq \frac{1}{2} \underset{I}{\operatorname{Min}} f(x) \leq \frac{4.942}{2}$ is satisfied. By iteration one can improve the result. Three iterations give

$$-0.100326 \leq x(0) \leq -0.100317 .$$

APPLICATIONS OF FUNCTIONAL ANALYSIS

10. Existence of Solutions

There is a long list of papers which generalize existence and uniqueness proofs for nonlinear integral equations, and which will not be cited here.

a) Classical results

Many of the classical results are collected and listed in a table in Tricomi [57], p. 210. This list refers to the equation

$$(10.1) \qquad u(x) + \int_B K(x,t)\, \varphi(t, u(t))dt = 0$$

or

$$(10.2) \qquad u + KFu = 0$$

with the operators

$$(10.3) \qquad Kz = \int_B K(x,t)\, z(t)dt, \quad Fw = \varphi(x, w(x)) \, .$$

Let K belong to the class L_2 and φ be continuous and let us introduce

$$(10.4) \qquad \phi(t, u) = \int_0^u \varphi(t, v(t))dv \, .$$

The existence of at least one continuous solution $u(x)$ can be proved under the additional conditions (table in Tricomi [57]):

1) φ satisfies for all $u \in R^1$ and all $x \in B$ with certain non-negative constants c_1, c_2

$$(10.5) \qquad |\varphi(x, \underline{u})| \le c_1 |u| + c_2$$

27

or

2) K continuous and ϕ bounded below or satisfying with certain nonnegative, not too large constants c_3, c_4

(10.6) $$\phi(x,u) \geq -(c_3 u^2 + c_4)$$

for instance $\varphi(x,u) = \sin u$ satisfies 1) and $\varphi(x,u) = e^u, u^3$ satisfies 2) but $\varphi(x,u) = u^2$ doesn't satisfy 2).

b) Theory of MDO

The theory of monotonically decomposible operators gives existence results in many cases in which other theories fail, as mentioned in Section 6.

c) Minty-Zarantonello Monotonicity

In Section 6 syntone operators of a certain type were introduced. Another kind of monotonicity was defined by Zarantonello [60] and Minty [62]. One can introduce for a real Banach space R, the dual space R^* and produce $<u,v>$ for $v \in R$, $u \in R^*$ and call the operator F in (10.2) monotone if

(10.7) $<v-w, F(v) - F(w)> \geq 0$ for all $v, w \in R^*$.

In the above considered case of a single equation (10.1) one can take

(10.8) $$<v,w> = \int_0^1 v(t) F(t, w(t)) dt$$

for systems of equations, v and V vectors with components v_j, F_j (j = 1,...,m) and the integrand is $\sum_{j=1}^m v_j F_j$. Then one can prove (Amann [69][70], Browder [70]) the existence of a solution U under the additional conditions

1) K is angle bounded, that means there exists a constant $\gamma \geq 0$ with

$$|<Ku,v> - <Kv,u>|^2 \geq \gamma^2 <Ku,u><Kv,v> \text{ for all } u,v \in R,$$

and $<Ku,u> \geq 0$ for all $u \in R$ (positive definiteness) is satisfied.

2) F is monotone in the sense of (10.7) and continuous. Amann [70] can replace these conditions by

1') K is angle bounded and completely continuous

2') F is continuous and bounded and it holds that

$<v,F(v)> \geq -\psi(\|v\|)$ with a function $\psi(z) = O(z^2)$ for $z \to \infty$.

Here it is not supposed that F is monotone. For the proofs Amann uses Schauder's theorem. He proves also convergence of the Galerkin-methods for getting approximate solutions.

11. Newton's Method and Related Methods

Ortega-Rheinboldt [70] give a survey on the methods of this section.

a) Consider the equation

(11.1) $$Fu = \theta$$

for an element u. The given operator F may map a set D of a Banach space R_1 into a Banach R_2 with zero element θ. If F has a Fréchet-derivative F' which is invertible, the very much used Newton procedure consists in determining a sequence of elements u_n, starting with an element $u_0 \in D$:

(11.2) $$u_{n+1} = u_n - F'^{-1}_n F_n \quad n = 0, 1, 2, \ldots$$

with the abbreviations $F_n = Fu_n$, $F_n^{(k)} = F^{(k)}(u_n)$. For the theory of the Newton-method see Kantorovitch-Akilov [64], Tapia [69], Rall [69], Ostrowski [70] and others.

b) Improvements

1) Döring [70] investigates improved formulas for $\delta_n = u_n - u_{n+1}$, for instance

$$\delta_n = F_n'^{-1} F_n - \frac{1}{2} F_n'^{-1} F_n'' (F_n'^{-1} F_n)(F_n'^{-1} F_n)$$

or (in symbolic form with quotients)

$$\delta_n = -\frac{2 F_n' F_n}{2F_n'^2 - F_n'' F_n}.$$

2) Dennis and Brown [68][69] investigate "Newton-like methods" (Secant method, method of Broyden [65], which has had great success in numerical work) and the nonlinear successive over-relaxation. They prove local convergence for a general class of related methods and convergence for some special methods; Dennis [70] develops a unified convergence theory for the different "Newton-like" methods.

3) A. Pasquali [70] describes a procedure of order 3, Figure 12

(11.3) $$u_{n+1} = u_n - F_n'^{-1}(u_n - \omega F_n'^{-1} F_n)F_n$$

with $0 < \omega \leq 1$; for $\omega = 0$ the procedure reduces to the classical Newton's method and for $\omega = \frac{1}{2}$ one has under some weak assumptions a procedure of order 3.

APPLICATIONS OF FUNCTIONAL ANALYSIS

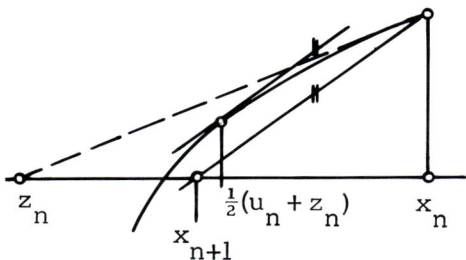

Figure 12

He solves approximately the Chandrasekhar-equation

(11.4) $$u(t) = 1 + \frac{1}{2} m\, u(t) \int_0^1 \frac{t}{s+t} u(s)ds = Tu(t)$$

He proves convergence for $0 < m < \frac{1}{2 \ln 2} \approx 0.72$. This example is also treated in Rall [69].

The operator Tu is syntone for $u(t) \geq 0$ and therefore the theory of MDO, No. 6 is applicable too. For $m = \frac{1}{2}$ one can use $v_0 = 0$, $v_1 = 1$, $w_0 = 2$, $w_1 \leq 1 + \ln 2 < 2$ and one has immediately existence of a solution and the bounds $1 \leq u(t) \leq 1 + \ln 2$, which can be improved easily.

4) Newton's method is convenient, if the derivatives $T^{(v)}$ and T'^{-1} are easy to get. This can be done often, if one has a finite system of nonlinear equations

(11.5) $$f_j(x_1, \ldots, x_n) = 0 \quad j = 1, \ldots, n$$

or if one is discretizing the given (nonlinear) differential or integral equations. In this way often nonlinear integral equations were solved approximately (for instance Anselone-Moore [66], Hammerstein-Equation and Application to Elasticity).

c) Avoiding derivatives

Many methods are developed which do not use values of the derivative F'. See for instance Schmidt-Leder [70], Dennis [70] and others. Hofmann [70] applies a modified regula falsi to equations in Banach spaces and he proves monotonicity principles for the regula falsi, which offer the possibility of bracketing the solutions in many cases. The principle in the following example may be illustrated by the symbolic sketch, Figure 13, of an equation $f(x) = 0$ for a single real variable x; in the sketch, the Newton-procedure, starting from x_0, gives x_1, x_2 and regula falsi starting from x_0, x_1 gives ξ_1 and in this example a solution x is bracketed between x_2 and ξ_1.

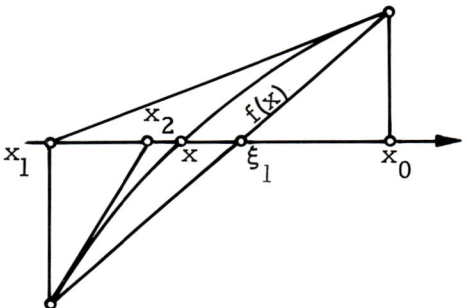

Figure 13

Numerical example: (I thank Mr. Hofmann for this example)

$$Tu(t) = 2u(t) - \frac{2}{3} - \frac{1}{3} t^2 - \int_0^t (t^2 - s^2) u^2(s) ds = 0 .$$

In the interval $0 \leq t \leq 1$ one gets the bounds

$$\frac{1}{3} + \frac{1}{6} t^2 \leq u(t) \leq \frac{1}{3} + \frac{2}{9} t^2 .$$

The lower bound is improved by Newton's method, the upper bound by regula falsi. The result leads to the much better bounds:

$$\psi(t) \le u(t) \le \psi(t) + e(t) \text{ in } 0 \le t \le 1 \text{ with}$$

$$\psi(t) = \frac{1}{3} + \frac{1}{6} t^2 + \frac{1}{27} t^3 + \frac{1}{135} t^5 + \frac{1}{972} t^6 + \frac{1}{1260} t^7 + \frac{1}{19440} t^8$$

$$+ \frac{1}{91854} t^9 + \frac{17}{45360} t^{10}$$

$$e(t) = \frac{1}{23328} t^8 + \left[\frac{9}{764400} - \frac{17}{45360} + 1.08 \cdot 10^{-5} \right] t^{10}$$

the maximum of the relative error is 2.2×10^{-5}.

Another procedure was suggested by Laasonen [69]. He works with two sequences $u_n, \bar{u}_n$ and supposes, that certain "divided differences" $F_{n\bar{n}}$ have an inverse $C_n = (F_{n\bar{n}})^{-1}$ in a Banach space, then his algorithm is

(11.6) $$\begin{cases} u_{n+1} = u_n - C_n F_n \\ \bar{u}_{n+1} = u_{n+1} - C_n F_{n+1} \end{cases}$$

as illustrated in the symbolic sketch, Figure 14.

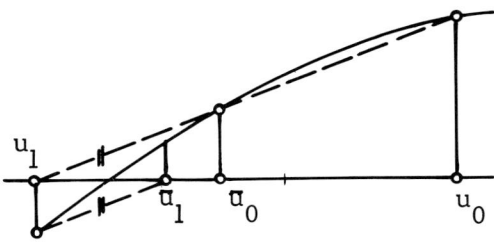

Figure 14

12. Method of Degenerate Kernels

a) Linear integral equations

For the linear equation (4.1), one very often replaces the given kernel $K(x,t)$ by a degenerate kernel

$$(12.1) \qquad \hat{K}(x,t) = \sum_{j=1}^{m} a_j(x) b_j(t) .$$

Then one gets the solution $\hat{u}(x)$ of

$$(12.2) \qquad \hat{u}(x) = \hat{f}(x) + \mu \int_B \hat{K}(x,t) \hat{u}(t) dt$$

by using the form

$$(12.3) \qquad \hat{u}(x) = \hat{f}(x) + \sum_{j=1}^{m} c_j a_j(x)$$

and solving a system of linear equations for the c_j. See for instance Tricomi [57], p. 55.

Generally, by comparing the original equation (4.1) with the perturbed equation (12.2), one is interested in an error bound for the error $\varepsilon(x) = \hat{u}(x) - u(x)$. Tricomi [57], p. 76 gives an estimation, and Kantorovitch-Krylov [56], p. 137 another, which uses only terms which are known.

__Theorem__: Suppose that one knows d, N, η, Γ_0 with

$$(12.4) \qquad \int_B |K(x,t) - \hat{K}(x,t)| dt \leq d \quad |f(x)| \leq N, \quad |f(x) - \hat{f}(x)| \leq \eta$$

for all $x \in B$

$$(12.5) \qquad \int_B |\hat{\Gamma}(x,t,\mu)| dt \leq \Gamma_0$$

APPLICATIONS OF FUNCTIONAL ANALYSIS

The perturbation should be small enough, so that

(12.6) $\quad d\,|\mu|\beta < 1 \quad \text{with} \quad \beta = 1 + |\mu|\Gamma_0$

then the error estimation holds

(12.7) $\quad |\varepsilon(x)| = |u(x) - \hat{u}(x)| \leq \dfrac{N\,|\mu|\,d\beta^2}{1 - |\mu|\,d\beta} + \eta\beta$.

Here $\hat{\Gamma}$ is the resolvent for the kernel $\hat{K}$ with

(12.8) $\quad \hat{u}(x) = \hat{f}(x) + \mu \int_B \hat{\Gamma}(x,t,\mu)\,\hat{f}(t)dt$.

The estimation becomes good if d is small; therefore we have in (12.4) to solve a Tschebyscheff-Approximation problem in several variables, (Collatz [70]).

The resolvent is computable, if $\hat{K}(x,t)$ is a degenerate kernel of the form (12.1). Especially in the case

(12.9) $\quad \hat{K}(x,t) = a(x)\,b(t)$

one gets

(12.10) $\quad \hat{\Gamma}(x,t,\mu) = a(x)\,b(t)\,[1 - \mu \int_B a(t)\,b(t)dt]^{-1}$.

b) Hammerstein-Equation (4.2)

Also in this case one can replace the kernel $K(x,t)$ by $\hat{K}(x,t)$ as in (12.1) and with (12.3) one gets for the unknown coefficients c_i a nonlinear system of equations. This procedure could also be applied for the Urysohn-Equation (4.3), but in this case occurs the additional difficulty that one doesn't know in which domain one has to carry out the Tschebyscheff-Approximation. Furthermore, in this nonlinear case no error bound seems to be known.

c) Bifurcation problems

The methods described in a) can also be applied with success in bifurcation problems but an error bound would be very desirable.
For the equation

$$(12.11) \qquad u(x) = \int_{-\infty}^{+\infty} K(x,t) [1 + \mu(u(t))^2] dt$$

with

$$K(x,t) = \frac{1}{1+x^2+t^2}$$

we look for the "endbifurcation", i.e. the greatest value $\tilde{\mu}$ of μ for which a solution exists. Replacing the kernel $K(x,t)$ by a degenerate kernel

$$\hat{K}(x,t) = \frac{1}{(a+bx^2)(a+bt^2)} \quad ,$$

one can solve the disturbed equation exactly in the form $\hat{u}(x) = \dfrac{c}{a+bx^2}$. [One has approximately $\hat{K}(x,t) = \dfrac{1.7484}{(1.3573+x^2)(1.3573+t^2)}$].

With

$$\gamma_j = \int_{-\infty}^{+\infty} \frac{dt}{(a+bt^2)^j}$$

36

one has to determine c from $c = \gamma_1 + \mu c^2 \gamma_3$, Figure 15. For $\tilde{\mu}$ one gets the approximate value $\hat{\mu} = \dfrac{1}{4\gamma_1 \gamma_3}$.

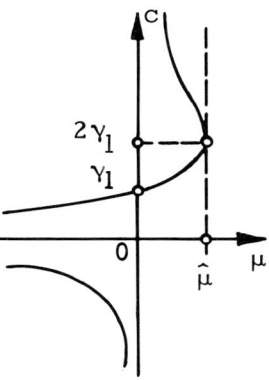

Figure 15

REFERENCES

H. Amann [69] Ein Existenz- und Eindeutigkeitssatz für die Hammersteinsche Gleichung in Banachräumen. Math. Z. 111 (1969), 175-190.

H. Amann [70] Zum Galerkin-Verfahren für die Hammersteinsche Gleichung. Arch. Rat. Mech. Anal. 35(1969), 114-121 and Arch. Rat. Mech. Anal. 37(1970), 33-47.

H. Amann [70] Existence Theorems for Equations of Hammerstein type, to appear soon.

P. M. Anselone [64] Nonlinear Integral equations. Proc. Advanced Seminar. University of Wisconsin Press, Madison, 1964. 378p.

P. M. Anselone-R. H. Moore [66] An extension of the Newton-Kantorovič Method for Solving nonlinear equations with an Application Elasticity, Journ. Math. Analysis and Applic. 13(1966), 476-501.

H. Bauer-H. S. Bear [69] The part metric in convex sets. Pacific J. Math. 30(1969), 15-33.

E. Bohl [69] Iteration und Fehlerabschätzung in Räumen mit einem archimedischen Kegel, Arch. Rat. Mech. Anal. 34(1969), 354-360.

E. Bohl [70] Die metrische Struktur von Räumen mit allgemeinerem Abstandsbegriff und ihre Verwendung bei der Behandlung nichtlinearer Probleme. Computing 5(1970), 189-199.

D. Boyd -J. S. W. Wong [69] On nonlinear contractions. Proc. Amer. Math. Soc. 20(1969).

L. E. J. Brouwer [12] Ueber die Abbildung von Mannigfaltigkeiten. Math. Ann. 71(1912), 97-115.

F. E. Browder [70] Nonlinear functional analysis and applications to nonlinear partial differential equations, Proceed. Internat. Congress of Mathematicians, Nice, Sept. 1970.

K. M. Brown-J. E. Dennis [68] On Newton-like Iteration Functions: General Convergence Theorems and a Specific Algorithm, Num. Math. 12(1968), 186-191.

H. Burchard [70] Splines with optimal knots, to appear.

G. L. Cain-M. Z. Nashed [70] Fixed points and stability for a sum of operators in locally convex spaces, to appear.

H. Caswell [70] A Note on the Addition of Time Lags to the Volterra Predator Prey Equations, to appear in Theor. Population Biology.

L. Collatz [52] Aufgaben monotoner Art. Arch. Math. 3(1952), 366-376.

L. Collatz [66] Functional Analysis und Numerical Mathematics, Acad. Press 1966, 473p.

L. Collatz [66a] Einige abstrakte Begriffe in der numerischen Mathematik (Anwendungen der Halbordnung). Computing 1(1966), 233-255.

L. Collatz [70] Applications of nonlinear optimization to approximation problems, in Abadie: Nonlinear and integer programming, North Holland Publ. Company (1970).

H. T. Davis [60] Introduction to nonlinear differential and integral equations, Dover, 1960, 566p.

J. E. Dennis [68] On Newton-like methods, Num. Math. 11 (1968), 324-330.

J. E. Dennis [70] Toward a unified convergence theory for Newton-like methods, to appear soon.

B. Döring [70] Ueber einige Klassen von Iterationsverfahren in Banachräumen, Math. Ann. 187(1970), 279-294.

A. Friedman [64] Partial Differential Equations of Parabolic Type, Prentice Hall, Englewood Cliffs, 1964.

W. Fulks and R. B. Guenther [69] A Free Boundary Problem and an Extension of Muskat's Model, Acta Mathematica, 122(1969), 273-300.

K. P. Hadeler [66] Einschließungssätze bei normalen und bei positiven Operatoren. Arch. Rat. Mech. Anal. 21(1966), 58-66.

A. Hammerstein [30] Nichtlineare Integralgleichungen nebst Abwendungen, Acta Math. 54(1930), 117-176.

G. H. Hardy - J. E. Littlewood - G. Polya [52] Inequalities, Cambridge 1952, 324p.

W. Hofmann [70] Die Regula falsi in Banachräumen, to appear in Computing.

N. C. Huang - E. H. Lee [66] Nonlinear Viscoelasticity for Short Time Ranges, Journ. Appl. Mechanics 1966, 313-321.

L. W. Kantorowitsch - G. P. Akilow [64] Funktionalanalysis in normierten Räumen, Berlin 1964, 622p.

L. W. Kantorowitsch - W. I. Krylow [56] Näherungsmethoden der höheren Analysis, Berlin 1956, 611p.

M. A. Krasnoselskij [55] Two remarks on the method of successive approximations. Us. Mat. Nauk 10(1955) 123-127 (Russian).

M. A. Krasnoselskij - R. J. Kachurovskij - P. P. Zabreiko [67] On a fixed point principle for operators in a Hilbert space (Russian). Funktional. Anal. 1 Prilozhen. 1(1967), 93-94. Transl. in Functional Analysis Applications 1(1967), 168-169.

R. Kussmaul - P. Werner [68] Fehlerabschätzungen für ein numerisches Verfahren zur Auflösung linearer Integralgleichungen mit schwach singulären Kernen, Computing 3(1968), 22-46.

P. Laasonen [69] Ein überquadratisch konvergenter iterativer. Algorithmus. Ann. Akad. Sci. Fenn. Ser. AI 405(1969), 9p.

L. A. Ljusternik - W. I. Sobolev [68] Elemente der Funktionalanalysis, Berlin 1968, 375p.

S. G. Michlin [69] Numerische Realisierung von Variationsmethoden, Berlin 1969, 343p.

G. L. Minty [62] Monotone (nonlinear) operators in Hilbert space, Duke J. 29(1962), 341-346.

N. I. Muschelischwili [65] Singuläre Integralgleichungen, Berlin 1965, 564p.

Mustaza [69] Parabolic differential equations, lecture at a congress Bucaresti, Romania 1969.

M. Z. Nashed - J. S. W. Wong [69] Some Variants of a Fixed Point Theorem of Krasnoselskij and Applications to Nonlinear Integral Equations, Journ. Math. Mech. 18 (1969), 767-778.

B. Noble [64] The numerical solution of nonlinear integral equations. Proc. Madison, see Anselone [64], 215-318.

J. M. Ortega - W. C. Rheinboldt [70] Iterative solution of nonlinear equations in several variables. Academic Press 1970, 572p.

A. Ostrowski [70] Theorie des Newtonschen Verfahrens, Vortrag Oberwolfach Juni 1970, Proc. Brosowski Martensen, Hochschultaschenbücher.

A. Pasquali [70] Some remarks on the convergence of an iterative higher order process, To appear soon.

W. Pogorzelski [66] Integral equations and their applications, Pergamon Press 1966, 714p.

Patricia M. Prenter [70] Polynomial operators and equations, this Proceedings.

L. B. Rall [61] Quadratic equations in Banach spaces, Rend. Circolo Matem. di Palermo 10(1961), 314-332.

L. B. Rall [69a] On the uniqueness of solutions of quadratic equations. SIAM Review 11(1969), 386-388.

L. B. Rall [69] Computational Solution of Nonlinear Operator Equations, John Wiley 1969, 224p.

L. B. Rall - R. A. Tapia [70] The Kantorovich Theorem and error estimates for Newton's method, to appear soon.

R. Reissig [69] Anwendung von Fixpunktsätzen auf das Problem der periodischen Lösungen bei nichtautonomen Systemen. DLR Forschungsbericht 69-53(1969), 63 S.

J. Schauder [30] Der Fixpunktsatz in Funktionenräumen. Studia Math. 2(1930), 171-182.

J. W. Schmidt - D. Leder [70] Ableitungsfreie Verfahren ohne Auflösung linearer Gleichungen, Computing 5(1970), 71-81.

J. Schröder [56] Das Iterationsverfahren bei allgemeinerem Abstandsbegriff, Math. Z. 66(1956), 111-116.

J. Schröder [59] Fehlerabschätzungen bei linearen Gleichungssystemen mit dem Brouwerschen Fixpunktsatz, Arch. Rat. Mech. Anal. 3(1959), 28-44.

H. Schwetlick [69] Lineare positive Operatoren und Fehlerabschätzungen bei Operatorgleichungen, Computing 4(1969), 345-358.

B. Sendov [70] Lecture on the conference on constructive function theory, Varna, Bulgaria, May 1970.

R. A. Tapia [69] The weak Newton method and boundary value problems. SIAM J. Numer. Anal. 6(1969), 539-550.

F. G. Tricomi [57] Integral equations, Interscience publishers 1957, 238p.

J. R. L. Webb [69] Fixed point theorems for nonlinear Semi-contractive Operators in Banach spaces, J. London Math. Soc. (2)1(1969), 683-688.

J. Weissinger [52] Zur Theorie und Anwendung des Iterationsverfahrens Math. Nachr. 8(1952), 193-212.

J. Werner [70] Einschließungssätze für periodische Lösungen der Lienardschen Dgl., Computing 5(1970). 246-252.

E. H. Zarantonello [60] Solving functional equations by contractive averaging. MRC Technical Summary Report No. 160, University of Wisconsin, Madison, 1960.

The Differentiation and Integration of Nonlinear Operators

R. A. TAPIA

CONTENTS

1. Introduction and Preliminaries 45
2. The Gâteaux and Fréchet Derivatives 48
3. Higher Derivatives 63
4. The Riemann Integral 69
5. The Metric Gradient of a Functional 76
6. Mean Value Theorems 92

1. Introduction and Preliminaries

There is no doubt that contemporary differentiation had its origin in the calculus of variations. Whenever possible we will take examples from classical variational theory.

Much of the material in Sections 2, 3 and 4 can be found, in one form or another, in the following four texts:

(i) Dieudonné [11];

(ii) Kantorovich and Akilov [27];

(iii) Rall [33];

(iv) Vainberg [40] .

The material in Sections 5 and 6 has not appeared in any text.
A real vector space will usually be denoted by X or Y. However, Euclidean n-space is denoted by R^n and the real line by R. When X is an inner product space we will denote the inner product by $<,>$. A function f defined from a subset D of a vector space X into a vector space Y is called an operator and is denoted by f: X → Y or f: D → Y. The set D is called the domain of f. Occasionally we write f(·) instead of f to emphasize that f is a function. The vector space of all operators from X into Y is denoted by [X, Y]. A real-valued operator, i.e., a member of [X, R] is called a functional. The operator f: X → Y is said to be

(i) homogeneous if $f(\alpha x) = \alpha f(x)$ and

(ii) additive if $f(x+y) = f(x) + f(y)$

for all $\alpha \in R$ and all x and y in the domain of f. An additive and homogeneous operator is said to be linear. By a nonlinear operator we mean any operator which is not necessarily linear. If X and Y are topological vector spaces, then the vector space of all continuous linear operators from X into Y is denoted by $L_1[X,Y]$. It is usual to denote $L_1[X,R]$ by X^* and call it the dual space of X. If X and Y are normed linear spaces and L is a linear operator from X into Y, then L is said to be bounded if there exists M > 0 such that $\|L(x)\| \leq M \|x\|$ for all $x \in X$. It is well known that L is continuous if and only if it is bounded. If we let $\|L\| = \inf\{M: \|L(x)\| \leq M\|x\|, x \in X\}$ then $L_1[X,Y]$ becomes a normed linear space. In particular X^* is a normed linear space whenever X is a normed linear space. The Hahn-Banach theorem for normed linear spaces asserts that for each $x \in X$ there exists $\delta \in X^*$ such that $\|\delta\|=1$ and $\delta(x) = \|x\|$. The reader unfamiliar with these notions is referred to Goffman and Pedrick [18].

DIFFERENTIATION AND INTEGRATION

Let $C_0^1[a,b]$ be the vector space of all real-valued functions which are continuously differentiable on the interval $[a,b]$ and vanish at the end points a and b. Suppose $f: R^3 \to R$ has continuous second partial derivatives with respect to all three variables. Consider the functional $J: C_0^1[a,b] \to R$ defined by

(1.1) $$J(y) = \int_a^b f(x, y(x), y'(x)) dx .$$

The simplest problem in the calculus of variations is essentially that of finding $y \in C_0^1[a,b]$ which minimizes J, i.e.,

(1.2) $$\text{minimize } J(y); \, y \in C_0^1[a,b] .$$

About 1790 Lagrange observed that if $y \in C_0^1[a,b]$ is a solution to Problem (1.2), then for each $\eta \in C_0^1[a,b]$, $\alpha(t) = J(y + t\eta)$ is a real-valued function of the real variable t which has a minimum at $t = 0$; hence $\alpha'(0) = 0$. Letting $\delta J(y; \eta)$ denote $\alpha'(0)$ we have

(1.3) $$\delta J(y;\eta) = \int_a^b [f_2(x,y(x),y'(x))\eta(x) + f_3(x,y(x),y'(x))\eta'(x)] dx ,$$

where f_i denotes the i-th partial derivative of f. Integrating by parts and recalling that $\eta(a) = \eta(b) = 0$ we obtain

(1.4) $$\delta J(y;\eta) = \int_a^b [f_2(x,y(x),y'(x)) - \frac{df_3(x,y(x),y'(x))}{dx}]\eta(x) dx .$$

Since $\delta J(y;\eta) = 0$ for all η it follows that

(1.5) $$\frac{df_3(x,y(x),y'(x))}{dx} - f_2(x,y(x),y'(x)) = 0; \, y(a) = y(b) = 0 .$$

47

That any solution to Problem (1.2) must be a solution of the boundary value problem (1.5) was known to Euler as early as 1744. Lagrange attributes equation (1.5) to Euler; Hilbert and others called it Lagrange's equation. It is presently called the Euler-Lagrange equation.

Remark. In this derivation Lagrange tacitly assumed that the integration by parts was valid, i.e., the integral of the left hand side of (1.5) exists. This is not immediately obvious unless we also know that y'' exists and is continuous.

Remark. Lagrange called $\delta J(y;\eta)$ the first variation of J. He called $\delta^k J(y;\eta) = \alpha^{(k)}(0)$ the k-th variation of J. Clearly

$$(1.6) \quad \delta^2 J(y;\eta) = \int_a^b [f_{22}(x,y,y') \eta(x)^2 + 2f_{23}(x,y,y') \cdot \eta(x)\, \eta'(x) + f_{33}(x,y,y')\eta'(x)^2] dx \, .$$

2. The Gateaux and Fréchet Derivatives

With others, we believe that the fundamental idea of the differential calculus should be the local approximation of operators by linear operators. Historically this idea has been completely obscured by the fact that for the real line, indeed for any one dimensional real vector space, there is a one-to-one linear correspondence between linear functionals on this space and real numbers; consequently the derivative of a real-valued function of a real variable was defined to be real-valued. This interpretation, void of intuitive meaning, led to the notion of the partial derivative, which in turn led to the gradient, from which came the ambiguous total derivative or total differential. The technique of defining the derivative from the gradient does not generalize for many reasons, not the least being that it is actually backwards. This will become clear in Section 5 when we consider the

DIFFERENTIATION AND INTEGRATION

definition given by M. Golomb for the gradient in Hilbert space.

Specifically, recall that for $f: R \to R$

$$(2.1) \qquad f'(\hat{x}) = \lim_{t \to 0} \left[\frac{f(\hat{x}+t) - f(\hat{x})}{t} \right]$$

and the line

$$(2.2) \qquad L(x) = f(\hat{x}) + f'(\hat{x})(x-\hat{x})$$

is a good approximation to $f(x)$ for x near $\hat{x}$.

If instead we consider $f: R^2 \to R$, then (2.1) no longer makes sense since we can not form $\hat{x} + t$, i.e., we can not add scalars and vectors. However, since $f(x) = f(x_1, x_2)$ where x_1 and x_2 are real we may consider the partial derivatives

$$\frac{\partial f(x)}{\partial x_1} = \lim_{t \to 0} \left[\frac{f(x_1+t, x_2) - f(x_1, x_2)}{t} \right], \text{ and}$$

$$\frac{\partial f(x)}{\partial x_2} = \lim_{t \to 0} \left[\frac{f(x_1, x_2+t) - f(x_1, x_2)}{t} \right].$$

We still do not know how to interpret $f'(x)(\eta)$ in (2.2) when x and η are in R^2. For real x and η we have

$$f'(x)(\eta) = f'(x) \cdot \eta ,$$

i.e., $f'(x)(\eta)$ is the product of the real numbers $f'(x)$ and η. Since the generalization of scalar multiplication is the inner product (scalar product) we take

$$f'(x)(\eta) = <d, \eta>$$

for some $d \in R^2$. Motivated by the real case, the obvious choice for d is the gradient vector

$$\nabla f(x) = (\frac{\partial f(x)}{\partial x_1}, \frac{\partial f(x)}{\partial x_2}) \ .$$

This leads to the definition

$$f'(x)(\eta) = <\nabla f(x), \eta> \ ,$$

which is usually written in the extremely confusing manner

$$df = \frac{\partial f}{\partial x_1} dx_1 + \frac{\partial f}{\partial x_2} dx_2 \ .$$

The linear functional $f'(x)$ or $df(x)$ is called the total differential or total derivative. This definition depends heavily on

(i)   the inner product in R^2,

(ii)   the natural basis in R^2, and

(iii)   the fact that f is real-valued;

hence it will only generalize to functionals in R^n.

However, let us return to $f: R \to R$ and not be so hasty this time. We have

$$f'(x)(\eta) = \lim_{t \to 0} [\frac{f(x+t) - f(x)}{t}] \cdot \eta \ .$$

Replacing t by $t\eta$ we obtain

$$f'(x)(\eta) = \lim_{t \to 0} \left[\frac{f(x+t\eta) - f(x)}{t\eta}\right] \cdot \eta$$

$$= \lim_{t \to 0} \left[\frac{f(x+t\eta) - f(x)}{t}\right].$$

This latter interpretation makes sense for $f: X \to Y$ where X need only be a vector space and Y a topological vector space. We have the following very natural definition.

Consider an operator $f: X \to Y$ where X is a vector space and Y is a topological vector space. Given x and η in X suppose

(2.3) $$Df(x)(\eta) = \lim_{t \to 0} \left[\frac{f(x+t\eta) - f(x)}{t}\right]$$

exists. Then $Df(x)(\eta) \in Y$ is called the Gâteaux derivative of f at x in the direction η and we say f is Gâteaux differentiable at x in the direction η. We say f is Gâteaux differentiable at x when f is Gâteaux differentiable at x in every direction. In this case the operator

$$Df(x): X \to Y$$

which assigns to each $\eta \in X$ the vector $Df(x)(\eta) \in Y$ is called the Gâteaux derivative of f at x.

The operator

(2.4) $$Df: X \to [X,Y]$$

which assigns to $x \in X$ the operator $Df(x) \in [X,Y]$ is called the Gâteaux derivative of f.

Remark. A few words of caution are in order. For historical reasons it is customary to say Gâteaux variation and not Gâteaux derivative; however I prefer the latter. The term Gâteaux differential is also used. Most authors speak of the Gâteaux derivative (variation, differential) of f at x in the direction η only when it has been previously established that f is Gâteaux differentiable at x, i.e., $Df(x)(\eta)$ exists for all η. This usage is very restrictive, see Propositions 2.3, 6.1, 6.2 and 6.3. Also many authors use the term Gâteaux derivative of f at x to mean $Df(x)$ is also a linear operator and use some other word when $Df(x)$ is not necessarily linear. We will not make this distinction and say linear Gâteaux derivative at x if necessary.

Remark. For the functional J given by (1.1) $DJ(y)(\eta) = \delta J(y;\eta)$ and $DJ(y)$ is a linear operator.

Remark. The vector space $[X,Y]$ in (2.4) is not a topological space and there is no obvious way to define a topology on it; consequently we cannot consider the Gâteaux derivative of the Gâteaux derivative. We could consider quantities analogous to Lagrange's higher variations, but for our purposes this would be a digression.

Example 2.1. If $f: R^n \to R$ and $e_1 = (1, 0, \ldots, 0), \ldots, e_n = (0, \ldots, 0, 1)$, then $x = (x_1, \ldots, x_n) = x_1 e_1 + \ldots + x_n e_n$; hence

$$Df(x)(e_i) = \lim_{t \to 0} \frac{f(x_1, \ldots, x_i+t, \ldots) - f(x_1, \ldots, x_n)}{t} = \frac{\partial f(x)}{\partial x_i},$$

i.e., the i-th partial derivative of f at x is the Gâteaux derivative of f at x in the direction e_i.

Example 2.2. Let $f: R^2 \to R$ be given by

$$f(x) = \frac{x_1 x_2}{(x_1^2 + x_2^2)}, \quad x = (x_1, x_2) \neq 0 \text{ and } f(0) = 0.$$

Then $Df(0)(\eta) = \lim_{t \to 0} \frac{1}{t} \frac{\eta_1 \eta_2}{(\eta_1^2 + \eta_2^2)}$ exists if and only if $\eta = (\eta_1, 0)$ or $\eta = (0, \eta_2)$.

Remark. This example shows that the existence of the partial derivatives is not a sufficient condition for the Gâteaux derivative to exist.

Example 2.3. Let $f: R^2 \to R$ be given by $f(x) = \frac{x_1 x_2^2}{x_1^2 + x_2^2}$, $x = (x_1, x_2) \neq 0$ and $f(0) = 0$. Then $Df(0)(\eta) = \frac{\eta_1 \eta_2^2}{\eta_1^2 + \eta_2^2}$

Remark. This example shows that the Gateaux derivative at a point is not necessarily a linear operator.

However, we do have the following result.

Proposition 2.1. The Gâteaux derivative of f at x is a homogeneous operator, i.e.,

$$Df(x)(\alpha\eta) = \alpha Df(x)(\eta), \text{ for } \alpha \in R.$$

Proof: Replacing t by $t\alpha$ in (2.3) we have

$$Df(x)(\eta) = \lim_{t \to 0} \left[\frac{f(x+t\alpha\eta) - f(x)}{t\alpha}\right] = \frac{1}{\alpha} Df(x)(\alpha\eta).$$

Remark. This proof used the fact that in a topological vector space scalar multiplication is continuous.

Proposition 2.2. If the functional $f: X \to R$ has a minimum or a maximum at $x \in X$ and $Df(x)$ exists, then $Df(x) = 0$.

Proof: If $\eta \in X$ is such that $Df(x)(\eta) > 0$, then for t

sufficiently small $\frac{f(x+t\eta) - f(x)}{t} > 0$; consequently $f(x+t\eta) > f(x)$ if $t > 0$ and $f(x+t\eta) < f(x)$ if $t < 0$. A similar argument can be used if $Df(x)(\eta) < 0$. This proves the proposition.

Example 2.4. Let $C[0,1]$ denote the vector space of real-valued functions which are continuous on the interval $[0,1]$. Consider $T: C[0,1] \to R$ defined by

$$T(y) = \int_0^1 [\frac{1}{2}(x+1)[y(x)]^2 - y(x)]dx .$$

If y is a constant function, say $y(x) = c$, then $T(c) = \frac{3}{4}c^2 - c$ which has a minimum when $c = \frac{2}{3}$ and $T(\frac{2}{3}) = -\frac{1}{3}$. Can we use Proposition 2.2 to improve on this? If y minimizes T, then by Proposition 2.2

$$DT(y)(\eta) = \int_0^1 [(x+1)y(x) - 1]\eta(x)dx = 0, \text{ for all } \eta .$$

Letting $\eta(x) = (x+1)y(x) - 1$ we see that $(x+1)y(x) - 1 = 0$ and $y(x) = \frac{1}{x+1}$. Also $T(\frac{1}{x+1}) = -\frac{1}{2}\log(2) \approx -.3466 < -\frac{1}{3}$.

The following proposition is very useful. In Section 6 we will prove a much stronger result.

Proposition 2.3. Let X be a vector space and Y a normed linear space. Consider an operator $f: X \to Y$. Given $x, y \in X$ suppose f is Gâteaux differentiable at each point of $\{x + t(y-x): 0 \le t \le 1\}$ in the direction $y - x$. Then for each $\delta \in Y^*$

(i) $\delta(f(y) - f(x)) = \delta[Df(x + \theta(y-x))(y-x)]$, for some $0 < \theta < 1$.

Also

(ii) $\|f(y) - f(x)\| \le \sup_{0 < \theta < 1} \|Df(x + \theta(y-x))(y-x)\|$.

Proof: Consider $g(t) = \delta(f(x + t(y-x)))$. Clearly $g'(t) = \delta[Df(x+t(y-x))(y-x)]$; hence by the ordinary mean value theorem $g(1) - g(0) = g'(\theta)$ for some $0 < \theta < 1$. This is equivalent to (i). Using the Hahn-Banach theorem, choose $\delta \in Y^*$ such that $\|\delta\| = 1$ and $\delta(f(y) - f(x)) = \|f(y) - f(x)\|$. Clearly (ii) now follows from (i).

Remark. Recall that in the definition of the Gâteaux derivative we require only the range space to have a topology defined on it. For many years mathematicians preferred not to introduce a topology on the domain space and when dealing with functionals did all the analysis using only the topological properties of the real line; consequently as fundamental and important a notion as convergence could not be adequately discussed. Eventually this trend changed, as it became apparent that any generality lost by introducing a norm or inner product was more than compensated for by being able to use the powerful functional analytic tools that were being developed.

Example 2.5. Let $f: R^2 \to R$ be given by $f(x) = \dfrac{x_1^3}{x_2}$, $x = (x_1, x_2) \neq 0$ and $f(0) = 0$. Then $Df(0)(\eta) = 0$ for all $\eta \in X$. Hence $Df(0)$ exists and is a continuous linear operator, but f is not continuous at 0.

We are quite confident that the reader will agree with us in saying that a satisfactory notion of differentiation should possess the property that differentiable functions are continuous. This leads us to the concept of the Fréchet derivative.

Consider $f: X \to Y$ where both X and Y are normed linear spaces. Given $x \in X$, if a linear operator $f'(x) \in L_1[X, Y]$ exists such that

(2.5) $\quad \lim\limits_{\|\Delta x\| \to 0} \dfrac{\|f(x + \Delta x) - f(x) - f'(x)(\Delta x)\|}{\|\Delta x\|} = 0$, $\Delta x \in X$,

then $f'(x)$ is called the Fréchet derivative of f at x. The operator

(2.6) $$f': X \to L_1[X, Y]$$

which assigns $f'(x)$ to x is called the Fréchet derivative of f.

Remark. The notation (2.6) does not imply that the domain of f' is all of X.

Remark. Contrary to the Gâteaux derivative at a point, the Fréchet derivative at a point is by definition a continuous linear operator.

Remark. We use the same terminology for the Fréchet derivative that we developed for the Gâteaux derivative; of course Gâteaux is replaced by Fréchet.

Remark. By (2.5) we mean, given $\varepsilon > 0$ there exists $\delta > 0$ such that

(2.7) $$\|f(x + \Delta x) - f(x) - f'(x)(\Delta x)\| \le \varepsilon \|\Delta x\|$$

for every $\Delta x \in X$ such that $\|\Delta x\| \le \delta$.

Proposition 2.4. If $f: X \to Y$ is Fréchet differentiable at x, then f is continuous at x.

Proof: Using the triangle inequality on (2.7) we obtain

$$\|f(x + \Delta x) - f(x)\| \le (\varepsilon + \|f'(x)\|) \|\Delta x\|.$$

The proposition clearly follows.

Remark. If $f'(x)$ exists, then by replacing Δx with $t \Delta x$ in (2.5) we have

$$\lim_{t \to 0} \left\| \frac{f(x + t \Delta x) - f(x)}{t} - f'(x)(\Delta x) \right\| = 0 ;$$

hence $Df(x)$ exists and $Df(x) = f'(x)$.

Proposition 2.5. The Fréchet derivative is unique.

Proof: This clearly follows from the previous remark.

Remark. If $Df(x)$ happens to be a bounded linear operator, then it is not difficult to show that $f'(x)$ will exist if and only if the convergence in (2.3) is uniform with respect to all η such that $\|\eta\| = 1$.

Remark. Observe that if $f: X \to Y$ is Fréchet differentiable, indeed if $Df(x) \in L_1[X,Y]$ for $x \in X$, then (ii) of Proposition 2.3 implies

$$\|f(y) - f(x)\| \leq \sup_{0 < \theta < 1} \|Df(x + \theta(y-x))\| \|y - x\|.$$

Example 2.6. Suppose $f: R^n \to R^m$ is Fréchet differentiable at x. Let us calculate $f'(x)(\eta)$. We have $f(x) = f(x_1, \ldots, x_n) = (f_1(x_1, \ldots, x_n), \ldots, f_m(x_1, \ldots, x_n))$. Since $\eta = \eta_1 e_1 + \ldots + \eta_n e_n$ where $e_1, \ldots, e_n$ is the natural basis for R^n, see Example 2.1, we need only calculate $f'(x)(e_i)$, then $f'(x)(\eta) = \eta_1 f'(x)(e_1) + \ldots + \eta_n f'(x)(e_n)$. Now

$$f'(x)(e_i) = \left(\frac{\partial f_1(x_1, \ldots, x_n)}{\partial x_i}, \ldots, \frac{\partial f_m(x_1, \ldots, x_n)}{\partial x_i} \right);$$

hence

$$f'(x)(\eta) = \begin{pmatrix} \frac{\partial f_1(x)}{\partial x_1} & \cdots & \frac{\partial f_1(x)}{\partial x_n} \\ \vdots & & \\ \frac{\partial f(x)}{\partial x_1} & & \frac{\partial f_m(x)}{\partial x_n} \end{pmatrix} \begin{pmatrix} \eta_1 \\ \vdots \\ \eta_n \end{pmatrix}.$$

As is well known $f'(x)$ turns out to be the continuous linear operator represented by the Jacobian matrix at x.

Remark. As Example 2.2 and Example 2.3 show $Df(x)$ may exist and not be the Jacobian matrix; also the Jacobian matrix may exist and $Df(x)$ not exist.

Proposition 2.6. Suppose $f: R^n \to R^m$ and $Df(x)$ exists. Then $Df(x)$ is represented by the Jacobian matrix at x if and only if $Df(x)$ is a linear operator.

Proof: Observe that in Example 2.6 we used only the fact that $f'(x)$ was linear.

Remark. A point worth mentioning is that the derivative (Gâteaux or Fréchet) always has its domain in the same space as the original operator, i.e., if $f: X \to Y$ then Df is also defined in X. However, since $Df: X \to [X,Y]$, we could consider $Df: X \times X \to Y$. With this interpretation f and Df will have the same range space. Historically when the derivative has been given this latter interpretation it is called the differential (Gâteaux or Fréchet) of f. Unfortunately even this interpretation is not standard among mathematicians.

The following three propositions are often very useful.

Proposition 2.7. Let X and Y be normed linear spaces. Suppose $f: X \to Y$ is Gâteaux differentiable in X. Suppose further that for fixed $x \in X$
(i) $Df(x)(\cdot): X \to Y$ is continuous at zero, and
(ii) $Df(\cdot)(\eta): X \to Y$ is continuous at x for each fixed $\eta \in X$.
Then $Df(x) \in L_1[X,Y]$, i.e., $Df(x)$ is a continuous linear operator.

Proof: By Proposition 2.1 $Df(x)$ is a homogeneous operator; hence $Df(0) = 0$. By (i) above there exists $r > 0$ such that $\|Df(x)(\eta)\| \le 1$ whenever $\|\eta\| \le r$. It follows that

$$\|Df(x)(\eta)\| = \left\| \frac{\|\eta\|}{r} Df(x)\left(\frac{r\eta}{\|\eta\|}\right) \right\| \le \frac{1}{r} \|\eta\|;$$

hence $Df(x)$ will be a continuous linear operator once we show it is additive. Consider $\eta_1, \eta_2 \in X$. Given $\varepsilon > 0$, by (2.3) there exists $\tau > 0$ such that

$$\|Df(x)(\eta_1+\eta_2) - Df(x)(\eta_1) - Df(x)(\eta_2)$$
$$- [\frac{f(x+t\eta_1+t\eta_2)-f(x)}{t}] - [\frac{f(x+t\eta_1)-f(x)}{t}] - [\frac{f(x+t\eta_2)-f(x)}{t}]\|$$
$$\leq 3\varepsilon,$$

whenever $|t| \leq \tau$. Hence

$$\|Df(x)(\eta_1+\eta_2) - Df(x)(\eta_1) - Df(x)(\eta_2)\|$$
$$\leq \frac{1}{|t|} \|f(x+t\eta_1+t\eta_2) - f(x+t\eta_1) - f(x+t\eta_2) - f(x)\| + 3\varepsilon.$$

By (i) of Proposition 2.3 and the Hahn-Banach theorem there exists $\delta \in Y^*$ such that $\|\delta\| = 1$ and

$$\|f(x+t\eta_1+t\eta_2) - f(x+t\eta_1) - f(x+t\eta_2) - f(x)\|$$
$$= \delta(f(x+t\eta_1+t\eta_2) - f(x+t\eta_1)) - \delta(f(x+t\eta_2) - f(x))$$
$$= t\delta(Df(x+t\eta_1+\theta_1 t\eta_2)(\eta_2)) - t\delta(Df(x+\theta_2 t\eta_2)(\eta_2))$$
$$= t\delta[Df(x+t\eta_1+\theta_1 t\eta_2)(\eta_2) - Df(x)(\eta_2)$$
$$\qquad + Df(x)(\eta_2) - Df(x+\theta_2 t\eta_2)(\eta_2)]$$
$$\leq |t| \|Df(x+t\eta_1+\theta_1 t\eta_2)(\eta_2) - Df(x)(\eta_2)\|$$
$$+ |t| \|Df(x)(\eta_2) - Df(x+\theta_2 t\eta_2)(\eta_2)\|, \quad 0 < \theta_1, \theta_2 < 1,$$
$$\leq 2|t|\varepsilon \text{ if } t \text{ is sufficiently small.}$$

It follows that

$$\|Df(x)(\eta_1 + \eta_2) - Df(x)(\eta_1) - Df(x)\eta_2\| \le 5\varepsilon ;$$

hence $Df(x)$ is additive. This proves the proposition.

Proposition 2.8. Let X and Y be normed linear spaces. Consider $f: X \to Y$. Suppose $Df: X \to L_1[X,Y]$ and Df is continuous at x. Then $f'(x)$ exists and f' is continuous at x.

Proof: By part (i) of Proposition 2.3 for any $\delta \in Y^*$

$$\delta[f(x+\eta) - f(x) - Df(x)(\eta)] = \delta[Df(x+\theta\eta)(\eta) - Df(x)(\eta)] .$$

By the Hahn-Banach theorem

$$\|f(x+\eta) - f(x) - Df(x)(\eta)\| \le \|Df(x+\theta\eta) - Df(x)\| \|\eta\| .$$

Since Df is continuous at x, given $\varepsilon > 0$ there exists $r > 0$ such that

$$\|f(x+\eta) - f(x) - Df(x)(\eta)\| \le \varepsilon \|\eta\|$$

whenever $\|\eta\| \le r$. This proves the proposition.

Proposition 2.9 (chain rule). Let X be a vector space and let Y and Z be normed linear spaces. Suppose

(i) $h: X \to Y$ is Gâteaux differentiable in X, and

(ii) $g: Y \to Z$ is Fréchet differentiable in Y.

Then $f = gh: X \to Z$ is Gâteaux differentiable in X and

$$Df(x) = g'(h(x)) Dh(x) .$$

DIFFERENTIATION AND INTEGRATION

If X is also a normed linear space and h is Fréchet differentiable in X, then f is Fréchet differentiable in X and

$$f'(x) = g'(h(x))\, h'(x) \; .$$

<u>Proof</u>: Given x and $\eta \in X$ let $y = h(x)$ and $\Delta y = h(x+t\eta) - h(x)$. Then

$$\frac{f(x+t\eta) - f(x)}{t} = \frac{g(y+\Delta y) - g(y)}{t}$$

$$= \frac{g'(y)(\Delta y) + g(y+\Delta y) - g(y) - g'(y)(\Delta y)}{t}$$

$$= g'(y)\frac{(h(x+t\eta) - h(x))}{t}$$

$$+ \frac{g(y+\Delta y) - g(y) - g'(y)(\Delta y)}{\|\Delta y\|} \cdot \frac{\|h(x+t\eta) - h(x)\|}{t} \; .$$

Hence

$$\left\| g'(h(x))\, h'(x)(\eta) - \left[\frac{f(x+t\eta) - f(x)}{t}\right] \right\|$$

$$\leq \|g'(h(x))\| \left\| h'(x)(\eta) - \left[\frac{h(x+t\eta) - h(x)}{t}\right]\right\|$$

$$+ \frac{\|g(y+\Delta y) - g(y) - g'(y)(\Delta y)\|}{\|\Delta y\|} \left\|\frac{h(x+t\eta) - h(x)}{t}\right\| \; .$$

Now, observe that if h is Gâteaux differentiable at x, then although h may not be continuous at x, we do have that h is continuous at x in each direction, i.e., $\|\Delta y\| \to 0$ as $t \to 0$. Also if h is Fréchet differentiable at x, then it is continuous at x and $\|\Delta y\| \to 0$ as $t \to 0$ uniformly for all η such that $\|\eta\| = 1$. This proves the proposition.

Remarks. If x is near x_0, i.e., $\|x - x_0\|$ is small, then (2.7) says that

$$\|f(x) - f(x_0) - f'(x_0)(x - x_0)\|$$

is nearly zero; hence

(2.8) $$f(x) \approx f(x_0) + f'(x_0)(x - x_0) \ .$$

If $f: R \to R$ we recognize (2.8) as the approximation of f by the line tangent to f at x_0. If $f: R^2 \to R$, then (2.8) becomes

$$f(x,y) \approx f(x_0, y_0) + \frac{\partial f(x_0, y_0)}{\partial x}(x - x_0) + \frac{\partial f(x_0, y_0)}{\partial y}(y - y_0) \ ;$$

which we again recognize as the approximation of f by the plane tangent to f at (x_0, y_0). It is now clear that for fixed x the Fréchet derivative of f at x is essentially a linear approximation to $f(x + \Delta x) - f(x)$.

In particular, consider a nonlinear operator $P: X \to X$. We wish to find $x \in X$ such that $P(x) = 0$. Given $x_0 \in X$ let $\Delta x = x - x_0$; our problem is now to find Δx such that $P(x_0 + \Delta x) = 0$; equivalently

$$P(x_0 + \Delta x) - P(x_0) = -P(x_0)$$

and if Δx is small, i.e., x_0 is near x, then $P'(x_0)(\Delta x)$ is a good linear approximation to $P(x_0 + \Delta x) - P(x_0)$; hence

$$P'(x_0)(\Delta x) \approx -P(x_0) \ .$$

If $P'(x_0)$ is invertible, i.e., there exists $P'(x_0)^{-1} \in L_1[X, X]$ such that $P'(x_0) P'(x_0)^{-1}(\eta) = \eta$ for all $\eta \in X$, then

$$\Delta x \approx -P'(x_0)^{-1} P(x_0)$$

DIFFERENTIATION AND INTEGRATION

and

$$x \approx x_0 - P'(x_0)^{-1} P(x_0) .$$

Repeating this procedure we arrive at Newton's method for approximating a zero of P, i.e., the construction of $\{x_n\}$ where

$$x_{n+1} = x_n - P'(x_n)^{-1} P(x_n), \quad n = 0, 1, 2, \ldots .$$

Newton's method uses the Fréchet derivative to replace a nonlinear problem by a sequence of linear problems. For applications of Newton's method to Problem (1.2) see [13], [36] and [37].

3. Higher Derivatives

Let X and Y be normed linear spaces. Suppose $f: X \to Y$ is Fréchet differentiable in an open subset of X. Recall that $f': X \to L_1[X, Y]$ and $L_1[X, Y]$ is a normed linear space. Consider f'', the Fréchet derivative of the Fréchet derivative of f. Clearly

$$f'': X \to L_1[X, L_1[X, Y]] .$$

In general let $L_n[X, Y]$ denote $L_1[X, L_{n-1}[X, Y]]$, $n = 2, 3, \ldots$. Then $f^{(n)}$, $n = 2, 3, \ldots$, the n-th Fréchet derivative of f is by definition the Fréchet derivative of $f^{(n-1)}$, the (n-1)-st Fréchet derivative of f. Clearly

$$f^{(n)}: X \to L_n[X, Y] .$$

Remark. Some generality could be gained by considering the Gâteaux derivative of the Fréchet derivative as a weaker form of higher derivative.

Remark. Clearly the operator f need not be Fréchet

differentiable in order to define the second derivative as the derivative of the derivative; it need only possess a bounded linear Gâteaux derivative at each point of some open set. However, this generality would lead to certain unpleasanties, see Example 2.5.

It is not immediately obvious how to interpret the elements of $L_n[X,Y]$. The following interpretation is very helpful. Recall that the Cartesian product of two sets U and V is by definition $U \times V = \{(u,v): u \in U, v \in V\}$. Also by U^1 we mean U and by U^n we mean $U \times U^{n-1}$, $n = 2, 3, \ldots$. Clearly U^n is a vector space in the obvious manner whenever U is a vector space.

An operator $K: X^n \to Y$ is said to be an n-linear operator from X into Y if it is linear in each of the n variables, i.e., for real α and β

$$K(x_1, \ldots, \alpha x_i' + \beta x_i'', \ldots, x_n)$$
$$= \alpha K(x_1, \ldots, x_i', \ldots, x_n) + \beta K(x_1, \ldots, x_i'', \ldots, x_n),$$
$$i = 1, \ldots, n.$$

The n-linear operator K is said to be bounded if there exists $M > 0$ such that

$$\|K(x_1, \ldots, x_n)\| \leq M \|x_1\| \|x_2\| \ldots \|x_n\|, \text{ for all}$$

$(x_1, \ldots, x_n) \in X^n$. The vector space of all bounded n-linear operators from X into Y becomes a normed linear space if we let $\|K\| = \inf\{M: \|K(x_1, \ldots, x_n)\| \leq M \|x_1\| \|x_2\| \ldots \|x_n\|, (x_1, \ldots, x_n) \in X^n\}$. This normed linear space we denote by $[X^n, Y]$.

Remark. Clearly by a 1-linear operator we mean a linear operator. Also a 2-linear operator is usually called a bilinear operator.

The following proposition shows that the spaces $L_n[X, Y]$ and $[X^n, Y]$ are essentially the same.

DIFFERENTIATION AND INTEGRATION

<u>Proposition 3.1</u>. The normed linear spaces $L_n[X,Y]$ and $[X^n, Y]$ are isometrically isomorphic.

<u>Proof</u>: We will construct

$$T_n : L_n[X,Y] \to [X^n, Y]$$

which is linear, one-one, onto and norm preserving. Clearly T_n^{-1} will have the same properties. Since $L_1[X,Y] = [X^1,Y]$, let T_1 be the identity operator. Assume we have constructed T_{n-1} with the desired properties. Define T_n as follows. For $W \in L_n[X,Y]$ let $T_n(W)$ be the n-linear operator from X into Y defined by

$$T_n(W)(x_1, \ldots, x_n) = T_{n-1}(W(x_1))(x_2, \ldots, x_n)$$

for $(x_1, \ldots, x_n) \in X^n$. Clearly $\|T_n(W)\| \leq \|W\|$; hence $T_n(W) \in [X^n, Y]$. Also T_n is linear. If $U \in [X^n, Y]$, then for each $x \in X$ let $W(x) = T_{n-1}^{-1}(U(x, \cdot, \ldots, \cdot))$. It follows that $W: X \to L_{n-1}[X,Y]$ is linear and $\|W\| \leq \|U\|$; hence $W \in L_n[X,Y]$ and $T_n(W) = U$. This shows that T_n is onto and norm preserving. Clearly a linear norm preserving operator must be one-one. This proves the proposition.

<u>Remark</u>. It follows that the n-th Fréchet derivative of $f: X \to Y$ is a bounded n-linear operator from X into Y.

Fréchet derivatives, especially higher derivatives, can be very difficult to calculate. The author has found the following technique most useful. Consider $f: X \to Y$. For $x, \eta \in X$ calculate $Df(x)(\eta)$ from (2.3). Then

(3.1) $\qquad \|Df(x)(\eta) - [\frac{f(x+t\eta) - f(x)}{t}]\| \to 0$ as $t \to 0$.

If $Df(x)(\eta)$ is not bounded and linear in η, then $f'(x)$ can not exist. If $Df(x)$ is a bounded linear operator and the convergence in (3.1) is uniform with respect to η such that $\|\eta\| = 1$, then $f'(x)$ exists and $f'(x) = Df(x)$.

It is often easier, in particular if we are also interested in establishing that f' is continuous at x, to show that $Df(z) \in L_1[X,Y]$ for all z in some neighborhood of x and that Df is continuous at x. The desired result then follows from Proposition 2.8.

Suppose now that we know $f': X \to L_2[X,Y]$. For x, $\eta_1, \eta_2 \in X$ let

$$B(x)(\eta_1, \eta_2) = \lim_{t \to 0} \left[\frac{f'(x+t\eta_1)(\eta_2) - f'(x)(\eta_2)}{t} \right].$$

Then

(3.2) $\left\| B(x)(\eta_1, \eta_2) - \left[\frac{f'(x+t\eta_1)(\eta_2) - f'(x)(\eta_2)}{t} \right] \right\| \to 0$ as $t \to 0$.

If f is twice Fréchet differentiable at x, then $B(x) = B(x)(\cdot, \cdot) \in [X^2, Y]$ and $f''(x) = B(x)$. In particular if $B(x)$ is not a bounded bilinear operator, then f can not be twice Fréchet differentiable at x. It should be clear that $B(x)$ may exist and even be a bounded bilinear operator and yet not even the Gâteaux derivative of f' at x exists. This is because a sequence of bounded linear operators which converges pointwise to a bounded linear operator need not converge in norm.

Suppose $B(x) \in [X^2, Y]$. If the convergence in (3.2) is uniform with respect to $\eta_2 \in X$ such that $\|\eta_2\| = 1$, then $B(x)$ is the Gateaux derivative of f' at x. If the convergence in (3.2) is uniform with respect to both $\eta_1, \eta_2 \in X$ such that $\|\eta_1\| = \|\eta_2\| = 1$, then $f''(x)$ exists and $f''(x) = B(x)$.

As before if for all z in a neighborhood of x $B(z) \in [X^2, Y]$, the convergence in (3.2) is uniform with respect to η_2 such that $\|\eta_2\| = 1$ and $B(z)$ is continuous at x, then $f''(x)$ exists, $f''(x) = B(x)$ and f'' is continuous at x.

When $f^{(n-1)}: X \to [X^{n-1}, Y]$ is known we have the obvious extension of this argument to

$$B(x)(\eta_1,\ldots,\eta_n) =$$

$$\lim_{t \to 0} \left[\frac{f^{(n-1)}(x+t\eta_1)(\eta_2,\ldots,\eta_n) - f^{(n-1)}(x)(\eta_2,\ldots,\eta_n)}{t} \right].$$

<u>Example 3.1.</u> Let us return to the simplest problem in the calculus of variations. We have already determined that if $J: C_0^1[a,b] \to R$ is given by (1.1), then for $y, \eta \in C_0^1[a,b]$

$$DJ(y)(\eta) =$$

$$\int_a^b [f_2(x,y(x),y'(x))\eta(x) + f_3(x,y(x),y'(x))\eta'(x)]dx \;;$$

hence $DJ(y)$ is a linear operator. Before we can discuss the Fréchet derivative of J we must introduce a norm on $C_0^1[a,b]$. Toward this end let

$$\|\eta\| = \max_{a \le x \le b} |\eta(x)| + \max_{a \le x \le b} |\eta'(x)|$$

for each $\eta \in C_0^1[a,b]$. In Section 5 we will introduce a different norm on $C_0^1[a,b]$. Recall that we have assumed f has continuous second partial derivatives; hence continuous first partial derivatives. It follows that

$$|DJ(y)(\eta)| \le$$

$$(b-a)[\max_{a \le x \le b}|f_2(x,y,y')| + \max_{a \le x \le b}|f_3(x,y,y')|]\|\eta\| \;;$$

hence $DJ(y)$ is a bounded linear operator. Also for $x, z \in C_0^1[a,b]$

$$\|DJ(y) - DJ(z)\| = \sup_{\|\eta\|=1} |DJ(y)(\eta) - DJ(z)(\eta)|$$

$$\leq (b-a)[\max_{a \leq x \leq b} |f_2(x,y,y') - f_2(x,z,z')| +$$

$$+ \max_{a \leq x \leq b} |f_3(x,y,y') - f_3(x,z,z')|].$$

Let $S = \{(u,v,w) \in R^3 : a \leq u \leq b, |v|, |w| \leq \|y\| + 1\}$. Clearly S is a closed and bounded subset of R^3 and therefore compact; hence f_2 and f_3 are uniformly continuous on S. Given $\varepsilon > 0$ there exists $\delta > 0$ such that

$$|f_2(u_1,v_1,w_1) - f_2(u_0,v_0,w_0)| \leq \frac{\varepsilon}{2(b-a)}, \text{ and}$$

$$|f_3(u_1,v_1,w_1) - f_3(u_0,v_0,w_0)| \leq \frac{\varepsilon}{2(b-a)}$$

whenever $(u_1,v_1,w_1), (u_0,v_0,w_0) \in S$ and $(u_1-u_0)^2 + (v_1-v_0)^2 + (w_1-w_0)^2 \leq \delta^2$. Hence if $\|z-y\| \leq \min(1, \delta^2)$ we have $\|Df(y) - Df(z)\| \leq \varepsilon$. This shows, by Proposition 2.8, that J' exists and J' is continuous.

A straightforward calculation gives

$$B(y)(\eta_1, \eta_2) = \lim_{t \to 0} \left[\frac{J'(y+t\eta_1)(\eta_2) - J'(y)(\eta_2)}{t} \right]$$

$$= \int_a^b [f_{22}(x,y,y')\eta_1(x)\eta_2(x) + f_{23}(x,y,y')\eta_1'(x)\eta_2(x)$$

$$+ f_{32}(x,y,y')\eta_1(x)\eta_2(x)' + f_{33}(x,y,y')\eta_1'(x)\eta_2'(x)]dx.$$

Clearly $B(y) \in [X^2, Y]$ and the convergence is uniform with respect to η_2 such that $\|\eta_2\| = 1$. Also

$$|B(y)(\eta_1,\eta_2) - B(z)(\eta_1,\eta_2)| \le (b-a) \|\eta_1\| \|\eta_2\| \cdot$$

$$\cdot \sum_{i,j=2}^{3} \max_{a \le x \le b} |f_{ij}(x,y,y') - f_{ij}(x,z,z')| \; ;$$

hence, arguing as above, J'' exists, $J''(y) = B(y)$ and J'' is continuous. Observe that $\delta^2 J(y;\eta) = J''(y)(\eta,\eta)$, see (1.6).

The previous argument generalizes in the obvious manner to show that if f in (1.1) has continuous partial derivatives of order n, i.e., is of class C^n, then $J^{(n)}$ exists and is continuous, i.e., J is of class C^n in $C_0^1[a,b]$.

4. The Riemann Integral

The material in this section is taken from the classical paper of L. M. Graves [23].

Consider $f: [0,1] \to Y$ where Y is a normed linear space. For P a partition of $[0,1]$, i.e., $P = \{[t_0,t_1],\ldots,[t_{n-1},t_n]\}$ where $0 = t_0 < t_1 < \ldots < t_n = 1$ and $|P| = \max_i \Delta t_i$ where $\Delta t_i = t_i - t_{i-1}$, let

$$S(P,f) = \sum_{i=1}^{n} f(t_i') \Delta t_i, \quad t_i' \in [t_{i-1}, t_i] \;.$$

Definition 4.1. Consider $J \in Y$. If for $\varepsilon > 0$ there exists $\delta > 0$ such that

$$\|J - S(P,f)\| \le \varepsilon$$

for any partition P with $|P| \le \delta$, then J is called the Riemann integral of f from 0 to 1 and denoted by

$$\int_0^1 f(t)dt \;.$$

Definition 4.2. Consider $f: X \to Y$ where X and Y are normed linear spaces. Given $x_0, x_1 \in X$ if

(4.1) $$\int_0^1 f(x_0 + t(x_1 - x_0))dt$$

exists, then it is called the Riemann integral of f from x_0 to x_1 and is denoted by

$$\int_{x_0}^{x_1} f(x)dx \ .$$

Remarks. If in Definition 4.2 $X = Y = R$, then making the change of variable $s = x_0 + t(x_1 - x_0)$ in (4.1) we have

$$\int_{x_0}^{x_1} f(x)dx = \frac{1}{(x_1-x_0)} \int_{x_0}^{x_1} f(s)ds$$

where $\int_{x_0}^{x_1} f(s)ds$ is the usual notion of the Riemann integral of a real-valued function of a real variable. Hence Definition 4.2, in this case, differs from the usual notion by the factor $(x_1 - x_0)$. This factor is readily apparent in equation (4.2).

Remark. If $f: X \to Y$ is continuous, then it will be uniformly continuous in $\{x_0 + t(x_1 - x_0): 0 \leq t \leq 1\}$. Hence if Y is complete, using an argument exactly the same as the one used for real-valued f, it follows that

$$\int_{x_0}^{x_1} f(x)dx$$

exists for all $x_0, x_1 \in X$.

<u>Proposition 4.1</u>. Consider $f: X \to Y$. Given $x_0, x_1 \in X$ suppose there exists $\phi: [0,1] \to R$ such that

$$\|f(x_0 + t(x_1 - x_0))\| \leq \phi(t) \text{ for } 0 \leq t \leq 1.$$

Then

$$\left\| \int_{x_0}^{x_1} f(x)dx \right\| \leq \int_0^1 \phi(t)dt,$$

provided these integrals exist.

<u>Proof</u>: Observe that $\|S(P,f)\| \leq S(P,\phi)$ for any partition P of $[0,1]$.

<u>Remark</u>. As a special case of Proposition 4.1 we have

$$\left\| \int_{x_0}^{x_1} f(x)dx \right\| \leq \int_{x_0}^{x_1} \|f(x)\|dx.$$

<u>Proposition 4.2</u>. If $f: X \to Y$ is continuously Fréchet differentiable and Y is complete, then

$$(4.2) \qquad f(x_1) - f(x_0) = \int_{x_0}^{x_1} f'(x)(x_1 - x_0)dx$$

for all $x_1, x_0 \in X$.

<u>Proof</u>: Let $g(t)$ denote $f(x_0 + t(x_1 - x_0))$ and $g'(t)$ denote $f'(x_0 + t(x_1 - x_0))(x_1 - x_0)$. Consider the partition $P_n = \{[0, 1/n], \ldots, [n-1/n, 1]\}$ of $[0,1]$, i.e., $t_i = (i-1)\frac{1}{n}$ and

$\Delta t_i = \frac{1}{n}$. Then

$$g(1) - g(0) - \sum_{i=1}^{n} g'(t_i) \Delta t_i = \sum_{i=1}^{n} [g(t_i + \Delta t_i) - g(t_i) - g'(t_i) \Delta t_i].$$

By the definition of the derivative given $\varepsilon > 0$ there exists N_0 such that for $n \geq N_0$

$$\|g(t_i + \Delta t_i) - g(t_i) - g'(t_i) \Delta t_i\| \leq \frac{\varepsilon}{2n};$$

hence

$$\|g(1) - g(0) - \sum_{i=1}^{n} g'(t_i) \Delta t_i\| \leq \frac{\varepsilon}{2} \quad \text{if} \quad n \geq N_0.$$

Also by the definition of the integral there exists N_1 such that

$$\|\sum_{i=1}^{n} g'(t_i) \Delta t_i - \int_0^1 g'(t) dt\| \leq \frac{\varepsilon}{2} \quad \text{if} \quad n \geq N_1.$$

It follows that

$$\|g(1) - g(0) - \int_0^1 g'(t) dt\| \leq \varepsilon, \quad \text{if} \quad n \geq \max(N_0, N_1).$$

This proves the proposition.

The following generalization of Proposition 4.2 is called Taylor's theorem.

Proposition 4.3. If $f: X \to Y$ is n times continuously Fréchet differentiable and Y is complete, then

DIFFERENTIATION AND INTEGRATION

$$f(x_1) = f(x_0) + \sum_{k=1}^{n-1} \frac{1}{k!} f^{(k)}(x_0)(x_1-x_0,\ldots,x_1-x_0) + R_n(x_0,x_1)$$

where

$$R_n(x_0,x_1) = \int_0^1 f^{(n)}(x_0 + t(x_1-x_0))(x_1-x_0,\ldots,x_1-x_0) \frac{(1-t)^{n-1}}{(n-1)!} dt.$$

Proof: For $n = 1$ this proposition is exactly Proposition 4.2. Now assume the proposition holds for $1,2,\ldots,n-1$. Let

$$h(t) = \frac{(1-t)^{n-1}}{(n-1)!} \quad \text{and}$$

$$g(t) = f^{(n-1)}(x_0 + t(x_1-x_0))(x_1-x_0,\ldots,x_1-x_0).$$

Then $R_{n-1}(x_0,x_1) = -\int_0^1 g(t) h(t) dt$. Using Proposition 4.2 on $g(t)\, h(t)$ we have

$$R_{n-1}(x_0, x_1) = -g(1)h(1) + g(0)h(0) + \int_0^1 g'(t)(h(t))dt$$

$$= \frac{1}{(n-1)!} f^{(n-1)}(x_0)(x_1-x_0,\ldots,x_1-x_0) + R_n(x_0,x_1).$$

This proves the proposition.

Remark. This proposition can be stated in much more generality, see [23].

Remark. If $f: X \to Y$ is n times continuously differentiable, then

$$\left\| f(x_1) - f(x_0) - \sum_{k=1}^{n-1} f^{(k)}(x_0)(x_1-x_0, \ldots, x_1-x_0) \right\|$$

(4.3)
$$\leq \sup_{0 \leq t \leq 1} \left\| f^{(n)}(x_0 + t(x_1-x_0)) \right\| \frac{\|x_1 - x_0\|^n}{n!} \, .$$

Remark. Consider $f: X \to Y$. If $\|f''(x)\| \leq K$ for all x in some convex set Ω, then from (4.3)

(4.4) $\quad \|f(x_1) - f(x_0) - f'(x_0)(x_1-x_0)\| \leq \dfrac{K}{2} \|x_1 - x_0\|^2$

for all $x_1, x_0 \in \Omega$. We would like the same result if instead of $\|f''(x)\| \leq K$ for all $x \in \Omega$ we only have

(4.5) $\quad \|f'(x) - f'(y)\| \leq K \|x-y\|, \qquad x,y \in \Omega \, .$

Clearly $\|f''(x)\| \leq K$ implies (4.5).
Using Proposition 2.3, if $\delta \in Y^*$, then

$$\delta[f(x_1) - f(x_0)] = \delta[f'(x_0 + \theta(x_1-x_0))(x_1-x_0)], \quad 0 < \theta < 1 \, ;$$

hence

$$\delta[f(x_1) - f(x_0) - f'(x_0)(x_1-x_0)] = [(f'(x_0 + \theta(x_1-x_0))$$

$$- f'(x_0))(x_1 - x_0)] \, .$$

Using the Hahn-Banach theorem to choose δ we have

$$\|f(x_1) - f(x_0) - f'(x_0)(x_1-x_0)\| \leq \|f'(x_0 + \theta(x_1-x_0)) - f'(x_0)\| \, \|x_1-x_0\|$$

$$\leq \theta K \|x_1 - x_0\|^2 \leq K \|x_1 - x_0\|^2$$

for $x_0, x_1 \in \Omega$. We have missed (4.4) by a factor of $\frac{1}{2}$ since we only knew that $0 < \theta < 1$ and not that $0 < \theta \leq \frac{1}{2}$.

Inequality (4.4) is very important in the proof of Kantorovich's theorem on Newton's method, see [10], [31] and [38]. That it follows from (4.5), obviously not in the above manner, is sufficiently important to state as a proposition.

<u>Proposition 4.4.</u> Consider f: X → Y where Y is complete. If

$$\|f'(x) - f'(y)\| \leq K \|x - y\|$$

for all x, y in some convex set Ω, then

$$\|f(x_1) - f(x_0) - f'(x_0)(x_1 - x_0)\| \leq \frac{K}{2} \|x_1 - x_0\|^2$$

for $x_0, x_1 \in \Omega$.

<u>Proof</u>: Since f' is continuous and Y is complete we have from Proposition 4.2

$$f(x_1) - f(x_0) = \int_0^1 f'(x_0 + t(x_1 - x_0))(x_1 - x_0) dt .$$

Hence, using Proposition 4.1,

$$\|f(x_1) - f(x_0) - f'(x_0)(x_1 - x_0)\|$$

$$= \|\int_0^1 [(f'(x_0 + t(x_1 - x_0)) - f'(x_0))(x_1 - x_0)] dt\|$$

$$\leq K \|x_1 - x_0\|^2 \int_0^1 t \, dt$$

$$= \frac{K}{2} \|x_1 - x_0\|^2 .$$

5. The Metric Gradient of a Functional

Suppose the partial derivative of $f: R^n \to R$ with respect to each variable exists at some point $x \in R^n$. Historically, the vector $\nabla f(x) = (\frac{\partial f(x)}{\partial x_1}, \ldots, \frac{\partial f(x)}{\partial x_n})$ has been called the gradient of f at x. The total differential is defined from the gradient by

$$df(x)(\eta) = <\nabla f(x), \eta>, \quad \eta \in R^n .$$

Observe that $df: R^n \times R^n \to R$ and $df(x): R^n \to R$ is a bounded linear functional. This, in essence, is the classical notion of differentiation of functions of several variables. The above notion does not generalize to arbitrary normed linear spaces, indeed to infinite dimensional Hilbert spaces, see Section 2. One of the difficulties being that the idea of partial derivative, as used in Euclidean n-space, involves the concept of a natural basis; which we do not have in more general spaces.

Notice that from Proposition 2.6 if f is Fréchet differentiable at x, then

$$f'(x)(\eta) = <\nabla f(x), \eta>, \quad \eta \in R^n ,$$

i.e., the total differential is the Fréchet derivative of f at x in the direction η. Also the Fréchet derivative at x is represented, via the inner product, by the gradient of f at x.

In 1934 F. Riesz showed that if δ is a continuous linear functional on a real Hilbert space H, then there exists $x_\delta \in H$ such that

$$\delta(\eta) = <x_\delta, \eta>, \quad \eta \in H .$$

It is immediate that x_δ must be unique and $\|\delta\| = \|x_\delta\|$. The previous remarks motivate the following definition.

<u>Definition 5.1</u>. Let H be a real Hilbert space and suppose the functional $f: H \to R$ is Fréchet differentiable at $x \in H$. Then the unique vector $\nabla f(x) \in H$ such that

(5.1) $\qquad f'(x)(\eta) = <\nabla f(x), \eta>, \quad \eta \in H$

is called the gradient of f at x. The operator $\nabla f: H \to H$ which assigns $\nabla f(x)$ to x is called the gradient of f.

<u>Remark</u>. This definition was first given by M. Golomb in 1934, see [21]. Clearly hints of a notion of gradient in spaces other than Euclidean n-space appear in the literature, especially in the earlier papers on the calculus of variations. In particular, Courant and Hilbert in 1922, see [8, p. 222], suggested a notion of the gradient of a functional for a particular class of functionals; however the notions of inner product and derivative were implied and not stated. It is very difficult to understand what they meant and it is the author's understanding that the situation is similar to Example 5.1.

<u>Remark</u>. It is clear that Definition 5.1 still makes sense if f only has a bounded linear Gâteaux derivative at x. However we will avoid this generality.

<u>Remark</u>. If H is merely an inner product space, i.e., not necessarily complete, then $\nabla f(x)$ in (5.1) may not exist. However when it does exist it will be unique and is called the gradient of f at x.

<u>Proposition 5.1</u>. If H is a real Hilbert space and $f: H \to R$, then f' is continuous at x if and only if ∇f is continuous at x.

<u>Proof</u>: If x and y are in the domain of f', then

$$\|f'(x) - f'(y)\| = \|\nabla f(x) - \nabla f(y)\|.$$

The proposition now follows from obvious considerations.

Remark. The Fréchet derivative of the gradient of f, for historical reasons, is called the Hessian of f and denoted by Hf. If $f: R^n \to R$, then $\nabla f = (\frac{\partial f}{\partial x_1}, \ldots, \frac{\partial f}{\partial x_n}): R^n \to R^n$; hence by Proposition 2.6

$$Hf(x)(\eta) = \begin{pmatrix} \frac{\partial^2 f(x)}{\partial x_1^2} & \cdots & \frac{\partial^2 f(x)}{\partial x_1 \partial x_n} \\ \vdots & & \vdots \\ \frac{\partial^2 f(x)}{\partial x_1 \partial x_n} & \cdots & \frac{\partial^2 f(x)}{\partial x_n^2} \end{pmatrix} \begin{pmatrix} \eta_1 \\ \vdots \\ \eta_n \end{pmatrix}.$$

Remark. Suppose $f: H \to R$ is twice Fréchet differentiable at $x \in H$. Then from (3.2) we have

$$f''(x)(\eta_1, \eta_2) = \lim_{t \to 0} [\frac{f'(x+t\eta_1)(\eta_2) - f'(x)(\eta_2)}{t}]$$

$$= \langle \lim_{t \to 0} [\frac{\nabla f(x+t\eta_1) - \nabla f(x)}{t}], \eta_2 \rangle$$

$$= \langle Hf(x)(\eta_1), \eta_2 \rangle.$$

Hence the bilinear functional $f''(x)$ is represented, via the inner product, by the Hessian of f at x. It is also true that $\|f''(x)\| = \|Hf(x)\|$, see [19].

The previous remarks clearly demonstrate the similarity between the derivative and the gradient in Hilbert space; for this reason many mathematicians often confuse the two, or at least fail to make a conscious distinction.

Definition 5.1 clearly overcomes the natural basis difficulty; however, as the following proposition shows, it is still closely related to the notion of a Hilbert space basis.

Proposition 5.2. Suppose H is a real Hilbert space and the functional $f: H \to R$ is Fréchet differentiable at $x \in H$. Then the following are equivalent

(i) $f'(x) = <\nabla f(x), \eta>$, $\eta \in H$

(ii) $\nabla f(x)$ maximizes $f'(x)(\eta)$ in $\{\eta: \|\eta\| \leq \|f'(x)\|\}$.

(iii) $\nabla f(x) = \sum_{\alpha \in A} f'(x)(x_\alpha) x_\alpha$, for any complete orthonormal set $\{x_\alpha : \alpha \in A\} \subset H$.

Proof. A straightforward application of the Cauchy-Schwarz inequality shows that (i) and (ii) are equivalent.
 Let $\{x_\alpha : \alpha \in A\}$ be a complete orthonormal subset of H. We know that for each $x \in H$, $x = \sum_{\alpha \in A} <x, x_\alpha> x_\alpha$ and the convergence is unconditional, see [18, p. 176]; hence

$$f'(x)(\eta) = f'(x)(\sum_\alpha <\eta, x_\alpha> x_\alpha)$$

$$= \sum_\alpha <\eta, x_\alpha> f'(x)(x_\alpha)$$

$$= <\sum_\alpha f'(x)(x_\alpha) x_\alpha, \sum_\alpha <\eta, x_\alpha> x_\alpha>$$

$$= <\sum_\alpha f'(x)(x_\alpha) x_\alpha, \eta> .$$

This shows that (i) and (iii) are equivalent.

Remark. The relationship between the gradient vector and the sum of partial derivatives is now clear. It is interesting that this sum is independent of the basis used. Strictly speaking we could use either (ii) or (iii) of Proposition 5.2 to define the gradient; however we feel (i) is the most natural. At any rate from either (i), (ii) or (iii) it is clear that the gradient is a consequence of the derivative and not vice-versa.

Remark. Part (ii) of Proposition 5.2 says that the functional f at x is increasing most rapidly in the direction of $\nabla f(x)$ and decreasing most rapidly in the direction of $-\nabla f(x)$. This observation leads to the gradient method for approximating a minimum of f.

Given $x_0 \in H$ construct the sequence

(5.2) $\qquad x_{n+1} = x_n - t_n \nabla f(x_n), \qquad n = 0, 1, 2, \ldots,$

choosing t_n to be the smallest nonnegative zero of

$$\phi_n(t) = f'(x_n - t \nabla f(x_n))(\nabla f(x_n)): R \to R .$$

Clearly $f(x_{n+1}) \le f(x_n)$ with equality if and only if $f'(x) = \nabla f(x) = 0$.

Example 5.1. Let us return to $J: C_0^1[a,b] \to R$ defined by (1.1). The obvious choice for an inner product in $C_0^1[a,b]$ is

(5.3) $\qquad <\eta_1, \eta_2> = \int_a^b \eta_1(x)\,\eta_2(x)\,dx .$

Hoping for the best we would have from (1.4)

$$J'(y)(\eta) = <f_2(x,y,y') - \frac{df_3(x,y,y')}{dx}, \eta(x)> ;$$

hence

(5.4) $\qquad \nabla J(y) = f_2(x,y,y') - \frac{df_3(x,y,y')}{dx} .$

However, with the norm induced by (5.3) J is not necessarily

Fréchet differentiable. Even when $J'(y)$ does exist we can not guarantee that (5.4) exists; hence we certainly can not guarantee that it is continuously differentiable. Also it is highly unlikely that (5.4) would vanish at $x = a$ and $x = b$. Clearly (5.3) was a bad choice. See Courant and Hilbert [8] and also (1.5). The space $C_0^1[a,b]$ is not complete with respect to the norm induced by (5.3).

<u>Example 5.2.</u>   Define an inner product on $C_0^1[a,b]$ by

(5.5) $$<\eta_1, \eta_2> = \int_a^b \eta_1'(x)\, \eta_2'(x)\, dx \ .$$

It is not difficult to see that $J'(y)$ exists and is given by (1.3). Integrating by parts in (1.3), this time in the opposite manner as before, we obtain

(5.6) $$J'(y)(\eta) = \int_a^b [f_3(x,y,y') - \int_a^x f_2(t,y,y')dt]\, \eta'(x) dx \ .$$

If $g(x) = f_3(x,y,y') - \int_a^x f_2(t,y,y')dt$, then clearly g is continuous on $[a,b]$. If g has an anti-derivative in $C_0^1[a,b]$, i.e., if there exists a constant c such that

(5.7) $$\int_a^x g(t)dt + c \in C_0^1[a,b]$$

we will be through and $\nabla J(y)$ will be given by (5.7). Since (5.7) must vanish when $x = a$, we have $c = 0$. However, we can not guarantee with $c = 0$ that (5.7) will vanish when $x = b$. Since $C_0^1[a,b]$ is not complete with respect to the norm induced by (5.5) it is possible that $\nabla J(y)$ does not exist.

However, observing that $\int_a^b \eta'(x)dx = \eta(b) - \eta(a) = 0$ we can write (5.6) as

$$J'(y)(\eta) = \int_a^b (g(x) - c)\eta'(x)dx$$

for any constant c. Again we ask does $g(x) - c$ have an anti-derivative in $C_0^1[a,b]$, i.e., for some c and d is

$$\int_a^x g(t)dt - c(x-a) + d \in C_0^1[a,b] \ .$$

Clearly $d = 0$ and $c = \frac{1}{(b-a)} \int_a^b g(t)dt$. It follows that if

(5.8) $$\nabla J(y) = \int_a^x g(t)dt - \frac{(x-a)}{(b-a)} \int_a^b g(t)dt \ ,$$

then

$$\nabla J(y) \in C_0^1[a,b] \text{ and } J'(y)(\eta) = <\nabla f(x), \eta> \ .$$

Remark. Observe that if y solves Problem (1.2), then $\nabla J(y) = 0$ and from (5.8) equation (1.5) will be satisfied without requiring y'' to exist.

The following material was developed in collaboration with M. Golomb and is taken from Golomb and Tapia [22]. The author acknowledges conversations with D. Cudia.

Let X be a normed linear space with dual X^* and second dual X^{**}. The natural embedding of X into X^{**} is denoted by χ. We also denote $\chi(x)$ by x^{**} for each $x \in X$. Let $S = \{x \in X: \|x\| = 1\}$ and $U = \{x \in X: \|x\| \leq 1\}$ with analogous definitions for S^*, U^*, X^{**} and U^{**}. By 2^X we mean the class of all subsets of X. A set-valued map T defined on X is said to be single-valued if $T(x)$ contains at most one point for each $x \in X$.

Definition 5.2. Let X be a normed linear space. For each $x \in X$ let $J_X(x)$ denote the set of $\delta \in X^*$ such that

(i) $\|\delta\| = \|x\|$ and

(ii) $\delta(x) = \|x\|^2$.

The map $J_X: X \to 2^{X^*}$ which assigns $J_X(x)$ to x is called the duality map of X.

Remark. The duality map has recently received a lot of attention in the study of monotone operators. It actually appears much earlier in the literature under the name of the extended spherical image map, see Klee [28] and Cudia [7].

Definition 5.3. Let X be a normed linear space. If $f: X \to R$ is Fréchet differentiable at $x \in X$, then by the metric gradient of f at x we mean

(5.9) $$\nabla f(x) = \chi^{-1} J_{X^*} f'(x) .$$

We call

$$\nabla f = \chi^{-1} J_{X^*} f' : X \to 2^X$$

the metric gradient of f.

Remark. Again some words of caution are necessary. As previously remarked, since the gradient and the derivative of a functional defined on a real Hilbert space behave much the same way many authors identify them via the correspondence 5.1. This is a consequence of, hence no worse than, the practice of identifying a real Hilbert space with its dual. However since this identification can not be done for an arbitrary normed linear space, these same authors take the term gradient to mean the Fréchet derivative of a functional defined on a normed linear space, see Vainberg [40, p. 54]. An objection to this interpretation is that when X is a Hilbert space we have

(5.10)   (i) $f'(x) \in X^*$
          (ii) $\nabla f(x) \in X$;

hence many useful consequences of (ii), no longer make sense (at least not without additional definitions) in normed linear spaces.

Notice that Definition 5.3 preserves (5.10). Recognizing this established use of the term gradient M. Golomb and I have decided to call (5.9) the metric gradient. This is not at all unreasonable since unlike the derivative the metric gradient depends on the norm, not just on the topology of the space in question, i.e., a functional defined on a normed linear space will always have the same derivative with respect to equivalent norms but not the same metric gradient. This is particularly true if we consider a finite dimensional space with different norms.

Proposition 5.3. Let X be a normed linear space. Then

(i) $J_X(x) \neq \phi$ for all $x \in X$.

(ii) $\chi(x) \in J_{X^*} J_X(x)$ for all $x \in X$.

(iii) J_X is homogeneous, i.e., $J_X(\alpha x) = \alpha J_X(x)$ for all $\alpha \in R$ and all $x \in X$.

(iv) J_X is monotone, i.e., if $\delta_x \in J_X(x)$ and $\delta_y \in J_X(y)$, then $(\delta_x - \delta_y)(x-y) \geq 0$.

(v) If X is reflexive, then J_X is onto, i.e., if $\delta \in X^*$, then there exists $x \in X$ such that $\delta \in J_X(x)$.

(vi) If X is strictly convex, then J_X is one-one, i.e., if $J_X(x) \cap J_X(y) \neq \phi$, then $x = y$.

(vii) If X^* is strictly convex, then J_X is single-valued.

Proof: A straightforward application of the Hahn-Banach theorem proves part (i). Parts (ii) and (iii) follow from Definition 5.2. If $\delta_x \in J_X(x)$ and $\delta_y \in J_X(y)$, then

$$(\delta_x - \delta_y)(x-y) \geq (\|x\| - \|y\|)^2 \geq 0 .$$

This establishes part (iv). Part (v) follows from the fact that for each $\delta \in X^*$

$$\delta \in J_X(x^{-1} J_{X^*}(\delta)) .$$

How, if $\delta \in J_X(x) \cap J_X(y)$, then

$$\|\delta\|^2 = \alpha \delta(x) + (1-\alpha) \delta(y) \leq \|\alpha x + (1-\alpha) y\| \|\delta\| \leq \|\delta\|^2 ;$$

part (vi) clearly follows. Similarly if $\delta_1, \delta_2 \in J_X(x)$, then

$$\|x\|^2 = \alpha \delta_1(x) + (1-\alpha) \delta_2(x) \leq \|\alpha \delta_1 + (1-\alpha) \delta_2\| \|x\| \leq \|x\|^2$$

which proves part (vii).

Remark. From part (ii) above whenever J_X and J_{X^*} are single-valued we have the following factorization of x

(5.11) $$x = J_{X^*} J_X .$$

Remark. Clearly if both X and X^* are reflexive and strictly convex, as in the case of $L_p(\Omega)$ ($1 < p < \infty$), then both J_X and J_{X^*} will be one-one, onto and single-valued, therefore invertible. Hence by (5.11)

$$X^{-1} = J_X^{-1} J_{X^*}^{-1},$$

and (5.9) can be written as

(5.12) $$\nabla f(x) = J_X^{-1} f'(x).$$

In the case of Hilbert space we know that $J_X(x) = <x, \cdot >$; hence (5.12) gives

$$f'(x) = J_X(\nabla f(x)) = <\nabla f(x), \cdot >$$

which is exactly (5.1).

Proposition 5.4. Let X be a normed linear space. Suppose $f: X \to R$ is Fréchet differentiable at $x \in X$. Then $\nabla f(x)$ is a closed, bounded and convex subset of X. Also if X is reflexive, then $\nabla f(x)$ is not empty and if X is strictly convex, then $\nabla f(x)$ contains at most one point.

Proof: The proposition follows directly from Proposition 5.3.

Remark. In Hilbert space J_X is a norm-preserving, invertible linear operator; hence ∇f depends linearly and continuously on f'. Certainly Hilbert space can not be the only space where ∇f depends linearly on f'? The following beautiful characterization of inner product spaces answers this question.

Proposition 5.4. Let X be a normed linear space. Then J_X is linear if and only if X is an inner product space.

Proof: It is clear that if X is an inner product space, then J_X is linear. Now assume J_X is linear. If $x, y \in X$, then

$$\|x+y\|^2 = J_X(x+y)(x+y) = \|x\|^2 + J_X(x)(y) + J_X(y)(x) + \|y\|^2,$$

DIFFERENTIATION AND INTEGRATION

and

$$\|x-y\|^2 = J_X(x-y)(x-y) = \|x\|^2 - J_X(x)(y) - J_X(y)(x) + \|y\|^2 ;$$

hence

$$\|x+y\|^2 + \|x-y\|^2 = 2\|x\|^2 + 2\|y\|^2 \quad \text{and} \quad X$$

is an inner product space.

Remark. It is quite surprising, especially since χ^{-1} is always linear and J_{X^*} is always homogeneous and monotone, that we have linear dependence of ∇f on f' only when X is a Hilbert space (inner product space).

Proposition 5.5. Suppose X is a normed linear space and the functional $f: X \to R$ is Fréchet differentiable at $x \in X$. Then the following are equivalent

(i) $\delta \in \nabla f(x)$

(ii) δ maximizes $f'(x)(\eta)$ in $\{\eta \in X: \|\eta\| \leq \|f'(x)\|\}$.

Proof. The proof follows from Definitions 5.2 and 5.3 and the fact that χ^{-1} is norm preserving.

Compare Proposition 5.5 with Proposition 5.2.

Remark. We now have the same motivation for the gradient method in normed linear spaces that we had in Hilbert space, see equation 5.2.

Conditions which guarantee the continuity of $\nabla f: X \to X$ are very important. Let us consider this problem.

The following two definitions are taken from Cudia [7] (slightly altered but equivalent).

Definition 5.4. The normed linear space X is said to be weakly uniformly convex in each direction if given $\varepsilon > 0$ and $g, h \in S^*$ there exists $\delta > 0$ such that

$$|g(\tfrac{x+y}{2})| > 1-\delta \implies |h(x-y)| < \varepsilon ,$$

for $x,y \in S$.

Definition 5.5. The normed linear space X is said to be weakly uniformly convex if given $\varepsilon > 0$ and $g \in S^*$ there exists $\delta > 0$ such that

$$|g(\tfrac{x+y}{2})| > 1-\delta \implies \|x-y\| < \varepsilon ,$$

for $x,y \in S$.

Remark. Clearly a weakly uniformly convex space is weakly uniformly convex in each direction. Also any uniformly convex space, therefore any L_p space ($1 < p < \infty$) is weakly uniformly convex.

Proposition 5.5. Suppose X is weakly uniformly convex in each direction and $f: X \to R$ is Fréchet differentiable. Then ∇f is single-valued. Also, if $f': X \to X^*$ is continuous at $x \in X$, then $\nabla f: X \to X$ is weak continuous at $x \in X$, i.e., if $x_n \to x$ in the norm of X and $\nabla f(x_n)$, $\nabla f(x)$ exist, then $\nabla f(x_n) \to \nabla f(x)$ weakly in X.

Proof. We first show J_{X^*} is single-valued. Suppose $g \in S^*$ and $\eta_1, \eta_2 \in J_{X^*}(g)$. Since $\chi(U)$ is weak* dense in U^{**}, there exist $\{x_\alpha\}, \{y_\alpha\} \subset U$ such that $x_\alpha^{**} \to \eta_1$ (weak*) and $y_\alpha^{**} \to \eta_2$ (weak*). We have $1 = \|\eta_1\| \leq \underline{\lim} \|x_\alpha\| \leq 1$; hence no generality is lost by assuming $\|x_\alpha\| = 1$. Also $g(\tfrac{x_\alpha + y_\alpha}{2}) = g^{**}(\tfrac{x_\alpha^{**} + y_\alpha^{**}}{2}) \to g^{**}(\tfrac{\eta_1 + \eta_2}{2}) = 1$. Therefore, for any $h \in S^*$ and any $\varepsilon > 0$, by Definition 5.4, eventually $|h(x_\alpha - y_\alpha)| < \varepsilon$ which implies $h^{**}(x_\alpha^{**} - y_\alpha^{**}) \to 0$; hence

DIFFERENTIATION AND INTEGRATION

$(x_\alpha^{**} - y_\alpha^{**}) \to 0$ (weak*) and $\eta_1 = \eta_2$. It follows, since J_{X^*} is homogeneous, that J_{X^*} is single-valued on all of X^*.

Now assume $x_n \to x$ in the norm of X and $f'(x_n) \to f(x)$ in the norm of X^*. Assume further that $\nabla f(x_n)$ and $\nabla f(x)$ exist, i.e., are not empty. Denote $f'(x_n)$ by d_n and $f'(x)$ by d. There exists $M < +\infty$ such that $\|d_n\| \leq M$; hence $\|J_{X^*}(d_n)\| \leq M$. By weak* compactness there exists $\eta \in \{q \in X^{**}: \|q\| \leq M\}$ such that, passing to a subsequence if necessary, $J_{X^*}(d_n) \to \eta$ (weak*). Hence

(i) $|J_{X^*}(d_n)(d) - \eta(d)| = |d^{**}(J_{X^*}(d_n)) - d^{**}(\eta)| \to 0$, and

(ii) $|J_{X^*}(d_n)(d) - J_{X^*}(d)(d)| \leq |d^{**}(J_{X^*}(d_n)) - d_n^{**}(J_{X^*}(d_n))|$

$\qquad + |d_n^{**}(J_{X^*}(d_n)) - d^{**}(J_{X^*}(d))|$

$\qquad \leq M\|d_n - d\| + |\,\|d_n\|^2 - \|d\|^2\,| \to 0$.

It follows that

(i) $\quad \eta(d) = \|d\|^2$, and

(ii) $\quad \|\eta\| \leq \|d\|$;

hence $\eta = J_{X^*}(d)$. If the original sequence $\{J_{X^*}(d_n)\}$ does not converge to $J_{X^*}(d)$ in the weak* topology, then some weak* neighborhood of $J_{X^*}(d)$ excludes infinitely many members of $\{J_{X^*}(d_n)\}$; these excluded members clearly have a weak* cluster point which must be $J_{X^*}(d)$. This is a contradiction; hence $J_{X^*}(d_n) \to J_{X^*}(d)$ (weak*). It follows, from the definition of χ^{-1}, that $\nabla f(x_n) \to \nabla f(x)$ weakly in X.

Proposition 5.6. If X is strictly convex and reflexive, then X is weakly uniformly convex in each direction; hence Proposition 5.5 can be used.

Proof: Recall that X is strictly convex if $\|\frac{x+y}{2}\| < 1$ whenever $x, y \in U$ and $x \neq y$. Given $g, h \in S^*$ and $\varepsilon > 0$ assume for $n = 1, 2, \ldots$, there exist $x_n, y_n \in S$ such that

$$|g(x_n - y_n)| \geq \varepsilon \quad \text{and} \quad |h(\frac{x_n + y_n}{2})| \geq 1 - \frac{1}{n}.$$

Since X is reflexive U is weakly compact, therefore, passing to subsequences if necessary, there exist $x^*, y^* \in U$ such that $x_n \to x^*$ (weakly) and $y_n \to y^*$ (weakly). Clearly

$$|g(x^* - y^*)| \geq \varepsilon \quad \text{and} \quad |h(\frac{x^* + y^*}{2})| = 1;$$

hence $x^* \neq y^*$ and $\|\frac{x^* + y^*}{2}\| = 1$. This contradicts the strict convexity of X. Therefore for some n we must have

$$|g(x-y)| \geq \varepsilon \implies |h(\frac{x+y}{2})| \leq 1 - \frac{1}{n}$$

for all $x, y \in S$. This proves the proposition.

Proposition 5.7. Suppose X is weakly uniformly convex and $f: X \to R$ is Fréchet differentiable. Then ∇f is single-valued. Also if $f': X \to X^*$ is continuous at $x \in X$, then $\nabla f: X \to X$ is continuous at x.

DIFFERENTIATION AND INTEGRATION

Proof: Suppose $x_n \to x$ in the norm of X, $f'(x_n) \to f'(x)$ in the norm of X^* and $\nabla f(x_n)$ and $\nabla f(x)$ exist. By Proposition 5.5

$$\frac{f'(x)}{\|f'(x)\|}\left(\frac{\nabla f(x_n)}{\|\nabla f(x_n)\|}\right) \to 1 \ .$$

Hence by weak uniform convexity given $\varepsilon > 0$ there exists $\delta > 0$ such that if

$$\frac{1}{2}\frac{f'(x)}{\|f'(x)\|}\left(\frac{\nabla f(x_n)}{\|\nabla f(x_n)\|} + \frac{\nabla f(x_m)}{\|\nabla f(x_m)\|}\right) > 1 - \delta$$

then

$$\left\|\frac{\nabla f(x_n)}{\|\nabla f(x_n)\|} - \frac{\nabla f(x_m)}{\|\nabla f(x_m)\|}\right\| < \varepsilon \ .$$

It follows that $\{\nabla f(x_n)\}$ is a norm Cauchy sequence in X. Hence $J_{X^*}(f'(x_n)) \to J_{X^*}(f'(x))$ in the norm of X^{**} and $\nabla f(x_n) \to \nabla f(x)$ in the norm of X.

Remark. It is possible, using results of Cudia, Mazur and Smulian, to show that Propositions 5.5 and 5.7 are sharp, i.e., if the metric gradient is a single-valued (weak) continuous function of the derivative, then the space must necessarily be weakly uniformly convex (in each direction).

Example 5.3. Consider $f: L_p(\Omega) \to R$, $1 < p < \infty$. It is well known that

$$f'(x)(\eta) = \int_\Omega g\eta \, d\mu \ ,$$

for some $g \in L_q(\Omega)$ where $\frac{1}{p} + \frac{1}{q} = 1$, see [18,p. 145]. A straightforward application of Hölder's inequality produces

(5.13) $$\nabla f(x) = \|g\|_q^{1-q/p} \operatorname{sgn}(g) |g|^{q/p} .$$

Remark. Observe that in (5.13)

$$\|\nabla f(x)\|_p = \|g\|_q = \|f'(x)\| .$$

Also, if $p = 2$, then $\nabla f(x) = g$.

6. Mean Value Theorems

Propositions 6.1 and 6.3 are due to R. McLeod, see [29]. I have slightly modified both McLeod's statement and proof of these propositions. In particular McLeod states these propositions in more generality. Proposition 6.2 is taken from A. A. Goldstein [20].

For a set A contained in a topological vector space let $C(A)$ denote the closed convex hull of A and $\bar{C}(A)$ the closed convex hull of A.

Proposition 6.1. Let X be a vector space and Y a locally convex topological vector space. Consider $f: X \to Y$. Given $x, y \in X$ suppose f is Gâteaux differentiable at each point of $\{x + t(y-x): 0 \le t \le 1\}$ in the direction $y - x$. Let

$$D_f = \{Df(x+t(y-x))(y-x): 0 < t < 1\} .$$

Then

$$f(y) - f(x) \in \bar{C}(D_f) .$$

Proof: It follows directly from the Hahn-Banach theorem [39,

p. 143] that in a locally convex topological linear space any closed convex set is the same as the intersection of all closed half-spaces containing it. Let H be a closed half-space containing $\overline{C}(D_f)$, i.e., $H = \{\eta \in Y: \delta(\eta) \leq \alpha\}$ for some $\delta \in Y^*$ and $\alpha \in R$. If $h(t) = \delta(f(x+t(y-x)))$, then $h'(t) = \delta[Df(x+t(y-x))(y-x)] \leq \alpha$; hence $h(t) - \alpha t$ is decreasing in $(0,1)$ and $h(d) - h(c) \leq \alpha(d-c)$ for $0 < c \leq d < 1$. By continuity $h(1) - h(0) \leq \alpha$, equivalently $\delta[f(y) - f(x)] \leq \alpha$; hence $f(y) - f(x) \in H$. Since H was arbitrary it follows that $f(y) - f(x) \in \overline{C}(D_f)$. This proves the proposition.

Remark. If in Proposition 6.1 Y is a normed linear space, then by considering convex combinations we have inequality (ii) of Proposition 2.3. A. A. Goldstein observed that under certain conditions Proposition 4.1 can be used to obtain a lower bound for $\|f(y) - f(x)\|$.

Proposition 6.2. Let X be a vector space and Y a normed linear space. Consider $f: X \to Y$. Given $x, y \in X$ suppose f is Gâteaux differentiable at each point of $\{x + t(y-x): 0 \leq t \leq 1\}$ in the direction $y - x$. Let $f'(t)$ denote $Df(x + t(y-x))(y-x)$. Suppose further that there exist t_0, $q \in (0,1)$ such that

$$\|f'(t_0) - f'(t)\| \leq q \|f'(t_0)\|, \quad \text{for } t \in (0,1).$$

Then

$$(1-q) \|f'(t_0)\| \leq \|f(x) - f(y)\| \leq (1+q) \|f'(t_0)\|.$$

Proof: The right hand side of the inequality is straightforward. By the Hahn-Banach theorem there exists $\delta \in Y^*$ such that $\|\delta\| = 1$ and $\delta[f'(t_0)] = \|f'(t_0)\|$. For $t \in (0,1)$

$$\delta[f'(t)] = \delta[f'(t_0)] + \delta[f'(t) - f'(t_0)]$$

$$\geq \|f'(t_0)\| - \|f'(t) - f'(t_0)\|$$

$$\geq (1-q) \|f'(t_0)\|.$$

By considering convex combinations we have

$$(1-q) \|f'(t_0)\| \leq \delta(z) \leq \|z\| \quad \text{for all} \quad z \in \overline{C}(D_f).$$

The proposition now follows from Proposition 4.1.

Proposition 6.3. Let X be a vector space. Consider $f: X \to R^n$. Given $x, y \in X$ suppose f is Gâteaux differentiable at each point of $\{x + t(y-x): 0 \leq t \leq 1\}$ in the direction $y - x$. Suppose further that $Df(x+t(y-x))(y-x)$ is a continuous function of t for $0 < t < 1$. Then

(6.1) $$f(y) - f(x) = \sum_{i=1}^{n} \lambda_i \, Df(x + \theta_i(y-x))(y-x)$$

where $0 < \theta_i < 1$, $\lambda_i \geq 0$ and $\sum_{i=1}^{n} \lambda_i = 1$.

Proof: Let $g: R \to R^n$ be given by $g(t) = f(x+t(y-x)) - t\,Df(\frac{x+y}{2})(y-x)$. We will use Proposition 4.1 on g in the interval $[0,1]$. If we denote $Dg(t)(1)$ by $g'(t)$, then $g'(t) = Df(x+t(y-x))(y-x) - Df(\frac{x+y}{2})(y-x)$. Hence, by hypothesis, $g'(t)$ exists for $0 \leq t \leq 1$ and $g'(t)$ is a continuous function of t for $0 < t < 1$. Recall that $Dg = \{g'(t): 0 < t < 1\}$. Note that $0 \in D_g$. Let V be the intersection of all subspaces containing $C(D_g)$. Since V is finite dimensional it is closed; hence the closure of $C(D_g)$ in V is $\overline{C}(D_g)$ and the interior of $C(D_g)$ relative to V is the same as the interior of $\overline{C}(D_g)$ relative to V. Using Proposition 4.1 on g in $[0,1]$ we have $g(1) - g(0) \in \overline{C}(D_g)$. Let b be a boundary point of $\bar{C}(D_g)$ relative to V. There exists a continuous non-zero linear functional δ on V and a scalar α such that $\bar{C}(D_g) \subset \{x \in V: \delta(x) \leq \alpha\}$ and $b \in U = \{x \in V: \delta(x) = \alpha\}$, see [39, p.142]. Now, if $D_g \in U$, then $\bar{C}(D_g) \subset U$. We must have $\alpha = 0$, since $0 \in D_g$. It follows that U is a subspace containing $\bar{C}(D_g)$; hence $V \subset U$. This implies that δ is the zero functional on V, which is a contradiction; hence $D_g \not\subset U$. It follows that if $h(t) = \delta(g(t))$, then for $0 < t < 1$ $h'(t) \leq \alpha$ and there exists at least one t such that $h'(t) < \alpha$. As before by considering $h(t) - t\alpha$ for $0 < t < 1$ we have $\delta[g(1) - g(0)] < \alpha$; hence $g(1) - g(0) \neq b$. Since b was an arbitrary boundary point of $\bar{C}(D_g)$ relative to V we must have $g(1) - g(0)$ in the interior of $\bar{C}(D_g)$, therefore in the interior of $C(D_g)$ relative to V. It follows that $g(1) - g(0)$ can be expressed as a convex combination of at most $n+1$ points of $C(D_g)$, i.e.,

$$g(1) - g(0) = \sum_{i=1}^{n+1} \lambda_i g_i$$

where $g_i \in D_g$, $\lambda_i \geq 0$ and $\sum_{i=1}^{n+1} \lambda_i = 1$.

We now show that this convex combination only requires n terms. We need only prove the result for the special case $g(1) = g(0)$. The general result then follows by considering $p(t) = g(t) - t(g(1) - g(0))$. To see this, observe that $D_p = D_g - (g(1) - g(0))$; hence, if $0 = p(1) - p(0) = \sum_{i=1}^{n} \lambda_i p_i$, where $p_i \in D_p$ and $\sum_{i=1}^{n} \lambda_i = 1$, then $g(1) - g(0) = \sum_{i=1}^{n} \lambda_i g_i$, for some $g_i \in D_g$.

If any proper subset of $\{g_1, \ldots, g_{n+1}\}$ has a convex combination equal to zero we are through. Suppose not. In particular $\lambda_i g_i \neq 0$ for $i = 1, 2, \ldots, n+1$. Reordering if necessary we have $g_1 = g'(t_1)$ and $g_2 = g'(t_2)$ with $0 < t_1 < t_2 < 1$. Let $A = \{-g_2, \ldots, -g_{n+1}\}$ and let C be the cone subtended at the origin by $C(A)$, i.e., C consists of all points $\sum_{i=2}^{n+1} \mu_i g_i$ with $\mu_i \leq 0$, $i = 2, \ldots, n+1$. Since $\sum_{i=1}^{n+1} \lambda_i g_i = 0$ and $\lambda_1 g_1 \neq 0$ we have $g_1 \in C$. If $g_2 \in C$, then we would have a convex combination of $\{g_2, \ldots, g_{n+1}\}$ equal to zero; hence $g_2 \notin C$.

The set A is linearly independent. Suppose $\sum_{i=2}^{n+1} v_i g_i = 0$ and at least one v_i is not zero. Without loss of generality say $v_2 < 0$. We have

$$\lambda_1 g_1 + \sum_{i=2}^{n+1} (\lambda_i + t v_i) g_i = 0$$

for all real t. Let $E = \{t: \lambda_i + t v_i \geq 0, i = 2, \ldots, n+1\}$. Then $0 \in E$ since $\lambda_i \geq 0$, $i = 1, \ldots, n+1$. Also E is bounded above by $-\lambda_2 v_2^{-1}$. Let $t_0 = \sup E$. By continuity $t_0 \in E$ and $\lambda_i + t_0 v_i = 0$ for at least one $i \in \{2, \ldots, n+1\}$. Let

$$v = \lambda_1 + \sum_{i=2}^{n+1} (\lambda_i + t_0 v_i).$$

Then $v \geq \lambda_i > 0$ and

$$v^{-1} \lambda_1 g_1 + \sum_{i=2}^{n+1} v^{-1} (\lambda_i + t_0 v_i) g_i$$

is a convex combination of a proper subset of $\{g_1, \ldots, g_{n+1}\}$

which is equal to zero. This contradicts our assumption; hence A is linearly independent.

The linearly independent set A contains n elements; hence any element $x \in R^n$ has a unique representation of the form $x = \sum_{i=2}^{n+1} \alpha_i g_i$. Also x is a boundary point of C if and only if $\alpha_i \leq 0$, $i = 2, \ldots, n+1$ and $\alpha_i = 0$ for at least one i.

Let $B = \{\tau \in (0,1): g'(t) \in C \text{ for } t_1 \leq t \leq \tau\}$. From before $t_1 \in B$ and $t_2 \notin B$. Hence, if $b = \sup B$, then $t_1 \leq b \leq t_2$. The set C is closed and $g'(t)$ is continuous for $0 < t < 1$; hence $g'(b) \in C$. There exists a sequence $\{b_n\} < 1$ decreasing to b such that $g'(b_n) \notin C$. By continuity $g'(b)$ belongs to the closure of the complement of C; hence $g'(b)$ is a boundary point of C and there exists $\alpha_i \leq 0$ with at least one $\alpha_i = 0$ such that

$$g'(b) - \sum_{i=2}^{n+1} \alpha_i g_i = 0 .$$

Dividing by $1 + \sum_{i=2}^{n+1} \alpha_i$ gives the desired convex combination of n members of D_g equal to 0.

We have shown that

$$g(1) - g(0) = \sum_{i=1}^{n} \lambda_i g'(\theta_i) ,$$

where $0 < \theta_i < 1$, $\lambda_i \geq 0$ and $\sum_{i=1}^{n} \lambda_i = 1$. This is equivalent to (6.1) and the proposition follows.

REFERENCES

1. P. M. Anselone, editor, Nonlinear Integral Equations, University of Wisconsin Press, Madison, 1964.

2. H. A. Antosiewicz and W. C. Rheinboldt, Numerical analysis and functional analysis, pp. 485-517, in Survey of Numerical Analysis, John Todd, editor, McGraw-Hill, New York, 1962.

3. O. Bolza, Lectures on the Calculus of Variations, Dover, New York, 1961.

4. W. E. Bosarge, Jr. and P. L. Falb, Infinite dimensional multipoint methods and the solution of two point boundary value problems, Numer. Math., 14(1970), 264-286.

5. H. Cartan, Calcul Différentiel, Hermann, Paris, 1967.

6. L. Collatz, Functionalanalysis und Numerische Mathematik, Springer Verlag, Berlin, 1964.

7. D. F. Cudia, The geometry of Banach spaces. Smoothness, Trans. Amer. Math. Soc., 110(1964), 284-314.

8. R. Courant and D. Hilbert, Methods of Mathematical Physics Vol. 1, Interscience, New York, 1965.

9. J. W. Daniel, Approximate Minimization of Functionals, Prentice-Hall, New Jersey, 1971.

10. J. E. Dennis Jr., On the Kantorovich hypothesis for Newton's Method, SIAM J. Numer. Anal. 6(1969), 493-507.

11. J. Dieudonné, Foundations of Modern Analysis, Academic Press, New York, 1960.

12. N. Dunford and J. Schwartz, Linear Operators Part I, Interscience, New York, 1964.

13. F. D. Faulkner, Newton's method applied to the solution of classical problems in calculus of variations and differential games, Doctoral thesis, University of Michigan, Ann Arbor, 1969.

14. W. H. Fleming, Functions of Several Variables, Addison-Wesley, Reading, Mass., 1965.

15. M. Fréchet, La notion de différentielle dans l' analyse générale, Ann. Sci. Ecole Norm. Sup., 3(1925), 293-323.

16. D. M. Friedlen and M. Z. Nashed, A note on one-sided directional derivatives, Math. Magazine, 41(1968), 147-150.

17. R. Gâteaux, Sur les functionnelles continus et les functionnelles analytiques, Bull. Soc. Math. France, 50(1922), 163-167.

18. C. Goffman and G. Pedrick, First Course in Functional Analysis, Prentice-Hall, New Jersey, 1965.

19. A. A. Goldstein, Constructive Real Analysis, Harper and Row, New York, 1967.

20. _____, A note on the mean value theorem, Mathematics Research Center Report 1086, University of Wisconsin, Madison, 1970.

21. M. Golomb, Zur Theorie der nichtlinear Integral-gleichungen, Integral-gleichungssysteme und algemeinen Functional-gleichungen, Math. Z., 39(1934), 45-75.

22. M. Golomb and R. A. Tapia, The metric gradient of a functional, in preparation.

23. L. M. Graves, Riemann integration and Taylor's theorem in general analysis, Trans. Amer. Math. Soc., 29(1927), 163-177.

24. M. R. Hestenes, Calculus of Variations and Optimal Control Theory, John Wiley, New York, 1966.

25. T. H. Hildebrandt and L. M. Graves, Implicit functions and their differentials in general analysis, Trans. Amer. Math. Soc., 29(1927), 127-153.

26. L. V. Kantorovich, Functional analysis and applied mathematics, National Bureau of Standards Report 1509, Washington, D.C., 1952.

27. L. V. Kantorovich and G. P. Akilov, Functional Analysis in Normed Spaces, Pergamon Press, New York, 1964.

28. V. Klee, Convex bodies and periodic homeomorphisms in Hilbert space, Trans. Amer. Math. Soc., 74(1953), 10-43.

29. R. M. McLeod, Mean value theorems for vector valued functions, Proc. Ed. Math. Soc., 14(1964), 197-209.

30. M. Z. Nashed, Some remarks on variations and differentials, Amer. Math. Monthly, 73(1966), 63-76.

31. J. M. Ortega, The Newton-Kantorovich theorem, Amer. Math. Monthly, 75(1968), 658-660.

32. J. M. Ortega and W. C. Rheinboldt, Iterative Solutions of Nonlinear Equations in Several Variables, Academic Press, New York, 1970.

33. L. B. Rall, Computational Solution of Nonlinear Operator Equations, John Wiley, New York, 1969.

34. L. B. Rall and R. A. Tapia, The Kantorovich theorem and error estimates for Newton's method, Mathematics Research Center Report 1043, University of Wisconsin, Madison, 1970.

35. E. H. Rothe, Gradient mappings, Bull. Amer. Math. Soc., 59(1953), 5-19.

36. M. Stein, On methods of obtaining solutions of fixed end point problems in the calculus of variations, Doctoral thesis, University of California, Los Angeles, 1950.

37. R. A. Tapia, The weak Newton Method and boundary value problems, SIAM J. Numer. Anal., 6(1969), 539-550.

38. _____, The Kantorovich theorem for Newton's method, Mathematics Research Center Report 1007, University of Wisconsin, Madison, 1969.

39. A. E. Taylor, Introduction to Functional Analysis, John Wiley, New York, 1963.

40. M. M. Vainberg, Variational Methods For the Study of Nonlinear Operators, Holden-Day, San Francisco, 1964.

41. L. C. Young, Lectures on the Calculus of Variations and Optimal Control Theory, Saunders, Philadelphia, 1969.

42. E. M. Zarantonello, Solving functional equations by contractive averaging, Mathematics Research Center Report 160, University of Wisconsin, Madison, 1960.

Differentiability and Related Properties of Nonlinear Operators: Some Aspects of the Role of Differentials in Nonlinear Functional Analysis

M. Z. NASHED

Dedicated to Professor E. H. Rothe on the occasion of his 75th birthday

"If one seeks a guiding thread through his many papers, it is perhaps the extensions to general spaces of the ideas of elementary calculus, in order to answer basic questions arising in partial differential equations and integral equations..."

W. Kaplan

CONTENTS

Introduction 106

I. Some Aspects of Differential Calculus in Normed and Topological Spaces 109

    1. Differentials and Variations 109

        1.1. Gâteaux Variation and Gâteaux Differential 109

        1.2. Fréchet Differential 115

        1.3. Hadamard Differential 123

	1.4. Gradients	130
	1.5. Strong and Bounded Differentials and Related Notions	133
	1.6. Equidifferentiability.	135
	1.7. Differentiation Along a Subspace and Partial Differentiation	136
2.	Differentials and Variations: Revisited	140
	2.1. Differentials in Linear Topological Spaces .	140
	2.2. The Axiomatic or Descriptive Approach to Differentials	143
	2.3. (H,σ)-Differentiability and Some Specific Realizations	146
	2.4. (H,σ)-Continuity and Partial Derivatives. .	150
3.	Higher Order Differentials and Variations	151
4.	Higher Order Differentials in Linear Topological Spaces	166
5.	Mean Value Theorems in Banach and Linear Topological Spaces	171
6.	Taylor's Theorems and Formulas in Banach and Locally Convex Spaces	187
7.	Converses of Taylor's Theorems and Characterizations of Higher Order Differentiability	193
8.	Direct Higher Order Differentials and Difference Differentials. Riemann, Peano, Taylor, and Difference Differentials	199

9. Remarks on Differentiability in Complex Linear Topological Spaces 205

10. Remarks on Weak Differentials 210

II. Some Aspects of the Role of Differentials in Nonlinear Functional Analysis 214

11. The Role of Differentials in Extrema of Functionals: Infinitesimal Conditions for Continuity and Lower Semicontinuity in the Weak Topology . 214

12. Tangents and Cones of Tangents 233

13. Gradient Mappings. Integrability and Hyperbolic Integrability 237

14. Compact, Collectively Compact Operators and Differentiability 247

15. Monotone Operators and Convex Functionals: Differential Characterizations 254

16. Numerical Range and Spectrum of Some Classes of Nonlinear Operators. Spectral Radii and Points of Attraction 259

17. Asymptotic Derivatives, Quasibounded Operators and Related Applications 265

18. One-Sided and Logarithmic Derivatives and Norms 268

References 272

Introduction

From the days when the derivative was vaguely regarded as just a rate of change, a velocity, or a slope (to cite a few definitions!), to the present day of differential calculus in the settings of topological linear spaces and infinite dimensional manifolds, there is no single concept which has profoundly influenced the development of mathematical ideas as much as the concept of a differential. From the viewpoint of applications, differentials are perhaps the most useful and often essential tool for the analysis of a variety of nonlinear problems.

The purpose of this paper is to give a systematic exposition of several aspects of differential calculus in normed and topological linear spaces, and to provide glimpses into various settings in nonlinear functional analysis in which differentials play an important role. The paper consists of two parts. The table of contents gives a general idea of the topics considered.

Historically the beginnings of the theory of differentiation in infinite dimensional spaces go back to Volterra [67] who in 1887 introduced the concept of the variational derivative, and to Hadamard [46-48], Fréchet [34, 24-31], Gâteaux [36, 37], Lévy [55, 56] and others (see references under B in I at the end of this paper). While the theory of differentiation in normed spaces was being developed in the early part of this century, there was also another group of mathematicians concerned with the extensions of other aspects of calculus to nonlinear problems in function spaces. Graves [42] extended Taylor's theorem to normed spaces, Hildebrandt and Graves [44] generalized the implicit function theorem, Kerner [52] introduced integrability conditions of abstract fields and generalized "Stokes' theorem" to Hilbert spaces. Golomb [41] formulated the notion of gradient in function spaces and Goldstine [275,276] gave a "multiplier rule" in abstract spaces. Rothe [64, 342, 343] studied topological properties of gradient mappings and their role in extrema of functionals. Topological methods in nonlinear problems were also developed by Birkhoff and Kellogg, Leray and Schauder, Nemytskii,

THE ROLE OF DIFFERENTIALS

Tikhonov, Krasnoselskii and others (see [295, 296]). Variational methods were developed in the work of Hammerstein, Golomb, Lichtenstein, Liusternik, Rothe, Sobolev and others (see [11], [337]). The 1920's and early part of the 1930's witnessed considerable interest in nonlinear problems in the setting of function spaces as indicated by the appearance of the celebrated paper of Birkhoff and Kellogg on invariant points in function spaces, the work of Graves and Hildebrandt, the development of differential calculus and the Leray-Schauder theory. Indeed it may be said that the interest in that period was in nonlinear rather than linear functional analysis. A year after the notion of a linear topological space was given, several definitions of differentials on such spaces were formulated by Michal and Paxson [110], Michal [106], and later by Hyers [96] and others. However a comprehensive theory for differentiation in topological spaces was not advanced until considerably later, as discussed in Section 2.

Differentials play at least three significant roles in the theory and applications of nonlinear functional analysis. The first and most familiar role is that of <u>local approximation</u> of a nonlinear map by a linear one; the sense of the approximation being determined by the notion of differential used. More generally, one may consider differentiation as a problem of <u>approximation</u> of a map $F: X \to Y$ <u>by a map from a given class</u> of operators. The sense of the approximation, which is a generalization of the usual order condition on the "remainder", is made precise by specifying the class of maps r which are considered "infinitesimal", i.e. $r(h)$ is small in comparison with h. All the differentials introduced in Sections 1 and 2 are examined from the type of representations they give for $F(x_0 + h) - F(x_0)$.

The differential is also important as a <u>tool</u> in nonlinear functional analysis. Often a certain property of a differentiable operator can be deduced from a related property of its derivative. Such a situation usually leads to amenable sufficient conditions for the operator to have that property. The setting and results of several sections in Part II of this paper are intended to illustrate this role of the differential in nontrivial situations.

The third role of the differential is in the <u>clue</u> it provides <u>for generalizations</u> of a certain class of differentiable operators having certain properties to a class of not necessarily differentiable operators having similar properties, and thereby providing a nonlinear analog of a class of linear operators having a certain property. For example a simple nonlinear analog of a nonnegative linear operator on a Hilbert space is a differentiable operator whose derivative is a nonnegative linear operator, which ultimately led to the notion of monotone operators. Another prototype of this process of generalization via properties of the derivative is the transition from a self-adjoint linear operator to a differentiable operator with a self-adjoint derivative and then to the general notion of a potential operator. We shall have several occasions in Part II to consider this process of generalization.

Part I deals with differential calculus <u>per se</u> in the settings of normed and linear topological spaces. The list of references for this part is reasonably complete relative to the topics considered. In contrast the bibliography in Part II is quite selective.

There are many measure-theoretic questions in the theory of nonlinear operators which have not been investigated. Such questions are suggested by functions of a real variable along the line of the references listed in Section H of the bibliography. Such investigations for nonlinear operators do not seem to have any applications at present. The theory of differentiation in linear topological spaces is fairly complete for practical purposes. It seems therefore that the main thrust of research in this area should now be directed toward applications of differential calculus to "hard" analysis in function spaces, as in [207]-[218]. A beautiful model for this is the paper of Donsker and Lions [211] which gives a profound interpretation of Volterra's variational derivative and demonstrates the power of this abstract notion in the formulation and solution of problems in concrete analysis.

For general background in functional analysis that is needed for this paper, we refer to Kantorovich and Akilov [8]; other useful references are given on p. 272.

THE ROLE OF DIFFERENTIALS

I. Some Aspects of Differential Calculus in Normed and Topological Spaces

1. Differentials and Variations

1.1. Gâteaux Variation and Gâteaux Differential

Let I be an open interval of the real line and X be a normed real space. The <u>derivative</u> of a mapping $\Phi: I \to X$ at $t_0 \in I$ is defined by $\lim_{t \to t_0} \dfrac{\Phi(t) - \Phi(t_0)}{t - t_0}$, if this limit exists, and is denoted by $\Phi'(t_0)$. The limit is to be understood in the sense of the norm in X, i.e., $\left\| \dfrac{\Phi(t) - \Phi(t_0)}{t - t_0} - \Phi'(t_0) \right\| \to 0$ as $t \to 0$. Φ is said to have a <u>weak derivative</u> at t_0 if there exists an element $u(t_0) \in X$ such that for each continuous linear functional ℓ on X,

$$\lim_{t \to t_0} \ell\left[\frac{\Phi(t) - \Phi(t_0)}{t - t_0} \right] = \ell[u(t_0)] .$$

The notion of a weak derivative will not be used in this paper, except in Section 10.

Let X and Y be normed real spaces and $\mathcal{U}$ an open subset of X. Let $x_0 \in \mathcal{U}$ and h be a fixed nonzero element in X. Since $\mathcal{U}$ is open there exists an interval $I = (-\tau, \tau)$ for some $\tau > 0$ such that if $t \in I$, then $x_0 + th \in \mathcal{U}$. If the mapping $\Phi: I \to X$ defined by $\Phi(t) = F(x_0 + th)$ has a derivative at $t = 0$, then $\Phi'(0)$ is called the <u>Gâteaux variation</u> of F at x_0 with increment h and is denoted by $\delta F(x_0; h)$, i.e.,

$$\delta F(x_0; h) = \frac{d}{dt} F(x_0 + th) \bigg|_{t=0} = \lim_{t \to 0} \frac{1}{t} \{F(x_0 + th) - F(x_0)\} .$$

Note that this equation may be used to define $\delta F(x_0; h)$ when X is any linear space, not necessarily normed. Clearly

$\delta F(x_0; h)$ may exist for some h, but fail to exist for others. However, $\delta F(x_0; h)$ is homogeneous in h of degree one, i.e., if $\delta F(x_0; h)$ exists for some $h \neq 0$, then for each real number λ, $\delta F(x_0; \lambda h)$ exists and is equal to $\lambda \delta F(x_0; h)$. From here on, unless otherwise stated, we shall take the Gâteaux variation to mean that $\delta F(x_0; h)$ exists for every $h \in X$. The Gâteaux variation, which was introduced for functionals by Gâteaux [36] in 1913, is a generalization of the notion of the directional derivative in calculus and of the notion of the first variation arising in the calculus of variations (see, for instance, [16], [17]). The existence of the Gâteaux variation at $x_0 \in \mathcal{U}$ provides a local approximation property in the following sense:

<u>Proposition 1.1.</u> Let $F: \mathcal{U} \subset X \to Y$. A necessary and sufficient condition for F to have a Gâteaux variation at $x_0 \in \mathcal{U}$ is that the following representation holds for every $h \in X$ for which $x_0 + h \in \mathcal{U}$:

$$(1.1) \qquad F(x_0 + h) - F(x_0) = H(x_0; h) + r(x_0; h)$$

where the map $h \to H(x_0; h)$ is homogeneous of degree one and

$$(1.2) \qquad \lim_{t \to 0} \frac{r(x_0; th)}{t} = 0$$

for each fixed h.

Note that if such a representation exists, it is unique and $\delta F(x_0; h) = H(x_0; h)$. The existence of $\delta F(x_0; h)$ implies the <u>directional</u> continuity of F at x_0, i.e.,

$$(1.3) \qquad \|F(x_0 + th) - F(x_0)\| \to 0 \text{ as } t \to 0 \text{ for fixed } h,$$

but does not necessarily imply that F is continuous at x_0. This is tantamount to saying that (1.3) does not necessarily hold uniformly with respect to h on the bounded set $\{h: \|h\| = 1\}$. Note also that the operator $h \to \delta F(x_0; h)$ is not necessarily linear nor continuous in h.

Example 1.1. Let $f(x_1, x_2) = x_1^2(1 + \frac{1}{x_2})$ for $x_2 \neq 0$ and $f(x_1, 0) = 0$. Let $x = (x_1, x_2)$, $h = (h_1, h_2)$, $h_2 \neq 0$, and $F(x) = f(x_1, x_2)$. Then $\delta F(0; h) = \lim_{t \to 0} \frac{F(th)}{t} = \frac{h_1^2}{h_2}$ is neither continuous nor linear in h.

Example 1.2. Let F be any homogeneous mapping of degree $s \geq 1$, i.e., $F(\lambda x) = \lambda^s F(x)$. Then F has a first variation at 0 and $\delta F(0; h) = F(h)$.

Example 1.3. Let E be the class of all functions y such that y is continuous on [0,1], y' and y" are continuous on [0,c) and (c,1] where c is a fixed point in (0,1) and y', y" may have a jump discontinuity at $x = c$, and $y(0) = 0$. Let J be the functional

$$J[y] = \int_0^1 [y'(x)]^2 dx + y(c),$$

defined on the space E, where y(c) is not prescribed a priori. Then the first variation of J at $\hat{y} \in E$ is given by

$$\delta J[\hat{y}; h] = 2 \int_0^{c^-} \hat{y}'h' dx + 2 \int_{c^+}^1 \hat{y}'h' dx + h(c) = 2\hat{y}'(c^-)h(c) - 2 \int_0^{c^-} \hat{y}''(x)h(x) + 2\hat{y}'(x)h(x) \Big|_{c^+}^1 - 2 \int_{c^+}^1 \hat{y}''(x)h(x) dx + h(c)$$

for all $h \in E$. In particular if $\hat{y}$ minimizes the functional $J[y]$, then $\delta J[\hat{y}; h] = 0$ for all admissible h, in this case for all $h \in E$. Choose $h \in E$ which vanishes on $[0, c]$. Then we get from the preceding equation

(1.4) $$\hat{y}'(1) h(1) - \int_{c^+}^1 \hat{y}''(x) h(x) dx = 0$$

for all h as chosen. Taking h in this class with $h(1) = 0$ we conclude that $\hat{y}'' = 0$ for $c < x < 1$. Similarly $\hat{y}'' = 0$ for $0 < x < c$. From (1.4) we then arrive at the so-called <u>natural boundary condition</u> $\hat{y}'(1) = 0$. The vanishing of the first variation for all $h \in E$ yields

$$2\hat{y}'(c^-) h(c) + 2\hat{y}'(1) h(1) - 2\hat{y}'(c^+) h(c) + h(c) = 0 .$$

Thus $\hat{y}$ satisfies the Euler equation

$$y''(x) = \delta(x-c) ,$$

where $\delta(x)$ is the Dirac generalized function, and the boundary conditions $y(0) = 0$, $2y'(c^-) - 2\hat{y}'(c^+) + 1 = 0$. Hence

$$\hat{y}(x) = \begin{cases} -\frac{1}{2} x & \text{for } 0 \leq x \leq c \\ -\frac{1}{2} c & \text{for } c \leq x \leq 1 \end{cases} .$$

Note that if F and G have a Gâteaux variation at x_0, then so does $T = \alpha F + \beta G$ for any real numbers α, β and $\delta T(x_0;h) = \alpha \delta F(x_0;h) + \beta \delta G(x_0;h)$. However the chain rule for the differentiation of a composite function fails in general.

<u>Example 1.4.</u> Let $f: \mathbb{R} \to \mathbb{R}^2$ be defined by $f(t) = (t, t^2)$ and $g: \mathbb{R}^2 \to \mathbb{R}$ be defined by $g(x,y) = x$ if $y = x^2$, and 0 otherwise. Then $(g \circ f)x = x$, f is differentiable at 0, g has a Gâteaux variation at $(0,0)$ which is equal to zero, but $(g \circ f)'(0) = 1$ and not 0.

Things can even get worse: The composite $g \circ f$ of two functions each of which has a variation may not have a variation!

<u>Example 1.5.</u> Let (r, θ) be the polar coordinates in the plane and define $f: \mathbb{R} \to \mathbb{R}^2$ by $r(t) = t$, $\theta(t) = t$, and $g: \mathbb{R}^2 \to \mathbb{R}$ by $g(r, \theta) = \frac{r^2}{\theta^3}$ for $0 < \theta < 2\pi$ and 0 for $\theta = 0$. Then f has a total differential at 0 and g has a Gâteaux variation

at $(0,0)$ but $g \circ f: \mathbb{R} \to \mathbb{R}$ is discontinuous at 0.

Several mathematicians, beginning with Gâteaux himself, found conditions ensuring that the Gâteaux variation is linear. Daniell [21] in 1919 obtained a sufficient condition in the case of the space $C[0,1]$ of all continuous functions: If $f: C[0,1] \to \mathbb{R}$ satisfies a Lipschitz condition $|f(x_1) - f(x_2)| \le M \max_{0 \le t \le 1} |x_1(t) - x_2(t)|$ and if $\delta f(x; h)$ exists in some neighborhood of $x_0 \in C[0,1]$ (in the maximum norm), then $\delta f(x_0; \cdot)$ is linear. Other sufficient conditions for the linearity (and continuity) of the Gâteaux variation in other classes of function spaces are also well known (see, for instance, Vainberg [11]). We now digress to state two theorems which guarantee linearity of the Gâteaux variation.

Theorem 1.1. If F has a Gâteaux variation $\delta F(x; \cdot)$ in some neighborhood W of x_0 and $\delta F(x; \cdot)$ is continuous at x_0, then $\delta F(x_0; \cdot)$ is a linear operator.

Proof: See Vainberg [11; Theorem 3.1].

Theorem 1.2. Suppose that F has a Gâteaux variation at x_0. A necessary and sufficient condition for $\delta F(x_0; h)$ to be linear and continuous in h is that F satisfies the following two conditions:

(a) To each h corresponds a $\delta(h)$ such that

$$|t| \le \delta \text{ implies } \|F(x_0 + th) - F(x_0)\| \le M \|th\|,$$

where M does not depend on h; and

(b) $\Delta^2_{th_1, th_2} F(x_0) = o(t)$, where

$$\Delta^2_{h_1, h_2} F(x_0) = F(x_0 + h_1 + h_2) - F(x_0 + h_1) - F(x_0 + h_2) + F(x_0).$$

For a proof see Vainberg [11; Theorem 3.2]. Other conditions guaranteeing linearity and continuity of the Gâteaux variation are given in Sections 1.2 and 1.3.

We say that F has a <u>Gâteaux differential</u> at x_0 if $\delta F(x_0; \cdot)$ is linear and continuous. In this case $\delta F(x_0; \cdot)$ is denoted by $DF(x_0)$ and is called the <u>Gâteaux derivative</u>.[†] Paul Lévy postulated the linearity of the Gâteaux variation in his book, "Leçons d'analyse fonctionelle" (1922); thus perhaps $DF(x_0; h)$ should be called the <u>Gâteaux-Lévy</u> differential. Note that the chain rule does not necessarily hold for Gâteaux derivatives (see Example 1.4). Note also that if $\delta F(x_0; \cdot)$ is additive, then $\delta F(x_0; h)$ is directionally continuous in h, i.e.,

(1.5) $$\lim_{\tau \to 0} \delta F(x_0; h + \tau k) = \delta F(x_0; h) .$$

An additive Gâteaux variation is a Gâteaux differential if and only if (1.5) holds uniformly with respect to k on the set $\|k\| = 1$.

Several notions of variations may be obtained also by varying the "difference quotient" in the definition of the Gâteaux variation. We introduce, for instance, the variation analogs of the Sindalovskii [137] and Murav'ev [216] derivatives. Let σ be an operator defined in a neighborhood of zero such that $\sigma(th) \to 0$ as $t \to 0$ for each fixed h. We define the <u>Sindalovskii variation</u> by $\lim t^{-1}\{F(x - \sigma(th)] - F[x - \sigma(th) - th]\}$ as $t \to 0$. Let G be any continuous operator. We define the <u>Murav'ev variation</u> of F (with respect to G) by $\lim_{t \to 0} t^{-1}\{F[x + th G(x)] - F(x)\}$. The Gâteaux variation is obtained as a special case of the Sindalovskii and Murav'ev variations. Other properties of these variations follow readily. For an excellent survey of various notions and properties of derivatives of a function of a real variable, see Bruckner and Leonard [123]. The various notions of a derivative of a function of a real variable may be used to define

---

[†] Some authors refer to the variation $\delta F(x; h)$ as the Gâteaux differential, and then use the phrase "linear Gâteaux differential" whenever $\delta F(x; \cdot)$ is linear.

analogous notions of variations of functionals or operators. A notion which is weaker than the Gâteaux variation is obtained if for instance the <u>Schwarz</u> derivative $\Phi^{[']}(0)$ is used instead of $\Phi'(0)$. (Recall that $\Phi^{[']}(0) = \lim\limits_{t \to 0} \frac{\Phi(t) - \Phi(-t)}{2t}$). We shall not pursue these variants in this paper.

1.2. Fréchet Differential

Let X and Y be normed real linear spaces and let $\mathcal{L}(X, Y)$ denote the space of all continuous linear operators on X to Y, with the usual norm.

<u>Definition 1.1.</u> A map $F: \mathcal{U} \to Y$, where $\mathcal{U}$ is an open subset of X, is said to be Fréchet differentiable at $x_0 \in \mathcal{U}$ if there exists a continuous linear operator $L(x_0): X \to Y$ such that the following representation holds for every $h \in X$ with $x_0 + h \in \mathcal{U}$,

(1.6) $\qquad F(x_0 + h) - F(x_0) = L(x_0)h + r(x_0; h)$

where

(1.7) $\qquad \lim\limits_{h \to 0} \frac{\| r(x_0; h) \|}{\| h \|} = 0$.

<u>Remark.</u> Historically, several mathematicians, besides Fréchet, also contributed to the formulation of the definition of a differential in this form. Until the end of the nineteenth century, even the notion of a differential of a function of several variables was not well formulated and only partial derivatives were considered. Stolz (1893), Pierpont (1905) and Young (1910) defined the differential of a function of several variables as follows: f is differentiable at $(x_1, \ldots, x_n)$ if $A_i = \frac{\partial f}{\partial x_i}$ exist at $(x_1, \ldots, x_n)$ and

$$f(x_1 + h_1, \ldots, x_n + h_n) - f(x_1, \ldots, x_n) = \sum_{i=1}^{n} A_i h_i + \sum_{i=1}^{n} \varepsilon_i h_i$$

where $\varepsilon_i \to 0$ as $\max(|h_1|,\ldots,|h_n|) \to 0$, $i = 1,\ldots,n$. This definition portrays of course the idea of linear approximation which was repeatedly stressed by Hadamard as a basis for any satisfactory definition of a differential. As a first step toward freeing this definition from the coordinates setting, Fréchet, who was a student of Hadamard, replaced $\sum_{i=1}^{n} \varepsilon_i h_i$ in the above representation by εD, where D is the "distance" between $(x_1,\ldots,x_n)$ and $(x_1+h_1,\ldots,x_n+h_n)$. For "distance", Fréchet used $D = \max(|h_1|,\ldots,|h_n|)$, or $D = \sqrt{h_1^2 + \ldots + h_n^2}$. In 1911, Fréchet [34] wrote: "The functional U_A has a differential at the point A_0 if there is a functional $V_{\Delta A}$ that is linear in the ΔA and differs from the increment of the functional U_A at A_0 by a quantity that is infinitely small in comparison with the distance between the arguments A_0 and $A_0 + \Delta A$." Fréchet obviously had in mind distance induced by a norm (all the examples he gave were in normed spaces, although such a notion was not yet formulated). The definition in quotation is unsatisfactory in linear metric spaces, for if we consider the space of real numbers with the metric $\rho(x,y) = |x-y|^{\frac{1}{2}}$, then the identity map has for its derivative at zero any real number λ. In 1925, Fréchet [25] arrived at a more precise definition of a differential which coincides with Definition 1.1. In this context, differentials appear in the early work of Graves and Hildebrandt [42], [44].

The uniqueness of the representation (1.6)-(1.7) and the implication relationships between the Fréchet and Gâteaux differentials follow from the following simple proposition.

<u>Proposition 1.2.</u> Let $\psi: X \to Y$, where X and Y are normed linear spaces. Then for any real number r, the following are equivalent:

(i) $\qquad \dfrac{\|\psi(h)\|}{\|h\|^r} \to 0$ as $h \to 0$

(ii) $\dfrac{\|\psi(\tau h)\|}{\tau^r} \to 0$ as $\tau \to 0$, $\tau \in \mathbb{R}$, $\tau \neq 0$, <u>uniformly with respect to</u> h on the bounded set $S = \{h: \|h\| = 1\}$.

We call the unique $L(x_0)h$ in Definition 1.1 the Fréchet differential of F at x_0 and denote it by $dF(x_0; h)$. The operator $F'(x_0) \in \mathcal{L}(X,Y)$ defined by $h \to dF(x_0; h)$ is called the <u>Fréchet derivative</u> of F <u>at</u> x_0; we write $dF(x_0; h) = F'(x_0)h$. If F has a Fréchet differential at each $x \in W$, then the map $F': W \to \mathcal{L}(X,Y)$ is called the Fréchet derivative of the operator F.

<u>Corollary 1.1.</u> An operator F is Fréchet differentiable at x_0 if and only if F is Gâteaux differentiable at x_0 and (1.2) holds uniformly with respect to h on the set $S = \{h: \|h\|=1\}$. In this case the two differentials coincide.

The Fréchet differential has the usual properties of the classical differential of a function of one or several variables. In particular, the chain rule holds. (Dieudonné [5]; this also follows as a special case of the corresponding assertion for Hadamard differential (Theorem 1.8)). Also the chain rule for $F \circ G$ holds if F has a Fréchet differential and G has a Gâteaux differential. Note that the chain rule may not hold for $F \circ G$ if F has a Gâteaux differential and G has a Fréchet differential (see Example 1.4). Fréchet differentiability of F at x_0 implies continuity of F at x_0. On the other hand, if we start with an operator F which is continuous at x_0, then the requirement of continuity of L in Definition 1.1 is redundant: the inequality

$$\|dF(x_0;h)\| \leq \|F(x_0+h) - F(x_0) - dF(x_0;h)\| + \|F(x_0+h) - F(x_0)\|$$

shows that $dF(x_0; \cdot)$ is continuous at zero and hence continuous everywhere.

<u>Remark.</u> The norm $\|\cdot\|'$ is said to be equivalent to the norm $\|\cdot\|$ on X if there exist positive numbers m and M such that
$$m\|x\| \leq \|x\|' \leq M\|x\|$$
for all $x \in X$. Obviously this defines an equivalence relation.

The definition of the Fréchet differential is given in terms of the norms on X and Y. However, it is evident that Fréchet differentiability is invariant under equivalent norms, i.e., if F is Fréchet differentiable at x_0 when the linear spaces X and Y are normed by $\|\cdot\|_X$ and $\|\cdot\|_Y$ respectively, then it is also Fréchet differentiable at x_0 when the spaces X and Y are normed by $\|\cdot\|_X'$ and $\|\cdot\|_Y'$, which are equivalent to $\|\cdot\|_X$ and $\|\cdot\|_Y$ respectively; the two differentials are equal. In the case of a finite dimensional space, all norms are equivalent; thus differentiability of mappings between finite dimensional normed spaces is independent of the norms used. Equivalent norms induce the same topology so that differentiability depends only on the topologies of X and Y in infinite dimensional spaces; this enables one to extend the definitions of Fréchet and Gâteaux differentials to linear topological spaces.

<u>Example 1.6.</u> Let $f(x_1, x_2) = \dfrac{x_1^3 x_2}{x_1^4 + x_2^2}$ if $(x_1, x_2) \neq (0,0)$ and 0 if $(x_1, x_2) = (0,0)$. Let $x = (x_1, x_2)$ and $F(x) = f(x_1, x_2)$. Then for $h = (h_1, h_2) \neq (0,0)$, $\delta F(0; h) = \lim\limits_{t \to 0} \dfrac{f(th_1, th_2)}{t} = 0$. If $(h_1, h_2) = (0,0)$, then the limit is also zero. Thus $\delta F(0;h)$ exists and is obviously continuous and linear in h so that F has a Gâteaux differential at 0. However in the representation (1.7), $r(0; h) = \dfrac{h_1^3 h_2}{h_1^4 + h_2^2}$ if $h \neq 0$, and $\lim\limits_{h \to 0} \dfrac{r(0, h)}{\|h\|} = \lim\limits_{(h_1, h_2) \to (0,0)} \dfrac{h_1^3 h_2}{(h_1^4 + h_2^2)(h_1^2 + h_2^2)^{\frac{1}{2}}}$ does not exist, and thus F is not Fréchet differentiable at 0. The same conclusion can be reached by noting that $\lim\limits_{\tau \to 0} \dfrac{r(0; \tau h)}{\tau}$ (which in this case is equal to zero for each <u>fixed</u> h) is not uniform with respect to h in the set $\{h: \|h\| = 1\}$.

<u>Example 1.7.</u> Consider the functional $J(x) = \int_0^1 f(t, x(t), x'(t)) dt$

where f has continuous and has continuous second partial derivatives. Let $C_0'[0,1]$, $C_1'[0,1]$ denote the space of all continuously differentiable functions on $[0,1]$, normed respectively by $\|x\|_0 = \max_{0 \le t \le 1} |x(t)|$ and $\|x\|_1 = \max_{0 \le t \le 1} \{|x(t)|, |x'(t)|\}$.

$\delta J(x;h)$ exists and $\delta J(x;h) = \int_0^1 (f_x h + f_{x'} h') dt$, which is linear in h. However it is not necessarily continuous in h on the space $C_0'[0,1]$. $\delta F(x; \cdot)$ is continuous on the space of twice continuously differentiable functions with the norm $\|x\|_0$. Yet in this space, $\delta F(x;h)$ is not necessarily a Fréchet differential since it may not even be continuous as the arc length functional $\int_0^1 \sqrt{1+[x'(t)]^2}\, dt$ shows. However, in the space $C_1'[0,1]$, J is Fréchet differentiable for each $x \in C_1'[0,1]$.

Example 1.8. Let $F: \mathbb{R}^n \to \mathbb{R}^m$, i.e., $y_i = F_i(x_1, \ldots, x_n)$, $i = 1, 2, \ldots, m$ relative to the standard bases. If F has a <u>linear</u> directional derivative at $a = (a_1, \ldots, a_n)$, i.e.

$$\lim_{t \to 0} \frac{F(a+th) - F(a)}{t}$$

exists and is linear in h, then the Gâteaux derivative is given by the linear transformation whose matrix is $[\frac{\partial F_i}{\partial x_j}]$ (evaluated at a). However the mere existence of the partial derivatives $\frac{\partial F_i}{\partial x_j}$ at a does not necessarily mean that $[\frac{\partial F_i}{\partial x_j}]$ is a Gâteaux derivative, and the formal expression $[\frac{\partial F_i}{\partial x_j}](h_1, \ldots, h_n)^T$ in this case is not necessarily a Gâteaux differential, i.e., it does not necessarily provide a linear approximation to $F(x_0 + h) - F(x_0)$ in the sense of (1.2). On the other hand, if there exists an

$r > 0$ such that for all x with $\|x-a\| \le r$, $\dfrac{\partial F_i}{\partial x_j}$ exist and are continuous at $x = a$, then F has a Fréchet differential at a. (See, for instance, [10], [12], [13]).

<u>Example 1.9</u>. Consider the operator T defined on the space $C[0,1]$ of all continuous functions by

$$T(x) = x(s) \int_0^1 (s^2 + t^2) x(t) dt .$$

Then $\delta T(x;h) = x(s) \int_0^1 (s^2 + t^2) h(t) dt + h(s) \int_0^1 (s^2 + t^2) x(t) dt .$

Thus the first variation is linear in h. It is easy to show that it is also continuous. This follows from

$$\|\delta T(x;h)\| = \max_{1 \le s \le 1} |\delta T(x;h)| =$$

$$\max_{1 \le s \le 1} \left| x(s) \int_0^1 (s^2 + t^2) h(t) + h(s) \int_0^1 (s^2 + t^2) x(t) dt \right|$$

$$\le \|x\| \|h\| \max_{1 \le s \le 1} \int_0^1 (s^2 + t^2) dt +$$

$$\|h\| \max_{0 \le s \le 1} \left| \int_0^1 (s^2 + t^2) x(t) dt \right| ,$$

which shows that $\delta T(x;h)$ is bounded, hence continuous. Similarly it is easy to show that

$$\|r(x;h)\| \le C \|h\|^2, \text{ where } C = \max_{1 \le s \le 1} \int_0^1 (s^2 + t^2) dt$$

and $r(x;h) = T(x+h) - T(x) - \delta T(x;h)$. Thus $\lim_{h \to 0} \dfrac{\|r(x;h)\|}{\|h\|} = 0$.

Example 1.10. Let $K(s,t)$ be a continuous real function for $0 \leq s, t \leq 1$. The functional

$$J(x) = \int_0^1 x^2(t)dt - \lambda \int_0^1 \int_0^1 K(s,t)\, x(s)\, x(t)ds\, dt - 2\int_0^1 x(t)y(t)dt$$

is defined on the space $C[0,1]$ with the norm $\|x\| = \max_{0 \leq t \leq 1} |x(t)|$. Here y is a fixed element in $C[0,1]$. The Gâteaux variation

$$\frac{d}{d\tau} J(x+\tau h)\bigg|_{\tau=0} = 2\int_0^1 x(t)h(t)dt - \lambda \int_0^1 \int_0^1 [K(s,t)+K(t,s)]x(s)h(t)dsdt$$

$$- 2\int_0^1 h(t)\, y(t)dt ,$$

is linear and continuous in h. Condition (1.7) is also satisfied, and thus J is Fréchet differentiable.

Note that the operator T and the functional J in the preceding two examples are quadratic; thus their Fréchet differentiability can be asserted and their differentials computed directly using standard differentiation rules. See also the companion papers of Prenter and Tapia in this volume.

In the calculus of several variables the existence and continuity of the partial derivatives is a sufficient condition for the existence of the total differential. The following theorem is the analogue of this result in normed spaces.

Theorem 1.3. Let X and Y be normed linear spaces, U an open subset of X, $F: U \to Y$. If F has a Gâteaux derivative $F'(x)$ which is continuous in x at x_0 in the operator topology, i.e. if the map $F': U \to \mathcal{L}(X,Y)$ is continuous at x_0, then F is Fréchet differentiable at x_0.

Proof: From the fundamental theorem of calculus (see p. 171 of this paper),

$$F(x_0+h) - F(x_0) = \int_0^1 \delta F(x_0 + th; h) dt$$

$$= \delta F(x_0; h) + \int_0^1 \{\delta F(x_0 + th; h) - \delta F(x_0; h)\} dt$$

for all $h \in X$ with $x_0 + h \in U$. But

$$\frac{1}{\|h\|} \| \int_0^1 \{\delta F(x_0+th;h) - \delta F(x_0;h)\} dt \| \leq \int_0^1 \|F'(x_0+th) - F'(x_0)\| dt$$

and the right hand side tends to zero as $h \to 0$, by assumption.

This proof is considerably shorter than the proof given in Vainberg [11; Theorem 3.3]. The next theorem follows from the proof of Theorem 1.3 and Proposition 1.2.

Theorem 1.4. Let $F: U \to Y$ have a Gâteaux differential at x_0 and suppose that $\delta F(x;h)$ exists in a neighborhood of x_0. If $\lim_{t \to 0} \|\delta F(x_0 + th; h) - \delta F(x_0; h)\| = 0$ uniformly with respect to $h \in X$ on $\{h: \|h\| = 1\}$, then F is Fréchet differentiable at x_0.

Theorem 1.4 generalizes a theorem of Marinescu [132]. Liusternik and Sobolev [9; Chapter 6, Section 3] showed that if $\delta F(x;h)$ is continuous in h and uniformly continuous in a neighborhood of x_0, then F is Fréchet differentiable at x_0.

We conclude this section by stating other sufficient conditions under which a Gâteaux variation is a Fréchet differential. See also Section 1.3, and references [129], [130], [142], [143] for other sufficient conditions.

Theorem 1.5 [129; Theorem 2]. Let E be a reflexive space and $F: E \to E$. If $\delta F(x;h)$ exists in a convex neighborhood of x_0 and if $\delta F(x;h)$ is jointly strongly continuous at $(x_0;h)$ for each $h \in E$, i.e. if $x_n \xrightarrow{W} x_0$ and $h_n \xrightarrow{W} h$ ($\xrightarrow{W}$ denotes weak convergence) imply $\delta F(x_n, h_n) \to \delta F(x,h)$. Then F has a Fréchet derivative at x_0.

Corollary. Let E be a finite dimensional normed space, $F: E \to E$. If $\delta F(x;h)$ exists for all x in a convex neighborhood of $x_0 \in E$ and $\delta F(x;h)$ is jointly continuous at (x_0, h), $h \in E$, then F has a Fréchet derivative at x_0.

Remark. Alexiewicz and Orlicz [122] gave an example of an operator F on a separable Banach space into itself, satisfying a Lipschitz condition, having everywhere a Gâteaux variation $\delta F(x, h)$ jointly continuous in x and h, and nowhere Fréchet differentiable. Another such example was given by Vainberg [142]. Thus strong joint continuity of $\delta F(x;h)$ cannot be replaced in Theorem 1.5 by joint continuity even if F satisfies a Lipschitz condition.

A mapping $F: X \to Y$ is called <u>completely compact</u> on a bounded set $K \subset X$ if F is uniformly continuous on K and compact (maps bounded sets into compact sets). The following property is equivalent to complete compactness of F: If $\{x_n\}$ and $\{y_n\}$ are sequences of K such that $\lim_{n \to 0} \|x_n - y_n\| = 0$, then there exist subsequences $\{x_{n_k}\}$, $\{y_{n_k}\}$ such that

$$\lim_{k \to \infty} F(x_{n_k}) = \lim_{k \to \infty} F(y_{n_k}) = u \in Y.$$

Theorem 1.6 [143; Theorem 3]. Let $F: X \to Y$ and $S_r = \{x \in X: \|x\| \leq r\}$ for $r > 0$. Suppose that the Gâteaux variation $\delta F(x;h)$ exists for every $x \in x_0 + S_r$. If $\delta F(x;h)$ is completely compact in $(x_0 + S_r) \times S_r \subset X \times X$, then F has a Fréchet derivative at $x_0 \in X$.

1.3. Hadamard Differential

In 1923 Hadamard published a note that contained the germ of a new definition of differential, which later motivated Fréchet to give in 1937 another definition of differential for functionals on the space of functions. We quote from [27]: "The functional $U[f]$ is differentiable at $f = f_0$ in the generalized sense of Hadamard if there is a functional

$W[df, f_0]$ linear in df such that if the function $f(t, \lambda)$ is differentiable with respect to λ at $\lambda = 0$ and $f(t, 0) = f_0(t)$, then the function $U[f(t, \lambda)]$ of λ is differentiable in λ for $\lambda = 0$ and

$$\frac{d}{d\lambda} U[f(t, \lambda)]\Big|_{\lambda=0} = W[\frac{df}{d\lambda}, f_0],$$

or, in the notation of "variations", $\delta U[f] = W[\delta f, f_0]$".

In the setting of normed spaces, we may formalize this as follows.

Definition 1.2. An operator $F: X \to Y$ is said to have an Hadamard differential at x_0 if there exists a continuous linear mapping $L: X \to Y$ such that for any continuous mapping $g: [0,1] \to X$ for which $g(0) = x_0$ and $g'(0^+)$ exists, the mapping $S(t) = F[g(t)]$ is differentiable at $t = 0^+$ and $S'(0^+) = Lg'(0^+)$. The mapping L, which is obviously unique, is called the <u>Hadamard derivative</u> of F at x_0 and $Lg'(0^+)$ is called the Hadamard differential.

An interesting property of this differential is that it is the weakest (most general) differential among all differentials which obey the chain rule and which reduce to the ordinary differential in the case of a real function of a real variable. (See Section 2.2).

Fréchet showed that in finite dimensional spaces, the Hadamard derivative coincides with the Fréchet derivative; in infinite dimensional spaces Hadamard differentiability does not necessarily imply Fréchet differentiability. Since Gâteaux differentiability at x_0 means that the composition of the map $\varphi: (-\tau, \tau) \to X$ defined by $\varphi(t) = x + th$ and the map F is differentiable at 0 for $h \in X$, it follows that Hadamard differentiability implies Gâteaux differentiability but not conversely.

Example 1.11. The function defined in Example 1.1 is Gâteaux differentiable, but is not Hadamard differentiable at $(0,0)$. Defining $g(t) = (t, t^2)$, we get $f[g(t)] = t$, $(f \circ g)'(0) = 1$, while $L[g'(0)] = 0$.

Example 1.12. The functional $f: L_2[0, \pi] \to \mathbb{R}$ defined by $f(x) = \int_0^\pi \sin x(t) dt$ is everywhere Hadamard differentiable but is not Fréchet differentiable at any point $x \in L_2[0, \pi]$ (see [120] for details).

Example 1.13. Define the functional f on the space ℓ_2 as follows: If $x = t e_k$, where $e_k = (0, \ldots, \underbrace{0, 1, 0}_{k}, \ldots)$, let $f(x) = t^{1 + \frac{1}{k}}$ and $f(x) = 0$ for other $x \in \ell_2$. Then f is Hadamard differentiable at zero, but it is not Fréchet differentiable at zero.

Theorem 1.7. If $F: U \to Y$ is Fréchet differentiable at x_0, then it is Hadamard differentiable at x_0.

Proof: Let $F'(x_0)$ be the Fréchet derivative of F at x_0 and let $g: (-\tau, \tau) \to X$ be any continuous function such that $g(0) = x_0$ and $g'(0)$ exists. Then

$$\frac{1}{t} \{F(g(t)) - F(x_0)\} = F'(x_0)[\frac{g(t) - g(0)}{t}] + \frac{1}{t} r(x_0; g(t) - x_0).$$

As $t \to 0$, $\frac{g(t) - g(0)}{t} \to g'(0)$ and

$$\frac{1}{t} r(x_0; g(t) - g(0)) = \frac{r(x_0; g(t) - g(0))}{\|g(t) - g(0)\|} \frac{\|g(t) - g(0)\|}{t} \to 0.$$

Hence $\frac{1}{t} r(x_0; g(t) - x_0) \to 0$ as $t \to 0$, and thus $(F \circ g)'(0)$ exists since $F'(x_0)$ is a continuous operator.

Theorem 1.8. Let $F: U \to Y$, $x_0 \in U$ and $y_0 = F(x_0) \in W \subset Y$. If F is Hadamard differentiable at x_0 and $G: Y \to Z$ is Hadamard differentiable at y_0, then the composite map $H = G \circ F$ is Hadamard differentiable at x_0 and

$$DH(x_0; h) = DG[y_0; DF(x_0; h)],$$

or equivalently,

$$(GF)'(x_0) = G'(F(x_0))F'(x_0).$$

Proof: Let U_0 be an open neighborhood of x_0 such that $H: U_0 \to Z$, and $\varphi: (-\tau, \tau) \to U_0$ be a continuous map such that $\varphi(0) = x_0$ and $\varphi'(0)$ exists.
 Since F is Hadamard differentiable at x_0, the map $A = F \circ \varphi$ is differentiable at zero and

$$A'(0) = DF[x_0; \varphi'(0)].$$

Also $A(0) = F[\varphi(0)] = F(x_0) = y_0$ and $A: (-\tau, \tau) \to W_0$ where W_0 is an open neighborhood of y_0. Thus by the Hadamard differentiability of G at y_0, the map $B = G \circ A$ is differentiable at zero and

$$B'(0) = DG[y_0; A'(0)] = DG[y_0; DF(x_0; \varphi'(0))].$$

On the other hand, $B = (G \circ F) \circ \varphi = H \circ \varphi$ and since $DG[y_0; DF(x_0; \cdot)]$ is linear, this shows that H is Hadamard differentiable and

$$DH(x_0; \cdot) = DG[y_0; DF(x_0; \cdot)].$$

In particular if L is a bounded linear operator on Y to Z and $F: X \to Y$ is Hadamard differentiable, then $G(x) = LF(x)$ is Hadamard differentiable and $G' = LF'$.
 The next theorem characterizes Hadamard differentiability in terms of a linear local approximation property.

Theorem 1.9. The operator $F: U \to Y$ is Hadamard differentiable at $x_0 \in U$ if and only if there exists $L(x_0) \in \mathcal{L}(X, Y)$

such that the representation (1.6) holds where for each continuous $g: (-\tau, \tau) \to U$ for which $g(0) = x_0$ and $g'(0)$ exists, we have

$$\lim_{t \to 0} \frac{r(x_0; g(t))}{t} = 0 \ .$$

This theorem is an immediate consequence of the definition of the Hadamard derivative. A more significant characterization of Hadamard differentiability is given by the following theorem [119] which avoids the functions g.

<u>Theorem 1.10.</u> A necessary and sufficient condition for F to be Hadamard differentiable at x_0 is that the representation (1.6) holds where

$$\lim_{t \to 0} \frac{r(x_0; th)}{t} = 0$$

uniformly with respect to $h \in S$ on each sequentially compact set $S \subset X$.

<u>Theorem 1.11.</u> A necessary and sufficient condition for F to be Hadamard (Fréchet) differentiable at x_0 is that for any positive integer n,

$$F(x_0 + t_1 h_1 + \ldots + t_n h_n) = F(x_0) + L(x_0)(t_1 h_1 + \ldots + t_n h_n)$$

$$+ \ r(t_1, \ldots, t_n; h_1, \ldots, h_n) \ ,$$

where $L(x_0) \in \mathcal{L}(X, Y)$ and for any sequentially compact (bounded) sets $S_1, \ldots, S_n \subset X$,

$$\frac{r(t_1, \ldots, t_n; h_1, \ldots, h_n)}{\|t\|} \to 0 \text{ as } \|t\| \to 0$$

uniformly in $h_i \in S_i$, $i = 1, \ldots, n$.

The following theorems give sufficient conditions for a Gâteaux variation to be linear and for a linear Gâteaux variation to be a Hadamard differential.

<u>Theorem 1.12</u>. A sufficient condition for a Gâteaux variation $\delta F(x_0; h)$ to be linear is that for all $h_1, h_2 \in X$, the function $\varphi(t_1, t_2) = F(x_0 + t_1 h_1 + t_2 h_2)$ of the real variables t_1, t_2 be Hadamard differentiable at $(0,0)$.

<u>Proof</u>: Assume that $\varphi(t_1, t_2)$ is Hadamard differentiable at $(0,0)$. Then, by Theorem 1.10

$$\varphi(t,t) = \varphi(0,0) + \varphi_1(0,0)t + \varphi_2(0,0)t + r(t,t) ,$$

where

$$\frac{r(t_1,t_2)}{\sqrt{t_1^2 + t_2^2}} \to 0 \text{ as } (t_1, t_2) \to (0,0) .$$

Or in terms of F,

$$F(x_0 + th_1 + th_2) = F(x_0) + t\delta F(x_0; h_1) + t\delta F(x_0; h_2) + r(t,t) .$$

On the other hand, by definition the Gâteaux variation

$$F(x_0 + th_1 + th_2) = F(x_0) + t\delta F(x_0; h_1 + h_2) + \gamma(t) ,$$

where $\frac{\gamma(t)}{t} \to 0$ as $t \to 0$. Thus equating the two expressions, we get

$$\delta F(x_0; h_1) + \delta F(x_0; h_2) - \delta F(x_0; h_1 + h_2) =$$

$$\lim_{t \to 0} \frac{1}{t} \{\gamma(t) - r(t,t)\} = 0 .$$

This proves the theorem.

THE ROLE OF DIFFERENTIALS

Theorem 1.13. Suppose that $F: U \to Y$ has a linear Gâteaux variation at $x_0 \in U$ and that F satisfies a Lipschitz condition on U, $\|F(x) - F(y)\| \leq M \|x-y\|$ for all $x, y \in U$. Then F is Hadamard differentiable at x_0.

Proof: Let g be a continuous mapping of a neighborhood of zero into U for which $g(0) = x_0$ and $g'(0)$ exists. Let $H = F \circ g$. Then

$$\|t^{-1}\{H(t) - H(0)\} - \delta F[x_0; g'(0)]\|$$

$$\leq \|t^{-1}\{F[g(t)] - F[x_0 + tg'(0)]\}\|$$

$$+ \|t^{-1}\{F[x_0 + tg'(0)] - F(x_0)\} - \delta F[x_0; g'(0)]\|$$

$$\leq Mt^{-1}\|g(t) - g(0) - tg'(0^+)\|$$

$$+ \|t^{-1}\{F[x_0 + tg'(0)] - F(x_0)\} - \delta F[x_0; g'(0)]\|,$$

from which it follows that $H'(0) = \delta F[x_0; g'(0)]$, and, since $\delta F[x_0; \cdot]$ is assumed to be linear, F is Hadamard differentiable at x_0.

Michal [106] gave a modified definition of Hadamard differentiability in linear topological space where the condition of differentiability of g at 0 in Definition 1.2 is replaced by its differentiability everywhere. It is not hard to show, however, that the two definitions are equivalent.

The Hadamard differential was studied in 1942 by Ky Fan [22] and Balanzat (1949) [78] in the setting of Fréchet spaces (linear metric spaces) and in 1959 by Madame de Foglio [87] in L-spaces (linear spaces with a definition of convergence). Balanzat (1960) considered the Hadamard differential for mappings between linear topological spaces. Some aspects of the Hadamard differential are also considered in [88], [5; pp. 151-152], [135, II], [72] and [73].

1.4. Gradients

Let $f: U \to \mathbb{R}$, $U \subset X$, and suppose that f is Fréchet differentiable at $x_0 \in U$. The the Fréchet derivative $f'(x_0)$ is an element of the dual space X^* of X. We call $f'(x_0)$ the <u>gradient</u> of f <u>at</u> x_0. If f is Fréchet differentiable on U, then the mapping $f': U \to X^*$ is called the <u>gradient</u> of f and denoted by grad f. A mapping $F: U \to X^*$ is called a <u>gradient mapping</u> if there exists a differentiable functional $f: U \to \mathbb{R}$ such that grad $f = F$ on U. Gradient mappings form the basis of variational methods for the study of nonlinear operator equations. Topological and analytic properties of gradient mappings have been extensively studied by Rothe [64], Vainberg [11], Krasnoselskii [296] and others. A gradient mapping is also called a <u>potential operator</u>, by analogy with classical potential theory and vector fields derived from scalar fields by the operation of gradient.

For example if f is a mapping on $\mathbb{R}^3$ into $\mathbb{R}$ which is Fréchet differentiable at $x = (x_1, x_2, x_3)$, then the differential of f at x with increment $h \triangleq (h_1, h_2, h_3)$ is given by $df(x; h) = \sum_{i=1}^{3} \frac{\partial f}{\partial x_i} h_i = \nabla f(x) \cdot h$, where $\nabla f =$ grad f. Thus ∇f assigns to each x a continuous linear functional $df(x; \cdot)$. In this example $df(x; h)$ is written as an inner product. Similarly, if X is any Hilbert space, then the gradient may be considered as a mapping from X into itself since the dual space of X may be identified with X. Furthermore, $df(x_0; h)$, being a continuous linear functional in h, can be uniquely represented as an inner product, i.e.,

(1.8) $$df(x_0; h) = <h, \text{grad } f(x_0)>.$$

The notion of gradient in function spaces was first introduced by M. Golomb in his early work on nonlinear integral equations [41] and appears informally also in Courant and Hilbert, Methods of Mathematical Physics, Vol. 1.

The representation in (1.8) may hold of course for some functionals on an inner product space which is not necessarily complete.

Gradients are defined in the same manner for functionals which have a Gâteaux derivative or a Hadamard derivative.

<u>Remark</u>. The gradient depends on the inner product used. Let $<\cdot,\cdot>$ and $[\cdot,\cdot]$ be two inner products defined on a linear space and which induce equivalent norms, and let f be Fréchet differentiable in the space $\{X, <\cdot,\cdot>\}$. Let grad f and Grad f denote the gradients of f relative to the inner proudcts $<\cdot,\cdot>$ and $[\cdot,\cdot]$ respectively. These gradients can be easily related if for instance,

(1.9) $$[x,y] = <Px,y>$$

where P is a self-adjoint positive linear operator, i.e., $<Px,x> > 0$ for $x \neq 0$. If the space is complete relative to the norm $\|\cdot\| = \sqrt{<\cdot,\cdot>}$, then it is also complete relative to the norm $\|\cdot\|' = \sqrt{[\cdot,\cdot]}$, and we get by definition of gradient,

(1.10) $$df(x;h) = [h, \text{Grad } f(x)] = <h, \text{grad } f> \quad \text{for all } h \in X.$$

But from (1.9)

(1.11) $$[h, \text{Grad } f] = <h, P \text{ Grad } f>.$$

From (1.10) and (1.11) we get

$$P \text{ Grad } f = \text{grad } f \quad \text{or} \quad \text{Grad } f = P^{-1} \text{grad } f.$$

This relation also holds even if the inner product spaces are not complete; a proof can be based on Lagrange's multipliers (see [147]). In particular, if X is finite dimensional, then any two inner products are related as in (1.9) for some symmetric positive matrix $P = [a_{ij}]$ and Grad f gives the gradient of f relative to the "ellipsoidal" distance $d(x,y) = \{\sum_{i,j=1}^{n} a_{ij}(y_j - x_j)(y_i - x_i)\}^{\frac{1}{2}}$, in terms of the gradient relative to the Euclidean distance.

Example 1.13. For the functional J in Example 1.10, we have

$$\text{grad } J(x) = 2x - \lambda \int_0^1 \{K(s,\cdot) + K(\cdot,s)\} x(s) ds - 2y.$$

Example 1.14. For any linear operator L on an inner product space X into itself, let $\phi(x) = \frac{1}{2} <Lx,x> - <x,y>$, where y is a fixed element in X. Then $\delta\phi(x;h) = \frac{1}{2}<Lh,x> - \frac{1}{2}<Lx,h> - <h,y>$. If L has an adjoint, then

$$\text{grad } \phi(x) = \frac{1}{2}(L + L^*)x - y.$$

In particular, if L is symmetric, then $\text{grad } \phi(x) = Lx - y$.

Example 1.15. Let $f(x) = \|x\|$ on a real inner product space X. Then

$$\delta f(x;h) = \lim_{t \to 0} \frac{1}{t}\{\|x+th\| - \|x\|\} = \lim_{t \to 0} \frac{1}{t} \frac{\|x+th\|^2 - \|x\|^2}{\|x+th\| + \|x\|}$$

$$= \frac{<h,x>}{\|x\|^2} \quad \text{and} \quad \text{grad } \|x\| = \frac{x}{\|x\|^2} \quad \text{for any } x \neq 0.$$

The gradient of a functional on a normed space X is an element of the dual space X^*. Thus f and grad f form a "dual" pair. This conforms with physical notions where, for example, momentum, strain, and current are considered to be dual to position, stress, and voltage, respectively. See various applications in the context of complementary variational principles in the paper of Robinson in this volume. Another discussion of the notion of gradient is given in the paper of Tapia in this volume. Several examples of gradients in function spaces are given in [133], [11], [64], [296], and others.

1.5. Strong and Bounded Differentials and Related Notions

Let $f: I \to \mathbb{R}$, where I is an open subset of $\mathbb{R}$. We say that f is strongly differentiable at $x_0 \in I$ if

$$\lim_{\substack{(x_1, x_2) \to (x_0, x_0) \\ x_1 \neq x_2}} \frac{f(x_2) - f(x_1)}{x_2 - x_1}$$

exists. This notion was first introduced by Peano [136] who felt it "portrayed the concept of the derivative used in the physical sciences more closely than does the usual derivative". This notion appears also in Leach [131], Esser and Shisha [124], Dieudonné [5], Cartan [4] and Nashed [135]. It is clear that this notion can be used in conjunction with each of the definitions of the differentials.

Definition 1.3. An operator $F: U \to Y$ is said to have a strong Fréchet differential at $x_0 \in U$ if there exists $L(x_0) \in \mathcal{L}(X,Y)$ such that

$$F(x) - F(y) = L(x_0)(x-y) + \psi(x_0; x, y) ,$$

where

$$\lim_{\substack{(x,y) \to (x_0, x_0) \\ x \neq y}} \frac{\|\psi(x_0; x, y)\|}{\|x - y\|} = 0 .$$

Obviously this definition is equivalent to saying that F is Fréchet differentiable at x_0 and that for each $\varepsilon > 0$, there exists $\delta > 0$ such that $r(x) = F(x) - F(x_0) - F'(x_0)(x - x_0)$ is ε-Lipschitzian for all x in $\{x: \|x - x_0\| \leq \delta\}$, i.e.,

$\|x - x_0\| \leq \delta$ and $\|y - x_0\| \leq \delta$ imply $\|r(x) - r(y)\| \leq \varepsilon \|x - y\|$.

Let $d^*F(x_0; h)$ denote the strong Fréchet differential of F at x_0. Let A denote the subset of U on which F is Fréchet differentiable and A^* the subset of A on which F is strongly Fréchet differentiable.

Theorem 1.14. (a) If F is strongly Fréchet differentiable at $x_0 \in U$, then for each $h \in X$,

$$\lim_{\substack{x \to x_0 \\ x \in A^*}} d^*F(x; h) = \lim_{\substack{x \to x_0 \\ x \in A}} dF(x; h) = d^*F(x_0; h) = dF(x_0; h)$$

whenever both limits exist.

(b) If F is strongly Fréchet differentiable at x_0, then F satisfies a Lipschitz condition in some neighborhood of x_0.

(c) Suppose F is Fréchet differentiable in some neighborhood of x_0. Then F is strongly Fréchet differentiable at x_0 if and only if F' is continuous at x_0.

Other properties of strong differentials are given in [135] and will appear elsewhere.

An operator F is said to have a <u>locally uniform</u> Fréchet differential $dF(x; h)$ on an open set U if F has a Fréchet differential on U and the remainder $r(x; h)$ is locally uniformly bounded, i.e. for each $\varepsilon > 0$ and each $x_0 \in U$, there exist $\delta(x_0; \varepsilon)$ and $\eta(x_0; \varepsilon)$ such that

$$\|h\| \leq \delta \quad \text{and} \quad \|x - x_0\| \leq \eta \quad \text{imply} \quad \|r(x; h)\| \leq \varepsilon \|h\|.$$

The Fréchet derivative F' is said to be <u>locally bounded</u> on U if every $x_0 \in U$ has a neighborhood in which $\|F'(x)\|$ is bounded. Vainberg [11] has shown that the Fréchet derivative F' is continuous on an open convex set Ω if and only if F has in Ω a locally uniform differential and F' is locally bounded in Ω. Combining this result with part (c) of Theorem 1.14, we get a necessary and sufficient condition for F to be strongly Fréchet differentiable on Ω.

An operator F is said to be uniformly Fréchet differentible on a set U, if

$$\lim_{h \to 0} \frac{\|r(x; h)\|}{\|h\|} = 0$$

uniformly with respect to $x \in U$.

For $r > 0$, let $S_r = \{x: \|x\| < r\}$. Let F be uniformly continuous in $S_{r+\alpha}$ where $\alpha > 0$. In order for F' to be uniformly continuous in S_r, it is necessary and sufficient that F has a uniform differential in S_r.

Finally we mention the concept of bounded differentials. A mapping $F: X \to Y$ is said to have at $x_0 \in X$ a <u>bounded differential</u> $BF(x_0; h)$ if for any given $\varepsilon > 0$ there exists $\delta > 0$ such that

$$|t| < \delta \text{ implies } \left\| \frac{1}{t}[F(x_0 + th) - F(x_0)] - BF(x_0; h) \right\| < \varepsilon$$

uniformly with respect to h on the set $S = \{h: h \in X, \|h\| = 1\}$, and $BF(x_0; h)$ is bounded on S, but is not necessarily linear.

It is not hard to show that if a mapping F of a Banach space X into itself has a bounded uniform differential at $x_0 \in X$, then $BF(x_0; h)$ is the Fréchet differential. The notion of bounded differentials was proposed by Suchomolinov [66a]. Ivanov [146] has shown that if X is a finite dimensional Banach space and f is a real function on X such that $\delta f(x_0; h)$ exists and f is Lipschitzian in a neighborhood of x_0, then f has a bounded differential at x_0. Under these conditions if moreover $\delta f(x_0; \cdot)$ is linear, then f has a Fréchet differential at x_0.

1.6. Equidifferentiability

Another notion that can be used in conjunction with each of the differentials defined in the preceding sections is the notion of equidifferentiability.

<u>Definition 1.4.</u> A family $\mathfrak{F}$ of operators on U into Y is called Fréchet equidifferentiable at $x_0 \in U$ if and only if each $F \in \mathfrak{F}$ is Fréchet differentiable at $x_0 \in U$ and

$$\lim_{h \to 0} \frac{\|F(x_0 + h) - F(x_0) - F'(x_0)h\|}{\|h\|} = 0$$

uniformly with respect to $F \in \mathfrak{F}$, i.e., for each $\varepsilon > 0$, there is $\delta > 0$, δ independent of $F \in \mathfrak{F}$, such that

$\|h\| < \delta$ implies $\|F(x_0 + h) - F(x_0) - F'(x_0)h\| < \varepsilon \|h\|$.

A family $\mathfrak{F}$ is Fréchet equidifferentiable on a set $S \subset U$, if it is equidifferentiable at each $x \in S$.

Gâteaux and Hadamard equidifferentiability can be defined in a similar way. Various implication relations can be given in an obvious manner.

1.7. Differentiation Along a Subspace and Partial Differentiation

In general the variation at a given point does not exist for every increment h. The case when the admissible increments (at a point x_0) form a subspace is important in problems of calculus of variations and boundary value problems, and motivates the notion of differentiation along a subspace.

Let X be a linear space, H a subspace of X, and Y a normed linear space. We say that $F: \Omega \subset X \to Y$ has a Gâteaux variation at $x_0 \in \Omega$ <u>along the subspace</u> H if $\delta F(x_0; h)$ exists for all $h \in H$. Similarly if H is also a normed space in addition to being a subspace of X, then we may define the Fréchet differential of F at x_0 along the subspace H. Note that if a map $F: X \to Y$ has a Gâteaux variation at x_0 along h, then it is Fréchet differentiable at x_0 along the one dimensional subspace spanned by h.

Differentiation along subspaces gives us a technical advantage at essentially no added expense. On the one hand, in many problems the derivative on the whole space does not exist, while the derivative along a subspace that is everywhere dense in the space exists. Such cases are encountered in two-point boundary value problems, variational problems, etc. For instance H may be taken as the subspace determined by the boundary conditions, or the class of admissible variations in calculus of variations problems, while X is the domain of the differential operator or the functional. On the other hand, differentiation along subspaces subsumes the notion of a partial derivative.

Let X be an open subset of the product space $\Pi = E_1 \times \ldots \times E_n$. Let $F: X \to Y$. The Fréchet <u>partial</u> differential at $u_1, \ldots, u_n$ of F with respect to x_i is defined in the usual way: there exists a bounded linear operator $L(u_1, \ldots, u_n; \cdot)$ such that for all $h_i \in E_i$ with

$$(u_1, \ldots, u_{i-1}, u_i + h_i, u_{i+1}, \ldots, u_n) \in X ,$$

$$F(u_1, \ldots, u_{i-1}, u_i + h_i, \ldots, u_n) - F(u_1, \ldots, u_n) =$$

$$= L(u_1, \ldots, u_n; h_i) + R(u_1, \ldots, u_n; h_i) ,$$

where

$$\frac{\|R(u_1, \ldots, u_n; h_i)\|}{\|h_i\|} \to 0 \text{ as } h_i \to 0 .$$

$L(u_1, \ldots, u_n; h_i)$ is called the <u>Fréchet partial differential</u> and is denoted by $d_i F(u_1, \ldots, u_n; h_i)$. F is said to <u>totally differentiable</u> if it is Fréchet differentiable considered as a mapping on $X \subset E_1 \times \ldots \times E_n$ into Y, that is if there exists an $L(u; h)$, $u = (u_1, \ldots, u_n) \in X$, $h = (h_1, \ldots, h_n) \in \Pi$, which is linear and continuous in h such that

$$\lim_{h \to 0} \frac{\|F(u_1+h_1, \ldots, u_n+h_n) - F(u_1 \ldots u_n) - L(u_1 \ldots u_n; h_1 \ldots h_n)\|}{\|h_1\| + \ldots + \|h_n\|} = 0 .$$

$L(u_1 \ldots u_n; h_1 \ldots h_n)$ is called the total Fréchet differential of F and is denoted by $dF(u_1 \ldots u_n; h_1 \ldots h_n)$.

An operator $F: X \subset \Pi \to Y$ which is totally differentiable at $u_1 \ldots u_n$ is partially differentiable with respect to each variable and its total differential is the sum of the differentials with respect to each of the variables. (Fréchet [26; p. 319], Dieudonné [5; p. 167]). If F is totally differentiable at each point of X, then a necessary and sufficient condition for $F': X \to \mathcal{L}(E_1 \times \ldots \times E_n; Y)$ to be continuous is that the partial derivatives $F'_i: X \to \mathcal{L}(E_i; Y)$, $i = 1, \ldots, n$, be continuous.

Similarly, the Gâteaux partial differential at $u_1 \ldots u_n$ of F with respect to x_i exists if and only if there exists a bounded linear operator $D_i F(u_1 \ldots u_n; \cdot): E_i \to Y$ such that

$$F(u_1, \ldots, u_{i-1}, u_i + h_i, u_{i+1}, \ldots, u_n) - Fu_1, \ldots, u_n$$
$$= D_i F(u_1, \ldots, u_n; h_i) + R(u_1, \ldots, u_n; h_i)$$

where

(1.12) $$\lim_{t \to 0} \frac{R(u_1, \ldots, u_n; th_i)}{t} = 0.$$

$D_i F(u_1, \ldots, u_n; h_i)$ is called the Gâteaux partial differential. The operator F is said to be totally Gâteaux differentiable at x_0 if F, considered as a mapping on $X \subset \Pi$ into Y, is Gâteaux differentiable at x_0. This means that

$$F(u_1 + h_1, \ldots, u_n + h_n) - Fu_1 \ldots u_n$$
$$= L(u_1 \ldots u_n; h_1 \ldots h_n) + R(u_1 \ldots u_n; h_1 \ldots h_n)$$

where L is a continuous linear operator in $h = h_1 \ldots h_n$, and

(1.13) $$\lim_{t \to 0} t^{-1} R(u_1 \ldots u_n; th_1, \ldots, th_n) = 0.$$

Clearly F has a Fréchet partial (total) differential at x_0 if and only if F has a Gâteaux partial (total) differential at x_0 and (1.12), (respectively (1.13)), holds uniformly with respect to $h_i \in E_i$ on the set $\|h_i\| = 1$ ($h_1 \ldots h_n$ on the set $\|h_1\| = \ldots = \|h_n\| = 1$).

Higher order partial derivatives are defined by induction. Note in particular that if $F: X \to Y$ is twice Fréchet (totally) differentiable at x_0, then the second order partial derivatives

$$\frac{\partial^2 F(x_0)}{\partial x_i \partial x_j} \in \mathcal{L}(E_i, E_j; Y), \qquad i,j = 1, \ldots, n$$

exist, and

(1.14) $$d^2 F(x_0; k_1, \ldots, k_n; h_1, \ldots, h_n) = \sum_{i,j=1}^{n} \frac{\partial^2 F(x_0)}{\partial x_i \partial x_j} k_i h_j.$$

Thus the second order derivative $F''(x_0)$ may be represented by the array $\left\{ \dfrac{\partial^2 F(x_0)}{\partial x_i \partial x_j} : i,j = 1, \ldots, n \right\}$. Since $d^2 F(x_0; h, k)$ is symmetric in h and k, it follows readily from (1.14) that

$$\frac{\partial^2 F(x_0)}{\partial x_i \partial x_j} = \frac{\partial^2 F(x_0)}{\partial x_j \partial x_i}, \qquad i,j = 1, \ldots, n.$$

Note that the symmetry of the mixed partial derivatives is a consequence of the existence of the second order total Fréchet differential. Continuity of the second order partial derivatives is not needed to reach this conclusion.

2. Differentials and Variations: Revisited

2.1. Differentials in Linear Topological Spaces

In Section 1 we presented some of the important definitions of differentials in the setting of normed linear spaces. Most of the theory of elementary calculus in infinite dimensional normed spaces has been completed by the 1940's and is thoroughly and elegantly presented in Dieudonné [5], Cartan [4]. Various aspects of differential calculus in normed spaces are given in Liusternik and Sobolev [9], Kantorovich and Akilov [8], Loomis and Sternberg [13], Rall [10], Vainberg [11], Nashed [134], [135], and others.

The status of differentiation in linear topological spaces until recently (1968) can be best described by the following quotations from [73], [72].

"By now a very peculiar situation has arisen. A considerable number of articles have accumulated that are devoted to the construction of differential calculus in linear topological spaces and are written approximately along the following lines: to begin with the author introduces a definition of the derivative of a map between two linear topological spaces which seems to him to be the most successful, and then he shows that this derivative has properties similar to those of an ordinary derivative. In many cases the authors of these articles know nothing at all about their predecessors and begin their constructions ab initio."

"There are at present more than a score of definitions of the derivative of a map from one linear topological space into another. These definitions are stated in what are superficially completely different ways and the authors who proposed new definitions have not as a rule concerned themselves with the relation between the new definitions and those already known. This leads to an impression of chaos."

From the numerous papers on differentiation in topological spaces which appeared before 1967 we single out the papers of Gil de Lamadrid [94,95], Sebastião e Silva [117, 118], Keller [99,100], and Sova [119,120], which we feel were significant partial contributions toward a unifying and satisfactory framework for differentiation in linear topological spaces.

In 1955 Gil de Lamadrid in his dissertation [94] (see also [95]) proposed and investigated a general and interesting approach to the definition of the derivative. Let X and Y be two linear topological spaces†. Let $\mathfrak{F}(X,Y)$ denote the set of all maps (not necessarily linear) from X to Y, and let $\mathfrak{T}$ be any (not necessarily linear) topology on $\mathfrak{F}(X,Y)$.

Definition 2.1. A map $F: X \to Y$ is said to be $\mathfrak{T}$-differentiable in the sense of de Lamadrid at x_0 if there exists a map $A \in \mathfrak{F}(X,Y)$ such that $F_t \to A$ as $t \to 0$ in the topology $\mathfrak{T}$, where for each $t \in \mathbb{R}$, $t \neq 0$, F_t is the map in $\mathfrak{F}(X,Y)$ defined by $F_t(h) = t^{-1}\{F(x_0 + th) - F(x_0)\}$.

In 1956, 1957, Sebastião e Silva [117, 118] independently of Gil de Lamadrid, constructed a theory of infinitesimal quantities ("little o's" of different order) and a corresponding theory of differentials. His work leads to two spectra of notions of differentials. We define only one of these spectra; the other does not coincide with ordinary derivative in the case of a function of a real variable, and is quite pathological.

Definition 2.2. Let β be a system of bounded sets in X. A map $r: X \to Y$ is said to be β-small if for each set $B \in \beta$,

$$\frac{r(th)}{t} \to 0 \text{ as } t \to 0 \underline{\text{uniformly}} \text{ for } h \in B.$$

Definition 2.3. A map $F: X \to Y$ is said to be β-differentiable at $x_0 \in X$ if there exists $L \in \mathfrak{L}(X,Y)$ such that

$$F(x_0 + h) - F(x_0) = Lh + r(x_0; h),$$

where r is β-small.

Note that this definition is a very interesting particular case of de Lamadrid's definition corresponding to the choice

---

† All spaces that we consider are assumed to be Hausdorff and all the scalars are real, unless otherwise stated.

of the class of linear topologies of uniform convergence on some system of bounded sets.

Keller, in his survey paper [99] and [100] considered some of the existing definitions and proved only partial connections among these definitions in the context of locally convex linear topological spaces. Sova in 1964 independently arrived at the definition of β-differentiability of Sebastião e Silva, and introduced the notion of compact differentiability (i.e. β-differentiability with respect to the system of all sequentially compact sets in X). He proved that this notion is equivalent to Hadamard differentiability and established interesting characterizations and properties of this notion of differentiability, which motivated a unifying definition of (H, β)-differentiability introduced later in [73].

Averbukh and Smolyanov, in their remarkable papers [72], [73], presented a unifying and thorough investigation of differentials in linear topological spaces and reduced all the definitions into a small number of classes, each consisting of equivalent definitions. Implications relationships and counter-examples are also provided to describe all the connections among the various classes.

We shall limit our presentation to two aspects of differential calculus in linear topological spaces. First we describe briefly the axiomatic approach and a descriptive characterization of a spectrum of definitions of differentials. Second we give an exposition of a particular case of de Lamadrid's definition as a unifying basis for a class of definitions. It is a minor modification of Definition 2.3 where differentiation is taken along a subspace and β may be taken to be _any_ system of subsets of X . It is our feeling that this definition, which subsumes Fréchet, Gâteaux, and Hadamard differentiability as particular realizations of the system β in the case of normed spaces, suffices for most of the applications of nonlinear functional analysis, particularly to variational problems, integral equations, and numerical analysis. Therefore we shall not dwell on other aspects of the "chaos" referred to in the quotation cited at the beginning of this section. For comparison purposes, a list of various definitions which have been proposed in linear topological spaces, along with a table of implications, may be found in [73, 72]. References [72]-[121] are devoted to calculus in topological spaces and related topics.

2.2. The Axiomatic or Descriptive Approach to Differentials

Let X and Y be two linear topological spaces. The problem of defining a derivative of a map $F: X \to Y$ is essentially a problem of <u>approximation</u> of F by a map from a given class of operators on X into Y. We denote the <u>class of approximating maps</u> by $\mathcal{G}(X, Y)$. (In most definitions $\mathcal{G}(X, Y)$ is usually taken as a subset of the space $\mathcal{L}(X, Y)$ of all continuous linear operators on X into Y). The sense of approximation, which is a generalization of the order condition on the "remainder", is made precise by specifying the class $\mathcal{R}(X, Y)$ of maps $r: X \to Y$ which are considered "infinitesimal", i.e. $r(h)$ is "small" in comparison with h. Thus the heart of the matter is the definition of "little o" for mappings between two linear topological spaces.

<u>Definition 2.4.</u> For given classes $\mathcal{G}(X, Y)$ and $\mathcal{R}(X, Y)$ the mapping $F: X \to Y$ is said to be $\mathcal{GR}$-differentiable at x_0 if there exist maps $A \in \mathcal{G}(X, Y)$ and $r \in \mathcal{R}(X, Y)$ such that

$$F(x_0 + h) - F(x_0) = A(h) + r(h).$$

The mapping A is called an $\mathcal{GR}$-derivative of F at x_0.

Note that a sufficient condition for such a representation to be unique is that the intersection of the sets $\mathcal{G}(X, Y)$ and $\mathcal{R}(X, Y)$ contains no nonzero element of the set of all maps on X into Y. As an example, if we take $\mathcal{G}$ to be the set of all homogeneous maps on X to Y and $\mathcal{R}$ to be the set of all maps $r: X \to Y$ such that $\frac{r(th)}{t} \to 0$ for each $h \in X$, then $\mathcal{G} \cap \mathcal{R} = \{0\}$ and the $\mathcal{GR}$-differential becomes the Gâteaux variation.

Different choices of the classes $\mathcal{G}(X, Y)$ and $\mathcal{R}(X, Y)$ give rise to different notions of derivatives which may not reduce to the usual derivative in the case of a real function of a real variable. Note that in general we have to allow some of the sets $\mathcal{R}(X, Y)$ to be empty in order to include in the setting of Definition 2.4 those notions of differentiability for maps on special classes of spaces, for example finite dimensional spaces, normed spaces, etc.

The setting of Definition 2.4 is too general to be useful. Therefore we shall restrict the classes $\mathcal{G}$ and $\mathcal{R}$ so that a greater interface is achieved with properties of classical differentiation. To this end it is natural to require some or all of the following properties:

(A). The derivative is unique (relative to fixed $\mathcal{G}$ and $\mathcal{R}$).

(B). The differentiation operation is linear.

(C). Every continuous linear operator is everywhere differentiable and its derivative coincides with itself; the derivative of any constant map is the zero map.

(D). If the domain of F is the real line, then the derivative coincides with the usual derivative of a function of a real variable.

(E). The derivative is a continuous linear map.

(F). The chain rule for the differentiation of composite functions holds.

Note that the Gâteaux variation satisfies properties (A)-(D) but not properties (E) and (F). The Gâteaux differential satisfies all the properties except (F). The Fréchet and Hadamard differentials satisfy all these properties.

It follows from properties D, E, and F that for each $h \in X$,

$$\lim_{t \to 0} \frac{r(th)}{t} = 0 \quad t \in \mathbb{R},$$

which also implies the uniqueness of the derivative.

With Averbukh and Smolyanov [73], we say that a method of differentiation is <u>quasiregular</u> if the classes of sets $\mathcal{G}, \mathcal{R}$ satisfy the following conditions for any linear topological spaces X, Y, Z.

1). $\mathcal{G}(X,Y)$ is the set $\mathcal{L}(X,Y)$ of all continuous linear maps of X to Y.

2). If $r \in \mathcal{R}(\mathbb{R},Y)$, then $\frac{r(t)}{t} \to 0$ as $t \to 0$.

3). If $g \in \mathcal{R}(Y,Z)$, $r \in \mathcal{R}(X,Y)$, $A \in \mathcal{L}(X,Y)$, and $B \in \mathcal{L}(Y,Z)$, then $B \circ r \in \mathcal{R}(X,Z)$ and $g \circ (r+A) \in \mathcal{R}(X,Z)$, where $\circ$ denotes composition.

4). All nonempty sets $\mathcal{R}(X,Y)$ are linear spaces under the natural algebraic operations.

5). $\mathcal{R}(X,Y) \cap \mathcal{G}(X,Y) = \{0\}$.

A quasiregular method of differentiability is called <u>regular</u> if

6). for each nonempty set $\mathcal{R}$, $r: \mathbb{R} \to Y$ with $\frac{r(t)}{t} \to 0$ as $t \to 0$ implies that $r \in \mathcal{R}(\mathbb{R}, Y)$.

We remark in passing that if we require all the sets $\mathcal{R}(\mathbb{R}, Y)$ to be nonempty, then an operation of differentiation is regular if and only if it satisfies properties B to F as listed above. Also if a mapping $F: X \to Y$ is differentiable at a given point by two different quasiregular methods, then the two derivatives at this point coincide.

A method of differentiation $\mathcal{G}\mathcal{R}$ is called stronger than another method $\widetilde{\mathcal{G}}\widetilde{\mathcal{R}}$ (or $\widetilde{\mathcal{G}}\widetilde{\mathcal{R}}$ is weaker than $\mathcal{G}\mathcal{R}$) if $\mathcal{G}(X,Y) \subset \widetilde{\mathcal{G}}(X,Y)$ and $\mathcal{R}(X,Y) \subset \widetilde{\mathcal{R}}(X,Y)$ for any linear topological spaces X,Y. In this case $\mathcal{G}\mathcal{R}$-differentiability implies $\widetilde{\mathcal{G}}\widetilde{\mathcal{R}}$-differentiability. If $\mathcal{G}(X,Y)$ is the set $\mathcal{L}(X,Y)$, then we refer to $\mathcal{G}\mathcal{R}$-differentiability simply as $\mathcal{R}$-differentiability.

We now show that the chain rule holds for every quasiregular method of $\mathcal{R}$-differentiability.

<u>Theorem 2.1.</u> Let $F: X \to Y$ and $G: Y \to Z$ and assume that F is differentiable at x_0 and G differentiable at $y_0 = F(x_0)$ in the sense of a given quasiregular $\mathcal{R}$-differentiation process. Then $G \circ F$ is also $\mathcal{R}$-differentiable at x_0 and

$$(G \circ F)'(x_0) = G'(y_0) \circ F'(x_0) \ .$$

<u>Proof</u>: Suppose without loss of generality that $x_0 = 0$ and $y_0 = 0$. Then $F = A + r$ and $G = B + s$, where $A \in \mathcal{L}(X,Y)$, $B \in \mathcal{L}(Y,Z)$, $r \in \Re(X,Y)$ and $s \in \Re(Y,Z)$. By properties 3) and 4) of a quasiregular method of differentiation, we have $B \circ r + s \circ (A+r) \in \Re(X,Z)$. On the other hand the composite map $G \circ F$ can be written in the form

$$G \circ F = (B+s) \circ (A+r) = B \circ A + B \circ r + s \circ (A+r) \ .$$

Thus $B \circ A \in \mathcal{L}(X,Z)$ and $B \circ r + s \circ (A+r) \in \Re(X,Z)$. Hence $G \circ F$ is $\Re$-differentiable at 0 and $(G \circ F)'(0) = B \circ A = G'(0) \circ F'(0)$.

2.3. (H, σ)-Differentiability and Some Specific Realizations

We now turn our attention to a specific choice of the class $\Re(X,Y)$ which seems to cover many interesting notions of differentiation.

Let X be a linear space, H, Y linear topological spaces, with H being also a linear subspace of X. Let σ be a system of subsets of H and β a system of bounded subsets of H. We assume that the union of sets in σ contains a neighborhood of $0 \in H$ and that β contains all the sets in H that consist of single elements. Let $\mathcal{L}(H,Y)$ denote the linear space of all continuous linear maps on H into Y and $\mathcal{L}_\beta(H,Y)$ the linear topological space obtained by endowing $\mathcal{L}(H,Y)$ with the topology of uniform convergence on sets of the system β. With Averbukh and Smolyanov [72] we make the following

<u>Definition 2.5.</u> The operator $F: X \to Y$ is said to be σ-differentiable at $x_0 \in X$ along H (briefly (H,σ)-differentiable at x_0) if there exists $L \in \mathcal{L}(H,Y)$, $L = L(x_0)$, such that

$$F(x_0 + h) - F(x_0) = Lh + r(x_0;h)$$

where

$$\lim_{\tau \to 0} \frac{r(x_0; \tau h)}{\tau} = 0 \quad \underline{\text{uniformly with respect to}} \ h \in S \ \text{for each} \ S \in \sigma.$$

The operator L is called the (H, σ)-<u>derivative</u> of F <u>at</u> x_0 and is denoted by $F'(x_0)$. Note that $F'(x_0)$ is a continuous linear operator on H and not on X.

Definition 2.5 is a specific realization of Definition 2.4 (modified to include differentiation along a subspace) where $\mathcal{G}(H,Y) = \mathcal{L}(H,Y)$, and $\mathcal{R}(H,Y)$ is the class of all maps $r: H \to Y$ such that $\frac{r(th)}{t} \to 0$ as $t \to 0$ uniformly with respect to $h \in S$ for each $S \in \sigma$.

Clearly if the operator F is (H, σ)-differentiable at x_0, then the first variation of F at x_0 exists along the subspace H.

If $F'(x)$ exist for all $x \in X$ (or a subset of X), then the map $F': x \to F'(x)$ of X into $\mathcal{L}_\beta(H,Y)$ is called the (σ, β)-<u>derivative along</u> H, or briefly the (H, σ, β)-derivative. If $\sigma = \beta$ then instead of (H, β, β) we write (H, β). When $X = H$, we say "σ-derivative" instead of "(H, σ)-derivative", and similarly for other abbreviations.

We say that F has a <u>uniform</u> (H, σ)-differential on a subset $\Omega \subset X$ if $\frac{r(x;th)}{t} \to 0$ as $t \to 0$ uniformly with respect to $h \in S$ and $x \in \Omega$ for each $S \in \sigma$.

Finally we consider four interesting choices of the system σ in Definition 2.5.

1. Let β_f denote the system of all <u>finite</u> subsets of H, and take $\sigma = \beta_f$ in Definition 2.5. Corresponding to this choice of σ, we recover the notion of Gâteaux differential along the subspace H.

2. Let β_c be the system of all <u>sequentially compact</u> subsets of H, and take $\sigma = \beta_c$. Then Definition 2.5 reduces in this case to the definition of the Hadamard differential (see Theorem 1.10).

3. Let β_b be the system of all <u>bounded</u> subsets of H. Then F is (H, β_b)-differentiable at x_0 if and only if there

exists $L \in \mathcal{L}(H,Y)$ such that

$$\lim_{t \to 0} \left\{ \frac{F(x_0+th) - F(x_0)}{t} - Lh \right\} = 0$$

uniformly with respect to $h \in S$ for each $S \in \beta_b$. In the case of normed spaces, this condition is tantamount to the following condition:

$$\lim_{h \to 0} \frac{\|F(x_0+h) - F(x_0) - Lh\|}{\|h\|} = 0 ,$$

i.e. to Fréchet differentiability of F at x_0. In view of this, corresponding to the choice of $\sigma = \beta_b$, we shall call (H, β_b)-differentiability Fréchet differentiability along the subspace H.

4. Let $\sigma = \{H\}$, i.e., σ is the system consisting of a single subset of H, the whole of H. In this case if $\lim_{t \to 0} \frac{r(th)}{t} = 0$ as $t \to 0$ uniformly for $h \in H$, then r is zero. Indeed, if $r(h_0) \neq 0$ for some h_0, then taking $h = \frac{h_0}{t}$, we have $\frac{r(th)}{t} = \frac{r(h_0)}{t}$, which does not tend to zero as $t \to 0$, which is a contradiction. Thus corresponding to the choice $\sigma = \{H\}$, only continuous linear maps are differentiable! This is the strongest of all quasiregular methods of differentiation.

One of the remarkable properties of Hadamard differentiability is that it is weaker than any quasiregular method of $\mathcal{R}$-differentiability, and hence it is the weakest notion of differential satisfying properties (A)-(E) for which the chain rule holds. We first prove two lemmas.

Lemma 2.1. Every quasiregular method of $\mathcal{R}$-differentiability is stronger than some regular method of $\mathcal{R}'$-differentiability, i.e., $\mathcal{R}' \supset \mathcal{R}$.

Proof: We let

$$\Re'(\mathbb{R}, Y) = \{r: \frac{r(t)}{t} \to 0 \text{ as } t \to 0\}$$

$$\Re_1(X,Y) = \{r: r = q \circ p,\ p \in \Re(X,\mathbb{R}),\ q \in \Re'(\mathbb{R},Y)\},$$

$$\Re_1'(X,Y) = \Re(X,Y) + \Re_1(X,Y),$$

and by induction,

$$\Re_n(X,Y) = \{r: r = q \circ p,\ p \in \Re(X,Z),\ q \in \Re_{n-1}'(Z,Y)\},$$

$$\Re_n'(X,Y) = \Re(X,Y) + \Re_n(X,Y).$$

It is obvious that

$$\Re(X,Y) \subset \Re_1'(X,Y) \subset \ldots \subset \Re_n'(X,Y) \subset \ldots.$$

Now we put

$$\Re'(X,Y) = \bigcup_{n=1}^{\infty} \Re_n'(X,Y).$$

Then $\Re'$ gives a regular method of differentiation which is weaker than $\Re$.

Lemma 2.2. Hadamard differentiability is weaker than any other regular differentiability.

Proof: Let $\Re$ be a regular method of differentiation and let $F: X \to Y$ be $\Re$-differentiable at $x_0 \in X$. Let $\varphi: \mathbb{R} \to X$ be differentiable at 0 and $\varphi(0) = x_0$. Since φ is $\Re$-differentiable by definition of a regular method of differentiability, the chain rule holds: $(F \circ \varphi)'(0) = F'(x_0) \circ \varphi'(0)$. Thus F is

149

Hadamard differentiable at x_0 and its Hadamard derivative is equal to the $\Re$-derivative $F'(x_0)$.

As a consequence of Lemmas 2.1 and 2.2 we have

<u>Theorem 2.2.</u> Hadamard differentiability is weaker than any other quasi regular method of $\Re$-differentiability.

2.4. <u>(H, σ)-Continuity and Partial Derivatives</u>

The map $F: X \to Y$ is called continuous at $x_0 \in X$ <u>along H</u> (briefly H-continuous at x_0) if the map $h \to F(x_0+h)$ of H into Y is continuous at zero. The map F is called σ-continuous at $x_0 \in X$ along H (briefly, (H,σ)-continuous at x_0) if $\lim_{t \to 0} F(x_0 + th) = F(x_0)$ uniformly with respect to $h \in S$ for each set $S \in \sigma$. If $H = X$, we write "σ-continuous" instead of "(H,σ)-continuous". It is not hard to show that if the map $F: X \to Y$ is H-continuous at x_0, then it is (H, β)-continuous at x_0 for any system β of bounded subsets of H, but not conversely. However if H is a metrizable linear topological space, and $F: X \to Y$ is (H, β_c)-continuous, then F is H-continuous at x_0. Thus if H is a metrizable linear topological space and the map $F: X \to Y$ is Hadamard differentiable along H at $x_0 \in X$, then it is H-continuous at x_0.

For $x_0 \in X$, and H a linear topological space which is also a subspace of the linear space X, we define an H-neighborhood of x_0 to be a set of the form $x_0 + U$, where U is a neighborhood of zero in the space H. A map $F: X \to Y$ is called H_1-continuously (H_2, σ, β)-differentiable at $x_0 \in X$ if it is (H_2, σ)-differentiable in some H_1-neighborhood of x_0 and the (H_2, σ, β)-derivative F' is H_1-continuous at x_0. Similarly, F is called (H_1, γ)-continuously (H_2, σ, β)-differentiable at $x_0 \in X$ if it is (H_2, σ)-differentiable at some H_1-neighborhood of x_0 and the (H_2, σ, β)-derivative F' is (H_1, γ)-continuous at x_0. It can be shown (using a mean value theorem) that if F is a map on a linear space X to a locally convex linear topological space Y and if F is continuously (H, β_f, β)-differentiable at x_0, then it is continuously (H, β)-differentiable at this point. This is essentially a generalization of Theorem 1.3.

Partial derivatives are defined as in the case of normed spaces. If $X = H_1 \oplus H_2 \oplus \ldots \oplus H_n$, then the partial derivative with respect to the kth variable of a mapping $F: X \to Y$ is simply the derivative along H_k of the map F; we denote it by F_k. If $F: X \to Y$, where $X = H_1 \oplus \ldots \oplus H_n = H$ is (H, σ)-differentiable at x_0, then the (H_i, σ_i)-derivatives $F_i(x_0)$ exist, where σ_i is the system of sets in H_i that are the intersections of H_i with all the sets in the system σ and $F'(x_0)h = \sum_{i=1}^{n} F_i(x_0)h_i$, where $h = h_1 + \ldots + h_n$, $h_i \in H_i$. The converse obviously is not true. However as in the case of finite dimensional spaces the converse holds under suitable continuity conditions on the partial derivatives. For instance if $X = H_1 \oplus H_2$, β the system of all subsets of X of the form $B_1 \times B_2$, where $B_i \in \beta_i$, and if for $i = 1, 2$, the (H_i, β_k, β_i)-derivative F_i is continuous at x_0, then F is β-differentiable at x_0.

3. Higher Order Differentials and Variations

The study of higher order Fréchet and Gâteaux differentials essentially involves the approximation in various senses of the difference $f(x_0 + h) - f(x_0)$ by abstract polynomials. This leads to consideration of various notions of continuity of mappings from a subset of a normed linear space to a space of multilinear operators, which we shall discuss in this section.

Let $E_1, E_2, \ldots, E_m$ and Y be normed real linear spaces and let Π denote the product space $E_1 \times E_2 \times \ldots \times E_m$ equipped with the usual product topology induced by the norms on E_i, $i = 1, 2, \ldots, m$. We write $x = (x_1, \ldots, x_m) \in \Pi$

as $x_1 \ldots x_m$ and let $\|x\| = \sup_i \|x_i\|$. We recall that an operator $L_m : \Pi \to Y$ is called multilinear or m-linear if it is separately additive and homogeneous in each of the variables. For a multilinear operator L_m and for each $x_1, \ldots, x_{k-1}, x_{k+1}, \ldots, x_m$ we let S_k be the linear operator $x_k \to L_m x_1 x_2 \ldots x_m$ of the space E_k into Y. We say that L_m is separately continuous if for each $x_1, \ldots, x_{k-1}, x_{k+1}, \ldots, x_m$, where $x_k \in E_k$, the operator S_k is continuous for $k = 1, \ldots, m$. L_m is uniformly separately continuous (bounded) in x_k if the linear operator S_k is continuous (bounded), uniformly on the set $\|x_1\| = \ldots = \|x_{k-1}\| = \|x_{k+1}\| = \ldots = \|x_m\| = 1$. Finally by a continuous multilinear operator we mean a multilinear operator which is jointly continuous in all the variables, that is, continuous on the product space Π. We shall need certain implications among these notions and characterizations of continuous multilinear operators which are stated in the following:

<u>Theorem 3.1.</u> Let L_m be a multilinear operator on Π into Y. Then the following implications hold among the following statements:

$$a \Longleftrightarrow b \Longleftrightarrow c \Longleftrightarrow d \Longleftrightarrow e \Longrightarrow f$$

$$f \Longrightarrow a \text{ if the space } \Pi \text{ is complete}$$

(a) L_m is continuous on Π.

(b) L_m is continuous at the point zero $\theta_1 \ldots \theta_m$.

(c) L_m is bounded on each bounded set of Π.

(d) L_m is bounded on Π, that is there exists a constant $M \geq 0$ such that for all $x_1 \ldots x_m \in \Pi$,

(3.1) $\|L_m x_1 \ldots x_m\| \leq M \|x_1\| \ldots \|x_m\|$.

(e) L_m is uniformly separately continuous (bounded) in x_k for some fixed k.

(f) L_m is separately continuous (bounded).

We omit the proof since the various implications of the theorem are either given in the literature or involve simple manipulations. The most important proposition in the theorem is the equivalence of statements (a) and (f) when the space Π is complete. For an exposition of continuous multilinear operators on normed spaces we refer to Dieudonné [5] and Hille and Phillips [6]. For computational aspects of such operators see Rall [10]. Bourbaki [1] and Dieudonné [84] give an account of the theory of multilinear operators on topological vector spaces, where it may be noted that (f) does not imply (a) even for locally convex topological vector spaces. However, the implication holds under weaker conditions than stated here, for example in the case of a bilinear operator which is separately continuous it suffices to take E_1 to be a metrizable barreled vector space, E_2 a metrizable vector space and Y any locally convex space, so that in particular joint continuity follows if the spaces E_i are Fréchet spaces. In this connection we also note that Bourbaki defines a notion of hypocontinuity (and equihypocontinuity) for bilinear operators which is intermediate between the notions of separately continuous and continuous operators. For normed spaces, this notion is equivalent to any of the statements of Theorem 3.1 if Π is a Banach space. These notions become important for the theory of higher order differentiability in vector spaces without a norm, which require modifications to be examined in Section 4. The general theory of continuous multilinear operators is closely connected with topological tensor products which are surveyed in [84].

Let $\mathcal{L}_m$ denote the vector space of all continuous multilinear operators on the product space Π into Y. The greatest lower bound of all constants M satisfying the inequality (3.1) is given by

$$\|L_m\| = \sup \{ \|L_m x_1 \ldots x_m\| : \|x_i\| = 1, i = 1, \ldots, m \}$$

and is a bonafide norm on the space $\mathcal{L}_m$. Furthermore the space $\mathcal{L}_m$ equipped with this norm is a Banach space if and only if Y is complete.

The inductive definitions of higher order Gâteaux and Fréchet differentials lead to consideration of the space $\mathcal{L}(E_1, \mathcal{L}(E_2, \ldots, \mathcal{L}(E_m, Y), \ldots))$, where $\mathcal{L}(X,Y)$ denotes the space of all continuous linear operators on X into Y. There is a canonical isometric isomorphism which identifies the space $\mathcal{L}_m(E_1, \ldots, E_m; Y)$ with the space $\mathcal{L}(E_1, \mathcal{L}(E_2, \ldots, \mathcal{L}(E_m, Y), \ldots))$. In the sequel we shall identify the corresponding elements of these spaces under this linear isometry.

Now let $E_1 = E_2 = \ldots = E_m = E$. Let σ be a permutation of the set $\{1, 2, \ldots, m\}$ and consider $x: \{1, 2, \ldots, m\} \to \Pi$. σ induces a linear transformation $P_\sigma : \mathcal{L}_m(\Pi; Y) \to \mathcal{L}_m(\Pi; Y)$ defined by $(P_\sigma L_m)x = L_m(x \circ \sigma)$, where $x \in \Pi$. The mean of an m-linear operator is defined by $\overline{L}_m x_1 \ldots x_m = \frac{1}{m!} \sum_\sigma (P_\sigma L_m) x$, where the sum is taken over all permutations of $\{1, 2, \ldots, m\}$. An m-linear operator is said to be symmetric if $E_1 = \ldots = E_m$ and $P_\sigma L_m = L_m$ for every permutation σ. The mean is always symmetric; a symmetric m-linear operator coincides with its mean. Note that $\|L_m\| = \|P_\sigma L_m\|$ for every σ.

This notion of symmetry should be distinguished from the weaker notion of symmetric m-linear operator used in exterior algebra where L_m is said to be symmetric if
$$\sum_\sigma s_\sigma P_\sigma L_m = 0, \text{ where } s_\sigma = 1 \text{ if } \sigma \text{ is even and } s_\sigma = -1$$
if σ is odd.

The study of differentials of order m leads to mappings from a subset X of a normed linear space to the space $\mathcal{L}_m(E_1 \times \ldots \times E_m; Y)$. We write $D: X \to \mathcal{L}_m$ and call $D(x; \ldots)$ a formal differential operator. $DF(x; h_1, \ldots, h_m)$ is called a formal differential form. Several notions of continuity and directional continuity may be defined for $D(x; \ldots)$ depending on the topology used for $\mathcal{L}_m$.

Definition 3.1. A differential operator is said to be <u>pointwise continuous</u> in x at x_0 if for any $h_1 \ldots h_m \in \Pi$, $\|D(x; h_1 \ldots h_m) - D(x_0; h_1 \ldots h_m)\|_Y \to 0$ whenever $\|x - x_0\| \to 0$.

This is equivalent to considering the space $\mathcal{L}_m$ in the topology of pointwise convergence.

Definition 3.2. A differential operator is said to be <u>jointly continuous</u> in the variables $x, h_1, \ldots, h_m$ at x_0 if for any $x \in X$, $k_1 \ldots k_m \in \Pi$, $\|D(x; h_1 \ldots h_m) - D(x_0; k_1 \ldots k_m)\|_Y \to 0$ whenever $\|x - x_0\| \to 0$ and $\|h_i - k_i\| \to 0$ for $i = 1, \ldots, m$. This is joint continuity in the classical sense.

Definition 3.3. A differential operator is said to be <u>continuous</u> in x at x_0 if it is continuous as a transformation from the set X to the space $\mathcal{L}_m$: $\|D(x; \ldots) - D(x_0; \ldots)\|_{\mathcal{L}_m} \to 0$ whenever $\|x - x_0\| \to 0$. This is the same as considering the space $\mathcal{L}_m$ in the topology of uniform convergence on bounded sets.

We remark that Definitions 3.1 and 3.3 are implicit in the work of Kerner [49] and Graves and Hildebrandt [44] respectively.

Notions of directional continuity may be defined similarly. For example,

Definition 3.2d. A differential operator is <u>jointly directionally</u> continuous at x_0 if for any fixed y, $\|D(x_0 + ty; h_1 \ldots h_m) - D(x_0; k_1 \ldots k_m)\|_Y \to 0$ whenever $t \to 0$ and $\|h_i - k_i\| \to 0$ for $i = 1, \ldots, m$. Definitions 3.1d and 3.3d will denote the analogs of Definitions 3.1 and 3.3 for directional continuity.

The implication relationships among these notions are stated in the following:

Theorem 3.2 (a) A differential operator $D(x; h_1 \ldots h_m)$ is continuous at x_0 if and only if it is pointwise continuous at x_0 uniformly with respect to $h_1 \ldots h_m$ on the set $\|h_1\| = \ldots = \|h_m\| = 1$.

(b) A differential operator is pointwise continuous at x_0 if and only if it is jointly continuous.

Furthermore, the continuity of $D(x; h_1 \ldots h_m)$ at x_0 in the sense of any of these definitions implies the existence of a positive number M and a neighborhood $N(x_0)$ of x_0 such that for all $x \in N(x_0)$,

(3.2) $$\|D(x; h_1 \ldots h_m)\|_Y \leq M \|h_1\| \ldots \|h_m\|.$$

Similar implication relations hold among Definitions 3.1d, 3.2d and 3.3d for directional continuity. The directional continuity of $D(x; h_1 \ldots h_m)$ at x_0 in the sense of any of these definitions implies the existence for fixed y of a positive number M and a neighborhood $N(0)$ such that for all $\tau \in N(0)$,

(3.3) $$\|D(x_0 + \tau y; h_1, \ldots, h_m)\|_Y \leq M \|h_1\| \ldots \|h_m\|.$$

The form $L_m x \ldots x$ obtained from a symmetric m-linear operator $L_m x_1 \ldots x_m$ by setting $x_1 = \ldots = x_m = x$ is called an abstract power and is denoted by $L_m x^m$. The induced operator is called a power operator of degree m. A power operator of degree m is continuous if and only if it is bounded on some sphere. If so, then it is bounded on each bounded set and satisfies a Lipschitz condition of order one uniformly on such a sphere. An abstract polynomial of degree m is an operator of the form $Q(x) = \sum_{k=1}^{m} L_k x^k$, where $L_k \in \mathcal{L}(E^k; Y)$. Note that there is no loss of generality in assuming that the multilinear operators L_k are symmetric since the polynomial form and its Fréchet derivatives are unchanged if each L_k is replaced by its mean $\bar{L}_k$.

For historical development of the notion of abstract polynomial in real a complex normed spaces we refer to Fréchet [30], [31], Gâteaux [36], Michal [60], Martin [58], Highberg [45] and Gavurin [38]. The paper of Prenter in this volume provides a thorough study of several quantitative and qualitative aspects of polynomials and polynomial equations in Banach spaces. See also Hille and Phillips [6], Liusternik and Sobolev [9], Rall [10], and Burýsek [145].

Let E be a linear space, Y a normed linear space and F a mapping on a subset E into Y. Let $x_0 \in X$ and $h \in E$, $h \neq 0$. If $\phi(t) = F(x_0 + th)$ has n^{th} order derivative at $t = 0$, then $\phi^{(n)}(0)$ is called the nth Gâteaux variation and is denoted by $\delta^n F(x_0; h)$. In this form, the definition is due to Lévy [56]. The notions of higher order variations passed through many stages of generalizations in the work of Gâteaux [36, 37], Lévy [55, 56] Graves and Hildebrandt [42, 43, 44], Kerner [51, 52] and others. Clearly $\delta^n F(x_0; h)$ is homogeneous in h of degree n. Every operator which is homogeneous of degree m has m^{th} order variation at 0 and $\delta^m F(0; h) = m! F(h)$.

Higher order variations can also be defined inductively. If $\delta F(x; h)$ exists in a neighborhood of x_0, then $\delta F(x; \cdot)$ is an element of the space of all homogeneous operators on E into Y. If for fixed h, $\delta F(x; h)$ has a first variation at x_0, then we say F has a second variation at x_0 and denote it by $\delta^2 F(x_0; h, k)$, i.e., $\delta^2 F(x_0; h, k) = \delta_x(\delta F(x_0; h); k) = \lim_{t \to 0} \frac{1}{t} \{\delta F(x_0 + tk; h) - \delta F(x_0; h)\}$. It then follows that

$$\delta^m F(x_0; h_1, \ldots, h_m) = \frac{\partial^m}{\partial t_1 \cdots \partial t_m} F(x_0 + \sum_{i=1}^{m} t_i h_i) \bigg|_{t_1 = \cdots = t_m = 0}.$$

Note that the order in which the increments $h_1, \ldots, h_m$ are taken is essential. (See Section 4 for connections between m^{th} order variation and m^{th} order difference and sufficient conditions under which $\delta^m F(x_0; h_1, \ldots, h_m)$ is symmetric). If $h_1 = \cdots = h_m = h$, then $\delta^m F(x; h, \ldots, h)$ coincides with $\delta^m F(x; h)$ defined earlier.

If $\delta F(x_0 + t\Delta x, h + s\Delta h)$, where Δx and Δh are fixed elements of E, has a total differential at $t = s = 0$, then we say that $\delta F(x; h)$ has a __total variation__ at (x_0, h), and denote it by $\delta \delta F(x_0, h; \Delta x, \Delta h)$. Clearly if $\delta F(x_0; h)$

has a total variation at (x_0, h), then

$$\delta\delta F(x_0, h; \Delta x, \Delta h) = \delta^2 F(x_0; h, \Delta x) + \delta F(x_0; h, \Delta h) .$$

Now we turn to higher order Fréchet differentials. Suppose that F has a Fréchet derivative on a subset X of E. Then $F': X \times E \to Y$ is continuous and linear on E. We give eight notions of second order Fréchet differentials of F.

<u>Definition 3.4.</u> If $dF(x; \cdot)$, considered as a mapping on X into $\mathcal{L}(E; Y)$, has a Fréchet derivative at x_0, then we denote this derivative by $F''(x_0)$ and call it the <u>second order Fréchet derivative</u> of F at x_0.

Accordingly $F''(x_0)$ is an element of the space $\mathcal{L}(E; \mathcal{L}(E; Y))$ which is isometrically isomorphic to $\mathcal{L}(E \times E; Y)$. We make the identification of these two spaces in the sequel so that $F''(x_0) \in \mathcal{L}(E \times E; Y)$. If $F''(x)$ exists on X, then F'' is a map from X to $\mathcal{L}(E \times E; Y)$.

We call $F''(x_0)hk$, for $h, k \in E$, the <u>second order Fréchet differential</u> of F at x_0 and denote it also by $d^2F(x_0; h, k)$. We then have

$$\lim_{k \to 0} \|k\|^{-1} \|dF(x_0 + k; \cdot) - dF(x_0; \cdot) - d^2F(x_0; \cdot, k)\|_{\mathcal{L}} = 0 ,$$

where $\|\cdot\|_{\mathcal{L}}$ denotes the norm in the space of all continuous linear operators on E into Y.

<u>Definition 3.5.</u> The operator F is said to have at x_0 a <u>second order pointwise</u> Fréchet differential if there exists a continuous bilinear operator $B(x_0; \cdot, \cdot)$ on E such that for each fixed $h \in E$,

(3.4) $\quad \lim_{k \to \theta} \|k\|^{-1} \|dF(x_0 + k; h) - dF(x_0; h) - B(x_0; h, k)\|_Y = 0 .$

We call B the second order pointwise Fréchet derivative and denote it by $\partial^2 F(x_0; \cdot\cdot)$.

Definition 3.6. The operator F is said to have at x_0 a second order <u>partial Fréchet</u> differential if $dF(x;h)$, considered as a map from $X \times E$ into Y, has at x_0 a partial Fréchet differential with respect to x. We denote the second order partial Fréchet differential by $d_x^2 F(x_0; h, k)$. Accordingly for each $h \in E$,

$$\lim_{k \to 0} \|k\|^{-1} \|dF(x_0 + k; h) - dF(x_0; h) - d_x^2 F(x_0; h, k)\|_Y = 0 \quad ,$$

where $d_x^2 F(x_0; h, k)$ is continuous and linear in k. Note that $d_x^2 F(x_0; h, k)$ is not required to be linear nor continuous in h. However, it turns out that it is automatically linear in h.

Definition 3.7. The operator F is said to have a second order <u>total</u> Fréchet differential at x_0, if $dF(\cdot, \cdot): X \times E \to Y$ has a total Fréchet differential at (x_0, h) for all $h \in E$. We denote this differential by $ddF(x_0, h; \Delta x, \Delta h)$. (Note that we do not abbreviate ddF to d^2F since we reserve the latter notation for the second order Fréchet differential in the sense of Definition 3.4).

Definition 3.8. The operator F is said to have at x_0 a second order <u>strong</u> Fréchet differential if there exists an operator $B(x_0; \ldots) \in \mathcal{L}(E \times E; Y)$ such that for each $\varepsilon > 0$, there exists $r > 0$ where

$$\|dF(y; \cdot) - dF(z; \cdot) - B(x_0; \cdot, y - z)\| \leq \varepsilon \|y - z\|$$

for each pair of elements y, z with $\|y - x_0\| \leq r$, $\|z - x_0\| \leq r$. The operator $B(x_0; \ldots)$ is called the second order strong Fréchet derivative at x_0 and is denoted by $d^2 F^*(x_0; ..)$.

The notions of strong pointwise differential and strong partial and total differentials may be similarly defined.

Before we state various implications among these differentials, we consider continuity properties of $dF(x; h)$

which are implied by the existence of each of these differentials. It is clear that if F has a second order Fréchet differential at x_0, then $F'(x)$ is continuous at x_0; equivalently the differential $dF(x;h)$ is continuous in x at x_0 in the topology of uniform convergence (Definition 3.3). The implications for the other differentials are stated in the following:

<u>Theorem 3.3.</u> (a) If F has a second order pointwise Fréchet differential at x_0, then $F'(x)$ is continuous at x_0 (Definition 3.3), hence also pointwise continuous at x_0.
  (b) If F has a second order partial Fréchet differential at x_0, then $dF(x;h)$ is jointly continuous at (x_0,h) (in the sense of Definition 3.2).
  (c) If F has a second order total Fréchet differential, then $dF(x;h)$ is jointly continuous at (x_0,h).
  (d) If F has a second order strong Fréchet differential, then $F'(x)$ satisfies a Lipschitz condition in some neighborhood of x_0.

<u>Theorem 3.4.</u> (a) F has a second order Fréchet differential at x_0 if and only if F has a second order pointwise differential at x_0 and (3.4) holds uniformly with respect to h on the set $\|h\|=1$. In this case the two differentials are equal.
  (b) F has a second order pointwise Fréchet differential at x_0 if and only if F has a second order partial Fréchet differential at x_0 and the latter is jointly continuous in h and k (or continuous in h and the space E is complete).
  (c) If $F(x)$ has a second order pointwise Fréchet differential at x_0, then $dF(x;h)$ is totally differentiable (Definition 3.7) at $(x_0;h)$ and the total differential is given by $\partial^2 F(x_0;h;\Delta x) + dF(x_0;\Delta h)$.
  (d) If F has a second order strong Fréchet differential at x_0, then F has a second order Fréchet differential at x_0.

For simplicity we have so far restricted our discussion to second order Fréchet differentials. Various notions of mth order differentials may be considered similarly. We call attention to the important fact that the mth order Fréchet derivative is automatically a <u>symmetric</u> m-linear operator, i.e. $d^m F(x_0; h_1, \ldots, h_m)$ is invariant under all permutations of $(h_1, \ldots, h_m)$. See [5; 8.12.2] or [4; Theorem 5.1.1].

We now consider various notions of mth order Gâteaux differentials. Suppose for $m \geq 2$ the Gâteaux differential of order $m-1$ has been defined as a continuous $(m-1)$-linear operator $D^{m-1} F(x; h_1 \ldots h_{m-1})$, $D^{m-1} F(x; \ldots) \in \mathcal{L}_{m-1}(E_1 \times \ldots \times E_{m-1}; Y)$ for each $x \in X$.

<u>Definition 3.4G.</u> If $D^{m-1} F(x; \cdot)$, considered as a mapping on X into $\mathcal{L}_{m-1}(E_1 \times \ldots \times E_{m-1}; Y)$ has a Gâteaux derivative at x_0, we denote this derivative by $D^m F(x_0; \cdot)$ and call it the mth order Gâteaux derivative of F at x_0. Thus $D^m F(x_0; \cdot) \in \mathcal{L}_m(R_1 \times \ldots \times E_m; Y)$. The mth order Gâteaux differential is denoted by $D^m F(x_0; h_1 \ldots h_m)$, $h_i \in E_i$. We have by definition:

$$\lim_{t \to 0} t^{-1} \| D^{m-1} F(x_0 + th_m; \cdot) - D^{m-1} F(x_0; \cdot)$$

$$- D^m F(x_0; \ldots, th_m) \|_{\mathcal{L}_{m-1}} = 0 .$$

<u>Definition 3.5G.</u> F is said to have at x_0 a <u>pointwise</u> Gâteaux differential of order m if F has a pointwise Gâteaux differential of order $m-1$ and there exists a continuous m-linear operator $L(x_0; \cdot)$ such that for each fixed $h_1 \ldots h_{m-1}$,

$$\lim_{t \to 0} t^{-1} \| D^{m-1} F(x_0 + th_m; h_1 \ldots h_{m-1}) - D^{m-1} F(x_0; h_1 \ldots h_{m-1})$$

$$- L(x_0; h_1 \ldots h_m) \| = 0 .$$

$L(x_0; \cdot)$ is called the mth order pointwise Gâteaux derivative and is denoted by $\partial_G^m F(x_0; \cdot)$.

The mth order partial Gâteaux differential (Definition 3.6G) and the mth order total Gâteaux differential (Definition 3.7G) may be defined in an obvious way and are denoted by $D_x D^{m-1} F(x_0; h_1 \ldots h_m)$ and $DD^{m-1} F(x_0; h_1 \ldots h_{m-1}; \Delta x, \Delta h_1, \ldots, \Delta h_{m-1})$ respectively. We remark that $D_x D^{m-1} F(x_0; h_1 \ldots h_m)$ is continuous and linear in h_m but is not required to be continuous or linear in $h_1, \ldots, h_{m-1}$. However if $D^{m-1} F(x_0; h_1 \ldots h_{m-1})$ is assumed to be linear and continuous in $h_1, \ldots, h_{m-1}$, then $D^m F(x_0; h_1 \ldots h_m)$ is automatically linear (but not necessarily continuous) in $h_1, \ldots, h_{m-1}, h_m$.

Definition 3.8G. F is said to have at x_0 a strong Gâteaux differential of mth order if there exists an operator $T(x_0) \in \mathcal{L}_m(E_1 \times \ldots \times E_m; Y)$ such that for each $\varepsilon > 0$, there exists $r > 0$, where

$$\|D^{m-1} F(x_0 + tk; \cdot) - D^{m-1} F(x_0 + sk; \cdot) - T(x_0)(\cdot; (t-s)k)\|$$

$$< \varepsilon |t - s|$$

for each $k \in E_m$ and each pair of numbers s, t with $|s| \leq r$ and $|t| \leq r$. $T(x_0)$ is called the mth order strong Gâteaux derivative and is denoted by $D^m F^*(x_0; \cdot)$.

The continuity implications of these notions are stated in the following theorem.

Theorem 3.5. (a) If $D^m F(x_0; h_1 \ldots h_m)$ exists, then $D^{m-1} F(x; h_1 \ldots h_{m-1})$ is directionally continuous (Definition 3.3d) at x_0.

(b) If $D^m F(x; h_1 \ldots h_m)$ has a pointwise Gâteaux differential at x_0, then $D^m F(x; h_1 \ldots h_m)$ is directionally continuous at x_0; hence also pointwise directionally continuous (Definition 3.1d).

(c) If $D^m F(x; h_1 \ldots h_m)$ has a partial Gâteaux differential at x_0, then $D^m F(x; h_1 \ldots h_m)$ is jointly directionally continuous (Definition 3.2d) at $(x_0; h_1 \ldots h_m)$.

(d) If $D^m F(x; h_1 \ldots h_m)$ has a total Gâteaux differential, at x_0, then $\hat{D}^m F(x; h_1 \ldots h_m)$ is jointly directionally continuous at $(x_0; h_1 \ldots h_m)$.

(e) If $D^m F(x, \cdot)$ has a strong Gâteaux differential at x_0, then for some $M \geq 0$,

$$\|D^m F(x_0 + tk; \cdot) - D^m F(x_0 + sk; \cdot)\|_{\mathcal{L}_m} \leq M |t - s|$$

for all t, s in some neighborhood of zero.

<u>Theorem 3.6.</u> (a) $D^m F(x; \cdot)$ has a Gâteaux differential at x_0 if and only if $D^m F(x; h_1 \ldots h_m)$ has a pointwise Gâteaux differential at x_0 and

$$(3.5) \quad \lim_{t \to 0} t^{-1} \{D^m F(x_0 + tk; h_1 \ldots h_m) - D^m F(x_0; h_1 \ldots h_m)\}$$

$$= \partial_G^{m+1} F(x_0; h_1 \ldots h_m k)$$

holds uniformly with respect to k on the set $\|k\| = 1$. Consequently, $D^m F(x; \cdot)$ has a Fréchet differential at x_0 if and only if (3.5) holds uniformly with respect to k and $h_1 \ldots h_m$ on the set $\|k\| = \|h_1\| = \ldots = \|h_m\| = 1$.

(b) $D^m F(x; h_1 \ldots h_m)$ has a pointwise Gâteaux differential at x_0 if and only if $D^m F(x; h_i \ldots h_m)$ has a Gâteaux partial differential at x_0 and the latter is jointly continuous in $h_1 \ldots h_{m+1}$.

(c) If $D^m F(x; h_1 \ldots h_m)$ has a pointwise Gâteaux differential at x_0, then $D^m F(x; h_1 \ldots h_m)$ is totally differentiable at x_0 and the total differential is given by

$$\partial_G D^m F(x_0; h_1 \ldots h_m; \Delta x) + D^m F(x_0; \Delta h_1, h_2 \ldots, h_m)$$

$$+ \ldots + D^m F(x_0; h_1, \ldots, \Delta h_m) .$$

(d) If $D^m F(x; \cdot)$ has a strong Gâteaux differential at x_0, then $D^{m+1} F(x; \cdot)$ exists.

Higher order Hadamard differentials may be defined inductively in a similar way. Higher order differentials of Gâteaux, Fréchet, and Hadamard are considered within a unifying framework and in the setting of linear topological spaces in Section 4. It is clear that existence of nth order Gâteaux or Fréchet differential in the sense of any definition given in this section implies the existence of the nth order variation. For a more detailed exposition of the material of this section and related aspects of higher order differentials, see [135].

We conclude this section by stating the chain rule and the analogue of Leibnitz's rule for higher order differentials. The proofs follow by (tedious) induction.

Let $F: X \to Y$ be n times Fréchet (Hadamard) differentiable at $x_0 \in X$, let $G: Y \to Z$ be n times Fréchet (Hadamard) differentiable at $y_0 = F(x_0)$. Then the composite map $T = G \circ F$ is n times Fréchet (Hadamard) differentiable at x_0 and

$$T^{(n)}(x_0)(h_1, \ldots, h_n) =$$

$$\sum_{k=1}^{n} \sum_{\sigma} G^{(k)}(y_0) (F^{(\ell_1)}(x_0) h_{i_1^1}, \ldots, h_{i_{\ell_1}^1}) \ldots (F^{(\ell_k)}(x_0) h_{i_1^k}, \ldots, h_{i_{\ell_k}^k}),$$

where the second sum is taken over all partitions σ of the set $\{h_1, \ldots, h_n\}$ into disjoint nonempty sets

THE ROLE OF DIFFERENTIALS

$\{h_{i_1^1}, \ldots, h_{i_{\ell_1}^1}\}, \ldots, \{h_{i_1^k}, \ldots, h_{i_{\ell_k}^k}\}$ with $\ell_1 + \ldots + \ell_k = n$,

$\ell_j > 0$, $i_1^j < i_2^j < \ldots < i_{\ell_j}^j$, $j = 1, 2, \ldots, k$, $i_1^1 < i_1^2 < \ldots < i_1^k$.

In particular for $X = \mathbb{R}$

$$T^{(n)}(x_0) = \sum_{k=1}^{n} \sum c(\ell_1, \ldots, \ell_k) G^{(k)}(y_0) (F^{(\ell_1)}(x_0), \ldots, F^{(\ell_k)}(x_0)),$$

where the second sum is taken over all solutions in positive integers of the equation $\ell_1 + \ldots + \ell_k = n$ such that $\ell_1 \leq \ell_2 \leq \ldots \leq \ell_n$ and

$$c(\ell_1, \ldots, \ell_k) = \frac{q!}{\prod_{i=1}^{k} \ell_i! \prod_{j=1}^{q} m_j!},$$

where m_1 is the number of $\ell_1, \ldots, \ell_k$ which are equal to 1, m_2 to 2, and so on, and $q = \ell_1 + \ldots + \ell_k$.

We finally consider the analogue of Leibniz's rule for the mth derivative of a "product" of two mappings. Let U be an open subset of E, $F: U \to Y_1$ and $G: U \to Y_2$ be m times (Gâteaux, Fréchet, or Hadamard) differentiable. Let I_2 be the identity continuous bilinear map on $Y_1 \times Y_2$ into Z, i.e., $I_2(u,v) = (u,v)$. Define FG by $(FG)(x) = I_2(F(x), G(x))$, for $x \in U$. Then FG is m times differentiable and for $k \leq m$, $x \in U$, we have

$$(FG)'(x) = \sum_{j=0}^{k} \binom{k}{j} (F^{(j)}(x), G^{(k-j)}(x)).$$

In particular, if $Y_1 = Y_2 = \mathbb{R}$, then the "product" is the ordinary product.

4. Higher Order Differentials in Linear Topological Spaces

There are several factors which make the transition of higher order differentiability from the setting of normed spaces to the setting of linear topological spaces less immediate than that of the first order differential. First we have several nonequivalent ways of defining polynomials on a topological space, while for normed spaces these definitions turn out to be equivalent. Second the theory of continuous multilinear operators is more complicated in topological spaces than the corresponding theory in normed spaces; in addition to the notions of continuous and separately continuous multilinear operators, there is an "intermediate" notion of a hypocontinuous multilinear operator. Finally there are complications that arise from the fact that in general topological spaces, "differentiability" does not always imply "continuity", which become more pronounced as one attempts to develop higher order chain rules.

We shall define higher order differentials inductively. To this end, let H_i $(i = 1, \ldots, n)$ be linear topological spaces which are subspaces of a linear space X, σ_i a system of subsets of H_i and β_i a system of bounded subsets of H_i. Each of the systems σ_i, β_i is assumed to contain all the one dimensional subsets of H_i. Let Y be a linear topological space, and let $\tilde{u}$ be a multilinear map of $H_1 \times \ldots \times H_n$ to Y. Define $u \cdot h_n \ldots h_1 \equiv \tilde{u}(h_1, \ldots, h_n)$ and, by induction, $u \cdot h_n \ldots h_{i+1}$ as the map $h_i \to u \cdot h_n \ldots h_i$ for $i = 1, \ldots, n-1$; also let u denote the map $h_n \to u \cdot h_n$.

A multilinear map $\tilde{u} \colon H_1 \times \ldots \times H_n \to Y$, $n \geq 2$, is said to be $(\beta_1, \ldots, \beta_{n-1})$-hypocontinuous if $u \cdot h_1 \ldots h_n \to 0$ as $h_i \to 0$ uniformly with respect to $h_j \in B_j$ for $i = 1, \ldots, n$, for all choices of the sets $B_j \in \beta_j$, $1 \leq j \leq i-1$, and all choices of the elements $h_k \in H_k$, $i+1 \leq k \leq n$.

A multilinear map $\tilde{u} \colon H_1 \times \ldots \times H_n \to Y$, $n \geq 2$, is said to be $(\sigma_1, \ldots, \sigma_n)$-continuous if $\tau \tilde{u}(h_1 \ldots h_n) \to 0$ as $\tau \to 0$ $\tau \in \mathbb{R}$, uniformly with respect to $h_i \in S_i$, for all choices of the sets $S_i \in \sigma_i$, $i = 1, 2, \ldots, n$.

Note that for $n = 1$, $(\sigma_1, \ldots, \sigma_n)$-continuity reduces to σ-continuity at zero of a linear map, and for $n = 2$, $(\beta_1, \ldots, \beta_{n-1})$-hypocontinuity is ordinary hypocontinuity in the sense of Bourbaki [1]; see also Schaefer [3] for a lucid exposition on hypocontinuous multilinear maps. It is easy to show that every continuous multilinear map is $(\beta_b, \ldots, \beta_b)$-hypocontinuous and $(\beta_b, \ldots, \beta_b)$-continuous, where β_b is the system of all bounded subsets of H_i. Also every $(\beta_1, \ldots, \beta_{n-1})$-hypocontinuous map is separately continuous and $(\beta_1, \ldots, \beta_{n-1}; \beta_b)$-continuous, but is not necessarily continuous. The linear space of all multilinear $(\beta_1, \ldots, \beta_{n-1})$-hypocontinuous maps is denoted by $\mathcal{L}_{\beta_1, \ldots, \beta_{n-1}}(H_1, \ldots, H_n; Y)$.

Let $\mathcal{L}(H, Y)$ denote the linear space of all continuous linear maps on H into Y and $\mathcal{L}_\beta(H, Y)$ the linear topological space obtained by endowing $\mathcal{L}(H, Y)$ with the topology of uniform convergence of sets of the system β. Put $\mathcal{L}_0 = Y$ and define by induction $\mathcal{L}_i = \mathcal{L}_{\beta_i}(H_i, \mathcal{L}_{i-1})$, $\mathcal{L}'_i = \mathcal{L}(H_i, \mathcal{L}_{i-1})$.

It can be shown that $\tilde{u}: H_1 \times \ldots \times H_n \to Y$, $n \geq 2$, is $(\beta_1, \ldots, \beta_{n-1})$-hypocontinuous if and only if $u \cdot h_n \ldots h_{i+1} \in \mathcal{L}'_i$ for $i = 1, \ldots, n-1$, and for all elements $h_k \in H_k$, $i+1 \leq k \leq n$, and $u \in \mathcal{L}'_n$. The space $\mathcal{L}_{\beta_1, \ldots, \beta_{n-1}}(H_1, \ldots, H_n; Y)$ is algebraically isomorphic to the space $\mathcal{L}'_n$.

Let $F: X \to Y$. We define the $(\sigma_1, \beta_1; \ldots; \sigma_n, \beta_n)$-derivative $F^{(n)}$ along the subspaces $H_1, \ldots H_n$ as the (H_n, σ_n, β_n)-derivative of the $(H_1, \sigma_1, \beta_1; \ldots; H_n, \sigma_n, \beta_n)$-derivative $F^{(n-1)}$. $F^{(n)}(x_0)$ is called the $(H_1, \sigma_1, \beta_1; \ldots; H_n, \sigma_n)$-derivative at x_0. We call this derivative the Fréchet (Hadamard, Gâteaux) $H_1, \ldots, H_n$-derivative if each of the systems σ_i, β_i consists of all bounded (sequentially compact, finite) subsets of H_i. When $\sigma_i = \sigma$ and $\beta_i = \beta$ for all i, we call $F^{(n)}$ the (H, σ, β)-derivative. Note that

$F^{(n)}(x_0) \in \mathcal{L}'_n$, whereas $F^{(n)}$ is a map on X to $\mathcal{L}_n$.

We remark that other inductive definitions of higher order differentials analogous to those given in the preceding section can be easily formulated. We shall not pursue this here.

The nth variation of a mapping F on a linear space X into a linear topological space can be defined by induction as usual:

$$\delta^n F(x; h_1, \ldots, h_n) =$$

$$\lim_{\substack{t \to 0 \\ t \in \mathbb{R}}} \frac{1}{t} \{\delta^{n-1} F(x + t h_n; h_1, \ldots, h_{n-1}) - \delta^{n-1} F(x; h_1, \ldots, h_{n-1})\}.$$

$\delta^n F(x; h_1, \ldots, h_n)$ is not necessarily symmetric in $h_1, \ldots, h_n$; the order in which the increments are taken is essential. It follows easily that

$$\delta^n F(x; h_1, \ldots, h_n) = \left. \frac{\partial^n F(x + t_1 h_1 + \ldots + t_n h_n)}{\partial t_n \ldots \partial t_1} \right|_{t_1 = \ldots = t_n = 0}.$$

If the map F is $(H_1, \sigma_1, \beta_1; \ldots; H_{n-1}, \sigma_{n-1}, \beta_{n-1}; H_n, \sigma_n)$-differentiable at x, then $\delta^n F(x; h_1, \ldots, h_n)$ exists and the map $(h_1, \ldots, h_n) \to \delta^n F(x; h_1, \ldots, h_n)$ is multilinear and $(\beta_1, \ldots, \beta_{n-1})$-hypocontinuous on $H_1 \times \ldots \times H_n$. On the other hand, every n^{th} β^n-derivative along the space H is symmetric.

We now examine the connection between higher order differences and variations. The nth difference of F at $x \in X$ is defined inductively by

THE ROLE OF DIFFERENTIALS

$$\Delta^1 F(x;h) = \Delta F(x;h) = F(x+h) - F(x)$$

$$\Delta^n F(x;h_1,\ldots,h_n) = \Delta^{n-1} F(x+h_n;h_1,\ldots,h_{n-1})$$
$$- \Delta^{n-1} F(x;h_1,\ldots,h_{n-1}) .$$

The nth difference is symmetric in $(h_1,\ldots,h_n)$ and

$$\Delta^n F(x;h) \equiv \Delta^n F(x;h,\ldots,h) = \sum_{k=0}^{n} (-1)^{n-k} \binom{n}{k} F(x+kh) .$$

It follows by induction that

(4.1) $\quad \delta^n F(x;h_1,\ldots,h_n) = \lim_{t_n \to 0} \cdots \lim_{t_1 \to 0} \dfrac{\Delta^n F(x;t_1 h_1,\ldots,t_n h_n)}{t_1 \cdots t_n} .$

That is, if one side of (4.1) exists, then so does the other and the two are equal. The equation

(4.2) $\quad \delta^n F(x;h_1,\ldots,h_n) = \lim_{(t_1,\ldots,t_n) \to (0,\ldots,0)} \dfrac{\Delta^n F(x;t_1 h_1,\ldots,t_n h_n)}{t_1 \cdots t_n}$

does not hold in general. However, if for some $x, h_1, \ldots, h_n$, both sides of (4.2) exist, then they are equal. If for some $x_0 \in X$ and all $(h_1,\ldots,h_n)$ in a subset A of $X \times \cdots \times X$, (4.2) holds, then it follows from the symmetry of the nth difference that $\delta^n F(x_0;h_1,\ldots,h_n)$ is symmetric on A. Note that this sufficient condition for the symmetry of the nth variation is considerably weaker than the continuity of the nth variation at x_0 along the subspace H; the latter condition being also sufficient for symmetry of the nth variation.

We now consider briefly various definitions of polynomials on a linear topological space. Let X be a linear space, Y, H, linear topological spaces with H being a subspace of X. Let $x_0 \in X$ and put $\mathfrak{m} = H + x_0$. A mapping $P: X \to Y$ may be considered to be a "polynomial" of degree n if it satisfies one of the following conditions (which are not necessarily equivalent and thus different notions of a polynomial arise).

P1. The $(n+1)^{st}$ β-derivative of P on H is zero at each point of $\mathfrak{m}$, and this is the lowest derivative with this property.

P2. For each $x \in \mathfrak{m}$ and each $h \in H$, the map defined by $P(x+th)$, $t \in \mathbb{R}$, is a polynomial in t of degree n with coefficients in Y, i.e.,

$$P(x+th) = \sum_{k=0}^{n} c_k(x;h) t^k, \quad c_n(x,h) \neq 0.$$

P3. For each $x \in \mathfrak{m}$ and $h_1, \ldots, h_{n+1} \in \mathfrak{m}$, the $(n+1)^{st}$ difference of P vanishes identically:

$$\Delta^{n+1} P(x; h_1, \ldots, h_{n+1}) = 0 \text{ but } \Delta^n P(x; h_1, \ldots, h_n) \neq 0.$$

P4. There is an element $y_0 \in Y$ and symmetric multilinear maps A_k of $\underbrace{H \times \ldots \times H}_{k}$ to Y, $k = 1, \ldots, n$, such that

$$P(x_0 + h) = y_0 + A_1(h) + A_2(h,h) + \ldots + A_n(h, \ldots, h),$$

with $A_n \neq 0$.

The implications relationships among these notions are as follow: (i) P3 $\Longleftrightarrow$ P4 $\Longrightarrow$ P2;

(ii) P1 $\Longrightarrow$ P2 if Y is locally convex;

(iii) P2 $\Longrightarrow$ P3 if P is (H, β_k)-differentiable on $\mathfrak{m}$;

(iv) P4 ⟹ P1 if the maps A_k are $(\underbrace{\beta, \ldots, \beta}_{k-1})$-hypo-continuous for $k = 1, \ldots, n$.

5. Mean Value Theorems in Banach and Linear Topological Spaces

The Riemann-Graves Integral

Let $F: \Omega \subset X \to Y$, where Ω is a convex subset of X, and X, Y are Banach spaces. Suppose that for any $x, x+h \in \Omega$, the mapping $g: [0,1] \to Y$ defined by $g(t) = F(x+th) = F[(1-t)x + tz]$, where $z = x + h$, is continuous on $[0,1]$. Then g is integrable in the Riemann-Graves sense, i.e. the Riemann integral of a function of a real variable with values in a normed space. We call the integral

$$\int_0^1 F[(1-t)x + tz]\,dt$$

the rectilinear integral of the function F on the line segment $\overline{xz}$ of X and denote it by $\int_{\overline{xz}} F(x)dx$.

The Primitive of a Differential and the Fundamental Theorem of Calculus

Let X and Y be Banach spaces and suppose that the variation of $F: X \to Y$ exists. Let π be a partition of the interval $[0,1]$, $0 = t_0 < t_1 < \ldots < t_n = 1$. Then

$$F(x + t_i h) - F(x + t_{i-1} h) = (t_i - t_{i-1})\delta F(x + t_{i-1}h; h)$$
$$+ (t_i - t_{i-1})\alpha(t_i - t_{i-1})$$

where

$$\alpha(t_i - t_{i-1}) \to 0 \text{ as } t_i - t_{i-1} \to 0.$$

Consider the sum

$$\sum_{i=1}^{n} \{F(x+t_i h) - F(x+t_{i-1} h)\} = F(x+h) - F(x)$$

$$= \sum_{i=1}^{n} (t_i - t_{i-1}) \delta F(x + t_{i-1} h; h) + \alpha_n .$$

If $\delta F(x+th;h)$ is continuous with respect to t, then as $\max_{1 \le i \le n} |t_i - t_{i-1}| \to 0$, the second sum in the preceding equality tends to the Riemann-Graves integral of $\delta F(x + th;h)$. Consequently, α_n has a limit α as $n \to \infty$,

(5.1) $\qquad \alpha = F(x+h) - F(x) - \int_0^1 \delta F(x + th;h) dt .$

We shall show that $\alpha = 0$. First consider the case when F is a functional. Then by the mean value theorem (see also (5.4) below)

$$F(x+th) - F(x) = t \delta F(x + \gamma th; h) \qquad 0 < \gamma < 1 .$$

Thus

$$F(x+t_i h) - F(x+t_{i-1} h) = (t_i - t_{i-1}) \delta F(x + \gamma_i (t_i - t_{i-1}) h; h)$$

and

$$\sum_{i=1}^{n} [F(x+t_i h) - F(x+t_{i-1} h)] = F(x+h) - F(x)$$

$$= \sum_{i=1}^{n} (t_i - t_{i-1}) \delta F(x + \gamma_i (t_i - t_{i-1}) h; h) .$$

Taking the limit as the norm of the partition goes to zero, we get

(5.2) $\qquad F(x+h) - F(x) = \int_0^1 \delta F(x + th; h) dt .$

Now consider the general case of a mapping $F: X \to Y$, and let $\ell \in Y^*$. Applying ℓ to both sides of (5.1) and using (5.2) for the functional $\ell(F)$, we obtain

$$\ell[F(x+h) - F(x)] - \ell[F(x)] = \ell[\int_0^1 \delta F(x+th;h)dt] + \ell(\alpha)$$

$$= \int_0^1 \ell \delta F(x+th;h)dt + \ell(\alpha) = \int_0^1 \delta \ell F(x+th;h)dt .$$

Thus $\ell(\alpha) = 0$ for every $\ell \in Y^*$; hence $\alpha = 0$, and the formula (5.2) is established in general. Note that (5.2) is what is often called the <u>fundamental theorem of calculus</u>, relating a differential to its primitive; it may also be considered as a mean-value theorem in integral form. It remains valid in the general setting of differentiation in topological spaces (Theorem 14 in [95]).

Remark. It follows from (5.2) that

(5.3) $\qquad \|F(x+h) - F(x)\| \leq \sup_{0 \leq t \leq 1} \|\delta F(x+th; \cdot)\| \; \|h\| .$

However in obtaining (5.3) from (5.2) it was assumed that $\delta F(x+th; \cdot)$ is continuous in t along the line segment $0 \leq t \leq 0$. This assumption is not necessary for the validity of (5.3). See Theorem 5.4.

Mean Value Theorems in Normed Spaces

Rolle's theorem and mean value theorem in <u>equality form</u> are both false in general for complex valued functions or mappings with values in a Banach space of dimension greater than one. For example, if $f(x) = e^{ix}$ on $[0, 2\pi]$, then $f(2\pi) - f(0) = 0$, while $|f'(x)| = 1$ for all x. (A restricted form of Rolle's theorem holds for a real valued differentiable function f defined on a relatively compact subset A of a finite dimensional space E, i.e., if under these conditions f is continuous on $\bar{A}$ and is equal to 0 on the boundary of A, then there exists a point $x_0 \in A$ such that $f'(x_0) = 0$).

The classical formulation of the mean value theorem (for real functions) in the form $f(b) - f(a) = f'(c)(b-a)$ is not very revealing and has no analogue for f with vector values; also it says nothing about the number c except that it lies between a and b, and one may as well write the mean value theorem in an inequality form: $|f(b) - f(a)| \leq \sup_{a < x < b} |f'(x)| |b-a|$. This form has analogues in infinite dimensional spaces.

We now present four types of mean value theorems:

1. "Soft" mean value theorems, sometimes called Lagrange's formulas;

2. mean value theorems in majorant form.;

3. strong and weak mean value theorems for vector valued functions;

4. geometric mean value theorems involving closed convex hulls of the set of values of the derivative.

Theorem 5.1. Let U be an open set of a normed space X and let f be a functional on U. Suppose that U contains the line segment joining $x, x+h \in U$, and that f has a Gâteaux variation at each point of this line segment. Then

(5.4) $f(x+h) - f(x) = \delta f(x + \tau h; h)$ for some $0 < \tau < 1$.

Relation (5.4), which is known as Lagrange's formula, is an immediate consequence of the mean value theorem for a function of a real variable applied to the function φ defined by $\varphi(t) = f(x+th)$.

Now let $F: U \to Y$, where Y is a normed linear space, and let ℓ be a linear functional on Y. Then $\ell \circ F$ is a functional on U. Application of Theorem 5.1 to $\ell \circ F$ yields

Theorem 5.2. Let $F: U \to Y$ have a Gâteaux variation at each point of the line segment joining $x, x+h \in U$. Then for each linear functional ℓ,

$$\ell[F(x+h)] - \ell[F(x)] = \ell[\delta F(x+\tau h; h)]$$

for some $\tau \in (0,1)$ (τ depends on ℓ).

Corollary 5.2. If the functional f has a gradient which is bounded on an open convex set Ω of a normed linear space X, then f satisfies a Lipschitz condition on Ω, i.e., for all $x_1, x_2 \in \Omega$,

$$|f(x_1) - f(x_2)| \leq M\|x_2 - x_1\| \text{ for some } M \geq 0.$$

Corollary. If $\delta f(x; h) = 0$ for all x in an open convex set Ω, then $f(x) =$ constant for all $x \in \Omega$.

Theorem 5.3. (Mean value theorem in majorant form). Let $a, b \in \mathbb{R}$, $a < b$, let Y be a normed linear space and let the continuous maps $\varphi: [a,b] \to Y$, $\psi: [a,b] \to \mathbb{R}$ have right derivatives $D^+\varphi(t)$ and $D^+\psi(t)$ at each point $t \in (a,b)$ (where $\lim_{h \to 0^+} \frac{1}{h}\{\varphi(t+h) - \varphi(t)\} = D^+\varphi(t)$) except possibly on a countable subset of (a,b). Suppose that

$$\|D^+\varphi(t)\| \leq D^+\psi(t) \text{ for } a < t < b.$$

Then $\|\varphi(b) - \varphi(a)\| \leq \psi(b) - \psi(a)$.

For a proof see Cartan [4, Théorème 3.1.3], Dieudonné [5], Loomis and Sternberg [13]. Note that the theorem remains valid if right derivatives are replaced by left derivatives on (a,b). The most immediate application of Theorem 5.3 is in the choice $\psi(t) = kt$, $k \geq 0$, which yields

Corollary 5.3. Let $\varphi: [a,b] \to Y$ be continuous and have a right derivative at each $t \in (a,b)$ except possibly on a countable subset. Suppose that $\|D^+\varphi(t)\| \leq k$ for some $k \geq 0$.

Then for any $t_1, t_2 \in [a,b]$, $\|\varphi(t_1) - \varphi(t_2)\| \le k |t_1 - t_2|$.

Corollary 5.3 yields immediately a mean value theorem for mapping on normed spaces.

Theorem 5.4. Let U be an open subset of a normed linear space X and $F: U \to Y$. If the line segment joining two points x_1 and x_2 in U lies in U and if F has a Gâteaux variation for all points on this line segment, then

$$\|F(x_2) - F(x_1)\| \le \|x_2 - x_1\| \sup\{\|\delta F((1-t)x_1 + tx_2; \cdot)\|: 0 \le t \le 1\}.$$

Proof: Take $\varphi(t) = F((1-t)x_1 + tx_2)$ in Corollary 5.3.

Corollary 5.4.1. Let Ω be an open convex subset of a normed linear space X and let $F: \Omega \to Y$ have a gradient which is bounded on Ω, i.e.,

$$\|F'(x)\| \le M \text{ for all } x \in \Omega.$$

Then F satisfies a Lipschitz condition on Ω.

Corollary 5.4.2. If $\delta F(x;h) = 0$ for all $x \in \Omega$ and all $h \in X$, where Ω is an open convex subset of X, then F is a constant function on Ω.

Theorem 5.4 admits a generalization to higher order differences. Let F be an n times differentiable mapping of an open subset U of X into Y. Let $x_0 \in U$, $h_i \in X$ ($1 \le i \le n$) be such that $x + \sum_{i=1}^{n} \lambda_i h_i \in U$ for $0 \le \lambda_i \le 1$, $1 \le i \le n$. We define by induction on k ($1 \le k \le n$)

$$\Delta^1 F(x_0; h_1) = F(x_0 + h_1) - F(x_0),$$

$$\Delta^k F(x_0; h_1, \ldots, h_k) = \Delta^{k-1} G_k(x_0; h_1, \ldots, h_{k-1}),$$

where

$$G_k(x) = F(x+h_k) - F(x) .$$

Theorem 5.5. Under the preceding assumptions,

(a) $\|\Delta^n F(x_0; h_1, \ldots, h_n)\| \le \|h_1\| \ldots \|h_n\| \sup\{\|F^{(n)}(z)\|: z \in \Gamma_n\}$,

where

$$\Gamma_n = \{z: z = x_0 + \sum_{i=1}^n \lambda_i h_i ; 0 \le \lambda_i \le 1\} , \text{ and}$$

(b) $\|\Delta^n F(x_0; h_1, \ldots, h_n) - d^n F(x_0; h_1, \ldots, h_n)\|$

$\le \|h_1\| \ldots \|h_n\| \sup\{\|F^{(n)}(z) - F^{(n)}(x_0)\|: z \in \Gamma_n\}$.

The proof is by induction. For $n = 1$, (a) follows from Theorem 5.4.

We now turn to some mean value theorems for continuous vector valued functions which are due to Aziz, Diaz, Mlak and Výborný. (See [148], [149], [151], [154]).

Theorem 5.6. Let $\varphi: [a,b] \to Y$, when Y is a normed space, a, b are finite real numbers with $b > a$. If φ is (strongly) continuous on $[a,b]$, then there is a number c, with $a < c < b$; and, together with c, either there is a decreasing sequence of real numbers b_k, such that $c < b_{k+1} < b_k < b$ for $k = 1, 2, \ldots$ and $\lim_{k \to \infty} b_k = c$, for which

$$\left\| \frac{\varphi(b) - \varphi(a)}{b-a} \right\| \le \left\| \frac{\varphi(b_k) - \varphi(c)}{b_k - c} \right\| ,$$

or else there is an increasing sequence of real numbers a_k, such that $a < a_k < a_{k+1} < c$ for $k = 1, 2, \ldots$ and $\lim_{k \to \infty} a_k = c$, for which

$$\left\| \frac{\varphi(b) - \varphi(a)}{b-a} \right\| \le \left\| \frac{\varphi(c) - \varphi(a_k)}{c - a_k} \right\| .$$

In particular, there is a number c, with $a < c < b$, such that

$$\left\| \frac{\varphi(b) - \varphi(a)}{b-a} \right\| \leq \limsup_{h \to 0^+} \left\| \frac{\varphi(c+h) - \varphi(c)}{h} \right\|$$

or

$$\left\| \frac{\varphi(b) - \varphi(a)}{b-a} \right\| \leq \limsup_{h \to 0^+} \left\| \frac{\varphi(c) - \varphi(c-h)}{h} \right\| .$$

Remark. If the function φ possesses a (finite) derivative then $\lim\limits_{k \to \infty} \dfrac{\varphi(b_k) - \varphi(c)}{b_k - c} = \lim\limits_{k \to \infty} \dfrac{\varphi(c) - \varphi(a_k)}{c - a_k} = \varphi'(c)$, and Theorem 5.6 implies that

(5.4) $$\left\| \frac{\varphi(b) - \varphi(a)}{b-a} \right\| \leq \| \varphi'(c) \|$$

which is the assertion of the mean value theorem in [148, p. 261].

On the other hand, if φ has both a finite right hand derivative and a finite left hand derivative (but are not necessarily equal), then Theorem 5.6 implies the existence of a number c between a and b such that either

$$\left\| \frac{\varphi(b) - \varphi(a)}{b-a} \right\| \leq \| D^+ \varphi(c) \| ,$$

or

$$\left\| \frac{\varphi(b) - \varphi(a)}{b-a} \right\| \leq \| D^- \varphi(c) \| .$$

Remark. In the case of a continuous <u>real</u> valued function, a slight modification in the proof of Theorem 5.6 (see [154]) leads to the following assertion: There is a number c, with $a < c < b$, such that either

$$\liminf_{h \to 0^+} \frac{\varphi(c+h) - \varphi(c)}{h} \leq \frac{\varphi(b) - \varphi(a)}{b-a} \leq \limsup_{h \to 0^+} \frac{\varphi(c) - \varphi(c-h)}{h}$$

or

$$\liminf_{h \to 0^+} \frac{\varphi(c) - \varphi(c-h)}{h} \leq \frac{\varphi(b) - \varphi(a)}{b-a} \leq \limsup_{h \to 0^+} \frac{\varphi(c+h) - \varphi(c)}{h}.$$

However this conclusion is weaker than the assertion of the mean value theorem of W. H. and G. C. Young [171, p. 10] concerning the Dini derivates, which states that there exists a number c, with $a < c < b$, such that either

$$\limsup_{h \to 0^+} \frac{\varphi(c+h) - \varphi(c)}{h} \leq \frac{\varphi(b) - \varphi(a)}{b-a} \leq \liminf_{h \to 0^+} \frac{\varphi(c) - \varphi(c-h)}{h}$$

or

$$\limsup_{h \to 0^+} \frac{\varphi(c) - \varphi(c-h)}{h} \leq \frac{\varphi(b) - \varphi(a)}{b-a} \leq \liminf_{h \to 0^+} \frac{\varphi(c+h) - \varphi(c)}{h}.$$

This observation naturally raises the question: Can the conclusion of Theorem 5.6 be strengthened so as to assert the existence of c, $a < c < b$, such that either

$$\left\| \frac{\varphi(b) - \varphi(a)}{b-a} \right\| \leq \liminf_{h \to 0^+} \left\| \frac{\varphi(c+h) - \varphi(c)}{h} \right\|,$$

or

$$\left\| \frac{\varphi(b) - \varphi(a)}{b-a} \right\| \leq \liminf_{h \to 0^+} \left\| \frac{\varphi(c) - \varphi(c-h)}{h} \right\| ?$$

The following theorem of Diaz and Výborný [154] shows that a stronger conclusion actually holds.

<u>Theorem 5.7.</u> Under the hypotheses of Theorem 5.6, there exists a number c, with $a < c < b$, such that either,

whenever both $h > 0$ and $a \leq c + h \leq b$, one has

$$\left\| \frac{\varphi(b) - \varphi(a)}{b - a} \right\| \leq \left\| \frac{\varphi(c + h) - \varphi(c)}{h} \right\| ;$$

or, whenever both $h > 0$ and $a \leq c - h \leq b$, one has

$$\left\| \frac{\varphi(b) - \varphi(a)}{b - a} \right\| \leq \left\| \frac{\varphi(c) - \varphi(c - h)}{h} \right\| .$$

L. J. Nicolescu [163, 164] extended the Mean Value Theorem 5.6 to mappings on a subset of $\mathbb{R}^2$ into a normed linear space Y. The role of the derivative is now played by the so-called bidimensional derivative. A mapping $\varphi: G \to Y$ of an open set in the plane into a normed linear space is called <u>bidimensionally continuous</u> at $(a, b) \in G$ if for any $\varepsilon > 0$ there exists a positive number δ such that $|h| < \delta$, $|k| < \delta$ imply

$$\|\varphi(a + h, b + k) - \varphi(a + h, b) - \varphi(a, b + k) + \varphi(a, b)\| < \varepsilon .$$

If there is an element $D\varphi(a, b) \in Y$ such that

$$\lim_{\substack{h \to 0 \\ k \to 0}} \left\| \frac{\varphi(a + h, b + k) - \varphi(a + h, b) - \varphi(a, b + k) + \varphi(a, b)}{hk} - D\varphi(a, b) \right\| = 0$$

then φ is said to be <u>bidimensionally differentiable</u> at (a, b).

Theorem 5.8. Let $\varphi: [a, b] \times [c, d] \to Y$. If φ is bidimensionally continuous at the points (a, c), (a, d), (b, d), (b, c) and is bidimensionally differentiable for all $(t, t') \in (a, b) \times (c, d)$, then there exists a point (ξ, η), $a < \xi < b$, $c < \eta < d$ such that

$$\left\| \frac{\varphi(a, c) - \varphi(b, c) - \varphi(a, d) + \varphi(b, d)}{(b - a)(d - c)} \right\| \leq \|D\varphi(\xi, \eta)\| .$$

Mean Value Theorems in Topological Spaces

We now turn to mean value theorems for mappings with values in topological spaces which are not necessarily normed. The mean value theorem mentioned above under type 4 will be included in this setting. Note first that Theorems 5.1 and 5.2 are obviously valid in topological spaces also.

We begin with an extension of Theorem 5.3 to topological linear spaces. The role of the norm in the majorant conditions of Theorem 5.3 is now played by gauge functionals. We recall that a gauge functional is a convex and positively homogeneous function.

Theorem 5.9. Let Y be a topological linear space. Let p be a continuous gauge functional on Y. Let f and g be continuous functions from [a,b] into Y and $\mathbb{R}$ respectively. Suppose there is a countable set K of [a,b] such that for each t in [a,b]∖K, f and g have right hand derivatives. Suppose also there is a set N of Lebesgue measure zero, such that $K \subseteq N \subseteq [a,b]$ and

$$p(D^+f(t)) \leq D^+g(t) \quad \text{for all} \quad t \in [a,b] \setminus N .$$

Then

$$p(f(d) - f(c)) \leq g(d) - g(c)$$

for all c, d, with $a \leq c \leq d \leq b$.

The idea of this theorem was suggested in [5, p. 156, problem 6] and was carried out in [159, Theorem B], which was crucial to the proof of several theorems in [159]. Two important cases of Theorem 5.9 occur when p is a continuous linear functional and when p is a norm or seminorm on Y .

For any set A in a linear topological space we let co A denote the convex hull of A (i.e. the intersection of all convex sets containing A) and $\overline{co}$ A the closure of the convex hull. We now state an important analogue of Theorem

5.4, which says that in a locally convex space the difference quotient belongs to the closed convex hull of the set of the derivatives. More precisely, we have

Theorem 5.10. Let K be a countable subset of $I = [a,b]$, where a and b are finite numbers. Let f be a continuous map of I into a locally convex space Y. If f is differentiable at all points of the set $I_0 = I \setminus K$, then

$$\frac{f(b) - f(a)}{b - a} \in \overline{co} \{f'(t): t \in I_0\} .$$

For a proof, see for instance McLeod [159] or Averbukh and Smolyanov [72].

This theorem yields immediately the following

Theorem 5.11. Let $F: X \to Y$, where X is a linear space and Y is a locally convex space. If F has a first variation along a given subspace H at each point of the line segment Γ joining x_0, $x_0 + h \in X$, for $h \in H$, except possibly on a countable subset K of Γ. Then

$$F(x_0 + h) - F(x_0) \in \overline{co} \{\delta F(x;h): x \in \Gamma_0 = \Gamma \setminus K\} ,$$

Theorem 5.12. Under the assumptions of Theorem 5.11, the following is true for each u at which $\delta F(u;h)$ exists:

$$F(x_0 + h) - F(x_0) - \delta F(u;h)$$
$$\in \overline{co} \{\delta F(x_0 + th;h) - \delta F(u;h): t \in [0,1] \setminus J\}$$

where J is a countable subset of $[0,1]$ on which $\delta F(x_0 + th;h)$ may fail to exist.

This theorem follows by applying Theorem 5.11 to the map $x \to F(x) - \delta F(u; x - x_0)$ defined on $x_0 + H$ and noting that the variation of this map is $\delta F(x;h) - \delta F(u;h)$.

Remarks

1. As in the case of normed spaces, the condition of

THE ROLE OF DIFFERENTIALS

existence of the derivatives in Theorems 5.10-5.12 can be replaced by right hand or left hand derivatives.

2. In the above theorems the countable set cannot be replaced by a set of measure zero. For example let T be the Cantor ternary set in $[0,1]$ and let f be the continuous function on $[0,1]$ defined as follows: $f(0) = 1$, $f(1) = 0$, $f(t) = \frac{1}{2}$ for $\frac{1}{3} < t < \frac{2}{3}$, $f(t) = \frac{3}{4}$ for $\frac{1}{9} < t < \frac{2}{9}$, $f(t) = \frac{1}{4}$ for $\frac{7}{9} < t < \frac{8}{9}$, etc. Then $f'(t) = 0$ for all $t \in [0,1] \setminus T$ and T is of measure zero, but Theorem 5.10 fails to hold.

3. The condition that Y be locally convex is essential for the validity of Theorems 5.10-5.12. Let $I = [0,1]$ and let Y be the space of all real measurable functions on I with the linear (but not necessarily locally convex) topology of convergence in measure. For each $\tau \in I$, let

$$y_\tau = \begin{cases} 1 & \text{for } 0 \leq t < \tau \leq 1, \\ 0 & \text{for } 0 \leq \tau \leq t \leq 1. \end{cases}$$

The derivative of the map F of I into Y defined by $F(\tau) = y_\tau$ is identically 0, but $F(1) - F(0) = y_1 - y_0 = 1$.

4. The <u>closure</u> of the convex hull in Theorems 5.10-5.12 cannot be replaced by the convex hull. Consider in the space ℓ_2 the sequence $a_k = \{1, \frac{1}{2}, \ldots, \frac{1}{2^k}, 0, 0, \ldots\}$, $k = 0, 1, 2, \ldots$, whose limit is $a = \{1, \frac{1}{2}, \frac{1}{4}, \frac{1}{8}, \ldots\}$. For $k = 0, 1, 2, \ldots$, let $t_k = 1 - \frac{1}{2^k}$ and define $f(t) = a_k + \frac{t - t_k}{t_{k+1} - t_k}(a_{k+1} - a_k)$ for $t \in [t_k, t_{k+1})$, and $f(1) = a$. Then f satisfies the conditions of Theorem 5.10, but

$$f(1) - f(0) \notin \operatorname{co} \{f'(t): t \in [0,1], t \neq t_k, k = 0, 1, \ldots\}.$$

5. Note that in the case of a mapping $f: [a,b] \to X$, where X is a normed space, the assertion of Theorem 5.10

$$f'(b) - f'(a) \in (b-a) \overline{\operatorname{co}} \{f'(t): t \in I_0\}$$

yields the <u>cruder</u> estimate

$$\|f'(b) - f'(a)\| \le (b-a) \sup\{\|f'(t)\|: t \in I_0\},$$

which is the content of Theorem 5.4.

6. McLeod [159] has shown that if Y is the finite dimensional space $\mathbb{R}^n$, then $\frac{f(b)-f(a)}{b-a}$ belongs to the convex hull of the set of derivatives $f'(t)$, $t \in I_0$, under the assumptions of Theorem 5.10. Algebraically this means that

$$f(b) - f(a) = (b-a) \sum_{k=1}^{n+1} \lambda_k f'(c_k)$$

for some $c_1, \ldots, c_{n+1}$. McLeod was able to show also that the number of terms in this sum can be reduced from $n+1$ to n under two additional assumptions: (i) the derivative $f'(t)$ exists for all $t \in [a,b] \smallsetminus M$, where M is a set having at most $n-1$ points; (ii) when $n \ge 2$, the derivative $f'(t)$ is continuous from the right at every point of $[a,b) \smallsetminus M$. Condition (i) cannot be dropped.

The following theorem generalizes the mean value theorem in the form of inequality (5.4) to topological linear spaces.

Theorem 5.13. Let Y be a topological linear space. Let f be a continuous function from $[a,b]$ into Y having a right hand derivative D^+f on $(a,b) \smallsetminus K$ where K is countable. Let p be a continuous gauge functional on Y. If N is any set of Lebesgue measure zero such that $K \subseteq N \subseteq (a,b)$, then there exists a number $c \in (a,b) \smallsetminus N$ such that

$$p[f(b) - f(a)] \le p[D^+f(c)](b-a).$$

We conclude our presentation of mean value theorems of differential calculus by stating a theorem due to Hartman and Wintner [157] which unifies several mean value theorems for the heat equation operator and other boundary value problems.

THE ROLE OF DIFFERENTIALS

Let V and W be subsets of a topological space X, and let V be dense in X. Let $\mathfrak{m}$ be a linear manifold of real or complex valued functions u which are defined and continuous on X, and let $\mathfrak{n}$ be a linear submanifold of $\mathfrak{m}$. For each fixed $u \in \mathfrak{m}$, let $L(u)$ be a function on V.

Theorem 5.14. Let the operator L satisfy the following conditions:

(i)   L is linear on $\mathfrak{m}$;

(ii) there exists a solution $u = u_0$ of the boundary value problem

(5.5) $\qquad L(u_0) = 1$ on V, where $u_0 \in \mathfrak{n}$;

(iii) if $u \in \mathfrak{n}$ and $L(u) \neq 0$ on V, then $u \neq 0$ on W. Further, let u be any function of $\mathfrak{m}$ for which there exists a solution v of the boundary value problem

(5.6) $\qquad L(v) = 0$ on V, where $v - u \in \mathfrak{n}$.

Then, corresponding to every $x \in W$, there exists at least one element φ of V satisfying

(5.7) $\qquad u(x) = v(x) + u_0(x) L[u(\varphi)]$

Proof: Since (iii) implies that the solution u_0 of (5.5) does not vanish on W, it follows that, corresponding to a fixed $x \in W$, there exists a number A such that

$$u(x) - v(x) - A u_0(x) = 0 .$$

On the other hand, the function $u - v - Au_0 \in \mathfrak{n}$, by (5.5) and (5.6) and vanishes at x. Hence (iii) implies that $L(u - v - Au_0)$ vanishes at some point $\varphi \in V$. Since L is linear, $L(v) = 0$ and $L(u_0) = 1$ imply that $L[u(\varphi)] - A = 0$, and hence (5.7).

185

In the applications of this theorem, V is usually either an open set or an open set and part of its boundary in $\mathbb{R}^n$, X the closure of V and W is either the interior of X or X itself, L a differential operator, $\mathfrak{m}$ the set of continuous functions on X for which L(u) is defined on V, and $\mathfrak{n}$ is the set of those functions $u \in \mathfrak{m}$ which satisfy a prescribed boundary condition. As an application of Theorem 5.14, let V be the open interval $0 < x < h$, $\mathfrak{m}$ the continuous functions u on $0 \leq x \leq h$ which possess continuous derivatives u', ..., $u^{(m-1)}$ on the half-open interval $0 \leq x < h$ and m^{th} order (not necessarily continuous) derivative $u^{(m)}$ on $0 < x < h$. Let $L(u) = u^{(m)}$, $W = (0,h]$ and $\mathfrak{n}$ the set of functions $u \in \mathfrak{m}$ satisfying $u(0) = u'(0) = \ldots = u^{(m-1)}(0) = 0$. Then it follows from Theorem 5.14, if $u \in \mathfrak{m}$, that there exists at least one number τ, $0 < \tau < h$, such that

$$u(h) = \sum_{k=0}^{m-1} \frac{u^{(k)}(0)}{k!} h^k + \frac{u^{(m)}(\tau)}{m!} h^m .$$

This fact was proved directly in [158]. Theorem 5.14 also includes as special cases Pólya's extension [164, 165] of Rolle's theorem in the case where L(u) is an ordinary differential operator, the mean value theorems of Blaschke [152] for the Laplacian, Zaremba's treatment [172] of solutions of Laplace's equation, as well as other results for differential operators.

Undoubtedly there are other results in the literature which can be considered as variants or extensions of the mean value theorem. The mean value theorem can be mean!

Finally we state a useful mean value theorem for integrals.

<u>Theorem 5.15</u>. Let X be a linear topological space, $\rho : [a,b] \to \mathbb{R}$, where a,b are finite. If ρ and the map $t \to \rho(t) f(t)$ of [a,b] to X are Riemann integrable and $\rho(t) \geq 0$ for $t \in [a,b]$ and $\int_a^b \rho(t) dt > 0$, then

$$\frac{\int_a^b \rho(t) f(t) dt}{\int_a^b \rho(t) dt} \in \overline{co} \{f(t): t \in [a,b]\} .$$

In particular, taking $\rho(t) \equiv 1$, Theorem 5.15 says that the mean value of a function f on a finite interval is in the closed convex hull of the values of the function on this interval.

6. Taylor's Theorems and Formulas in Banach and Locally Convex Spaces

We mention several forms of Taylor's theorems and formulas in Banach and locally convex topological linear space.

Let U be an open subset of a Banach space X, F a mapping on U into a Banach space Y.

<u>Theorem 6.1</u> (W. H. Young's form of Taylor's theorem). Let $F: U \to Y$ be n times Fréchet differentiable at a point $x_0 \in U$. Then

$$F(x_0 + h) = F(x_0) + F'(x_0)h + \ldots + \frac{1}{n!} F^{(n)}(x_0) h^n + r(x_0; h) ,$$

where

$$\lim_{h \to 0} \frac{\|r(x_0; h)\|}{\|h\|^n} = 0 ,$$

or equivalently,

$$\lim_{\tau \to 0} \frac{\|r(x_0; \tau h)\|}{\tau^n} = 0 \quad \text{uniformly with respect to } h \text{ in the set } \{h: \|h\| = 1\} .$$

This form of Taylor's theorem requires only differentiability of order $n-1$ in some neighborhood of x_0 and n^{th} order differentiability at x_0; it asserts only the asymptotic behavior of the remainder as $h \to 0$. The proof is by induction (see, for instance, Cartan [4; Theorem 5.6.3]). A similar theorem holds for Gâteaux or Hadamard differentiability (see Theorem 6.5 below). In particular, if F has n^{th} order Gâteaux variation at $x_0 \in U$, then $F(x_0 + h) = F(x_0) + \delta F(x_0;h) + \ldots + \frac{1}{n!} \delta^n F(x_0;h) + r(x_0;h)$, where

$$\lim_{\tau \to 0} \frac{\|r(x_0;\tau h)\|}{\tau^n} = 0 \ .$$

Theorem 6.2 (Taylor's theorem with integral remainder). Let $F: U \to Y$ and let F have an nth variation on U. Let x_0, $x_0 + h \in U$. If the line segment joining x_0 and $x_0 + h$ is contained in U (or in particular if U is convex), and if $\delta^n F(x_0 + th;h)$ is continuous in t on $[0,1]$, then

(6.1) $F(x_0 + h) = F(x_0) + \delta F(x_0;h)$

$$+ \frac{1}{2} \delta^2 F(x_0;h) + \ldots + \frac{1}{(n-1)!} \delta^{(n-1)} F(x_0;h)$$

$$+ \int_0^1 \frac{(1-t)^{n-1}}{(n-1)!} \delta^n F(x_0 + th;h) dt \ .$$

Note that the integral in (6.1) exists and is a Banach space valued integral in the Riemann sense; $\delta^k F(x_0;h) = \delta^k F(x_0; h_1, \ldots, h_k)$ for $h_1 = \ldots = h_k = h$. One way to prove Theorem 6.2 is to apply the Taylor formula (with integral remainder) for a real valued function on $[0,1]$ to the function φ defined on $[0,1]$ by $\varphi(t) = \ell F(x_0 + th)$, where ℓ is a continuous linear functional on Y. To complete the proof note that ℓ commutes with the integral operator in (6.1) and apply the corollary to the Hahn-Banach theorem which asserts that for every element u_0 of a Banach space E there exists an element $\ell \in E^*$ of norm 1 such that $\ell(u_0) = \|u_0\|$. An interesting direct proof of Theorem 6.2 (i.e. without assuming the validity of (6.1) for a real valued function of a real variable) can be given using the majorant form of the mean value theorem (see Dieudonné [5] or Cartan [4]). The first direct proof of Theorem 6.2 was given by Graves [42] using induction and the fundamental theorem of calculus (see (5.2) of this paper) applied to the function $t \to F(x_0 + th)$ on $[0,1]$. Note also that in Theorem 6.2 it suffices to require that $\delta^n F(x_0 + th; h)$, as a function of t on $[0,1]$, is bounded and its set of discontinuities is of measure zero. For a proof using this approach see Rall [10; Theorem 20.2].

Corollary 6.2. Let $F: U \to Y$ have an nth order Fréchet differential. Let $x_0, x_0 + h \in U$. If the line segment joining x_0 and $x_0 + h$ is contained in U and the map $\varphi: [0,1] \to F^{(n)}(x_0 + th)$ is bounded and its set of discontinuities is of measure zero, then

$$F(x_0 + h) = F(x_0) + F'(x_0)h + \frac{1}{2} F''(x_0)h^2$$
$$+ \cdots + \frac{1}{(n-1)!} F^{(n-1)}(x_0) h^{n-1}$$
$$+ \int_0^1 \frac{(1-t)^n}{(n-1)!} F^{(n)}(x_0 + th)h^n \, dt .$$

$(F^{(k)}(x_0)h^k$ is the value of the k-linear operator $F^{(k)} \in \mathcal{L}_k(X,Y)$ at $(h, \ldots, h))$.

The following theorem generalizes Lagrange's formula (Theorem 5.1) for functionals.

Theorem 6.3 (Taylor's theorem in Lagrange's form for functionals). Let $f: U \to \mathbb{R}$ have n^{th} order variation at each point of the line segment $\{x_0 + th: 0 \le t \le 1\} \subset U$, where x_0,

$x_0 + h$ are fixed elements in U. Then there exists a real number τ with $0 < \tau < 1$ such that

$$f(x_0 + h) = f(x_0) + \delta f(x_0; h) + \frac{1}{2} \delta^2 f(x_0; h) + \ldots + \frac{1}{(n-1)!} \delta^{n-1} f(x_0; h)$$

$$+ \frac{1}{n!} \delta^n f(x_0 + \tau h; h) .$$

The proof is a trivial application of Taylor's theorem for a real valued function of a real variable to the function $\phi: [0,1] \to \mathbb{R}$, defined by $\phi(t) = f(x_0 + th)$.

Theorem 6.3 is not valid for operators with values in Banach spaces. However its real essence, which is exhibited in the form of an inequality, is valid and is indeed a generalization of Theorem 5.4.

Theorem 6.4 (Taylor's theorem in Lagrange's form). Let $F: U \to Y$ be n times Gâteaux differentiable at every point of the line segment $\{x_0 + th: 0 \leq t \leq 1\} \subset U$, where x_0, $x_0 + h$ are fixed elements in U. Then

$$\left\| F(x_0 + h) - F(x_0) - \sum_{k=1}^{n-1} \frac{1}{k!} F^{(k)}(x_0) h^k \right\|$$

$$\leq \frac{1}{n!} \sup \{\|F^{(n)}(x_0 + th)\|: 0 < t < 1\} \|h\|^n .$$

Proof: Let $y = F(x_0 + h) - F(x_0) - \sum_{k=1}^{n-1} \frac{1}{k!} F^{(k)}(x_0) h^k$. Therefore y is a fixed element in the Banach space Y. By the corollary to the Hahn-Banach theorem alluded to earlier in this section, there exists a bounded linear functional ℓ on Y such that $\ell(y) = \|y\|$ and $\|\ell\| = 1$. We define $f(t) = \ell F(x_0 + th)$ for $0 \leq t \leq 1$. Then

$$f^{(k)}(t) = \ell(F^{(k)}(x_0 + th) h^k) \text{ for } k = 1, 2, \ldots, n .$$

Hence,

$$\ell(y) = \ell(F(x_0+h)) - \ell(F(x_0)) - \sum_{k=1}^{n-1} \frac{1}{k!} \ell(F^{(k)}(x_0)h^k)$$

$$= f(1) - f(0) - \sum_{k=1}^{n-1} \frac{1}{k!} f^{(k)}(0).$$

It then follows from the definition of ℓ and the usual Taylor's formula for real functions (a special case of Theorem 6.1) that

$$\ell(y) = \|y\| = \|F(x_0+h) - F(x_0) - \sum_{k=1}^{n} \frac{1}{k!} F^{(k)}(x_0)h^k\| =$$

$$\frac{1}{n!} f^{(n)}(\tau) = \frac{1}{n!} \ell(F^{(n)}(x_0+\tau h)h^n), \quad \text{for some } \tau \in (0,1).$$

Thus

$$\|y\| \le \frac{1}{n!} \|F^{(n)}(x_0+\tau h)h^n\|$$

$$\le \frac{1}{n!} \sup\{\|F^{(n)}(x_0+th)\|: 0 < t < 1\} \|h\|^n,$$

which completes the proof.

A direct proof of Theorem 6.3, which does not invoke the Taylor's theorem for real functions, can also be given based on majorants (see Cartan [4; Théorème 5.6.2]). Finally note that if we slightly strengthen the hypotheses of Theorem 6.4 by requiring in addition that the function $\varphi(t) = F(x_0+th)$ is bounded on $0 \le t \le 1$ and its set of discontinuities is of measure zero, then Theorem 6.4 follows easily from (5.2).

In the case $n = 1$, the generalization of the preceding theorem to locally convex spaces was stated earlier (Theorem 5.11); it asserts that the difference quotient is contained in the closed convex hull of the set of values of the first derivative. We expect the generalization for $n > 1$ to assert that the nth order difference is in the closed convex hull of the

n^{th} order differential. More precisely we have

<u>Theorem 6.4.</u> Let Y be a locally convex topological linear space, X a linear space and H a subspace of X and a topological linear space. If $F: X \to Y$ has n^{th} order Gâteaux variation along the subspace H at every point of the set

$$\Gamma = \{x_0 + \tau_1 h_1 + \ldots + \tau_n h_n, \ 0 \leq \tau_i \leq 1, \quad i = 1, \ldots, n\}$$

for $x_0 \in X$, $h_1, \ldots, h_n \in H$, then

$$\Delta^n F(x_0; h_1, \ldots, h_n) \in \overline{co} \{\delta^n F(x; h_1, \ldots, h_n): x \in \Gamma\}.$$

Note that Theorem 5.5 is a corollary to the preceding theorem.
Theorem 6.1 extends also to locally convex spaces, for various notions of (H, β)-differentiability discussed in Section 2.

<u>Theorem 6.5</u> (Taylor's theorem in Young's form in locally convex spaces). Let X, H, and Y be as in Theorem 6.4. Let β be a system of bounded sets in H containing all the bounded sets of dimension n and let $F: X \to Y$ be n times (H, β)-differentiable at $x_0 \in X$. Then

$$F(x_0 + h) - F(x_0) = \sum_{k=1}^{n} \frac{1}{k!} F^{(k)}(x_0) h^k + r(x_0; h)$$

where

$$\frac{r(x_0; \tau h)}{\tau^n} \to 0 \text{ as } \tau \to 0$$

uniformly with respect to $h \in B$ for each subset $B \in \beta$.
Finally Theorem 6.2 holds in the same form for a locally convex space Y. The integral in (6.1) has values in Y and is defined as usual (see, for instance, Dunford and Schwartz [2]).

7. Converses of Taylor's Theorems and Characterizations of Higher Order Differentiability

The existence of an mth order variation or differential of an operator F at a point x_0 provides a local representation of the operator in a neighborhood of x_0. Such a representation gives, in this neighborhood, an approximation in some sense of the operator by a sum of homogeneous operators or by a polynomial operator. The "converse" of Taylor's theorem is not necessarily true: For instance, a real function of a real variable can have an expansion such as

$$f(a+h) = f(a) + c_1 h + \ldots + c_n h^n + r_n(a;h)$$

where $\lim_{h \to 0} \dfrac{r_n(a;h)}{h^n} = 0$; and it does not, in general, follow that even the second derivative of f exists at a. This seems to have been recognized first by Graves [42] who gave a simple counterexample: $f(x) = x^3 \sin \frac{1}{x}$, $x \neq 0$, $f(0) = 0$, $n = 2$. Graves, however, did not investigate converses of Taylor's theorem. We give two types of theorems on converses of Taylor's theorem which also provide characterizations of higher order differentiability.

Theorem 7.1. Let X be a linear space and let Y be a normed space both over the field of real numbers. An operator $F: X \to Y$ has n^{th} Gâteaux variation $\delta^n F(x_0; h)$ at $x_0 \in X$ if and only if there exist mappings $P_i: X \to Y$, $P_i = P_i(x_0; \cdot)$, $i = 1, \ldots, n$, where P_k is homogeneous of degree k such that for each fixed $h \in X$, the function $r_n(t)$ defined by

$$r_n(t) = F(x_0 + th) - F(x_0) - P_1(x_0; th) - \ldots - P_n(x_0; th)$$

satisfies the following conditions

(i) $\qquad \lim_{t \to 0} \dfrac{r_n(t)}{t^n} = 0$,

(ii) the $(n-1)^{st}$ derivative $r_n^{(n-1)}(t)$ of $r_n(t)$ exists in some deleted neighborhood of zero,

and

(iii) $\displaystyle\lim_{t \to 0} \frac{r_n^{(j)}(t)}{t^{n-j}} = 0, \quad j = 1, \ldots, n-1$.

Theorem 7.2. Let X and Y be two normed linear spaces and let U be an open subset of X. An operator $F: U \to Y$ has an nth order Gâteaux differential at $x_0 \in U$ if and only if there exist $L_1, \ldots, L_n$ with $L_k \in \mathcal{L}_k(X,Y)$ $L_k = L_k(x_0)$ such that the function $\alpha_n(h)$ defined in a deleted neighborhood $\tilde{N}(0)$ of the zero element $0 \in X$ by

(7.1) $\alpha_n(h) = F(x_0 + h) - F(x_0) - L_1 h - \ldots - L_n h^n$

satisfies the following conditions:

(i) $\displaystyle\lim_{t \to 0} \frac{\|\alpha_n(th)\|}{t^n} = 0$;

(ii) $D^{(n-1)} \alpha_n(h; \cdot)$, the $(n-1)^{st}$ Gâteaux derivative of α_n, exists for all $h \in \tilde{N}(0)$;

and

(iii) $\displaystyle\lim_{t \to 0} \frac{\|D^{(j)} \alpha_n(th; \cdot)\|_{\mathcal{L}_j}}{t^{n-j}} = 0 \text{ for } j = 1, \ldots, n-1$.

In this case $L_j = \dfrac{1}{j!} D^j F(x_0; \cdot), \quad j = 1, \ldots, n$.

Theorem 7.3. Let X and Y be two normed linear spaces and let U be an open subset of X. An operator $F: U \to Y$

has an nth order Fréchet differential at $x_0 \in U$ if and only if there exist $L_k \in \mathcal{L}_k^s(X,Y)$ such that the function $\alpha_n(h)$ defined in a deleted neighborhood $\tilde{N}(0)$ of $0 \in X$ by (7.1) satisfies the following conditions:

(i) $\lim\limits_{h \to 0} \dfrac{\|\alpha_n(h)\|}{\|h\|^n} = 0$;

(ii) $d^{(n-1)}\alpha_n(h; \cdot)$, the $(n-1)^{st}$ Fréchet derivative of α_n, exists for all $h \in \tilde{N}(0)$;

(iii) $\lim\limits_{h \to 0} \dfrac{\|d^{(j)}\alpha_n(h; \cdot)\|_{\mathcal{L}_j}}{\|h\|^{n-j}} = 0, \quad j = 1, \ldots, n-1$.

Proof: Obviously the conditions (i), (ii), and (iii) are necessary and sufficient for $n = 1$. Assume that the conditions are sufficient for $n = s$ and suppose there exist $L_j \in \mathcal{L}_j^s(X,Y)$, $L_j = L_j(x_0)$, $j = 1, \ldots, s+1$ such that

(7.2) $F(x_0 + h) - F(x_0) = L_1 h + \ldots + L_{s+1} h^{s+1} + \alpha_{s+1}(h)$

with $\alpha_{s+1}(h)$ satisfying (i), (ii), and (iii). Since $\alpha_{s+1}(h)$ has Fréchet differential of order s for h in some deleted neighborhood of 0, we obtain by differentiating (7.2) with respect to h, writing $L_j h^j$ as $P_j(h)$,

$dF(x_0 + h; k) = L_1 + \ldots + dP_{s+1}(h;k) + d\alpha_{s+1}(h,k)$

where the $dP_j(h;k)$ is $(j-1)$-linear in h. Moreover, $\lim\limits_{h \to 0} dF(x_0 + h; k) = dF(x_0; k)$ and thus

(7.3) $dF(x_0 + h; k) - dF(x_0; k) = L_2 k + \ldots + dP_{s+1}(h;k) + d\alpha_{s+1}(h;k)$

Let $\gamma_s(h) = d\alpha_{s+1}(h;k)$. Then by the induction hypothesis $d^{s-1}\gamma_s(h;\cdot) = d^s\alpha_{s+1}(h;k,\cdot)$ exists and

$$\lim_{h \to 0} \frac{\|d^j\gamma_s(h;\cdot)\|}{\|h\|^{s-j}} = \lim_{h \to 0} \frac{\|d^{j+1}\alpha_{s+1}(h;k,\cdot)\|}{\|h\|^{s-j}}$$

$$= \lim_{h \to 0} \frac{\|d^i\alpha_{s+1}(h;k,\cdot)\|}{\|h\|^{s+1-i}} = 0 \quad \text{for} \quad j = i-1 = 1, \ldots, s,$$

and

$$\lim_{h \to 0} \frac{\|\gamma_s(h)\|}{\|h\|^s} = \lim_{h \to 0} \frac{\|d\alpha_{s+1}(h;k)\|}{\|h\|^s} = 0,$$

uniformly with respect to k on $\{k: \|k\| = 1\}$. Therefore $dF(x;\cdot)$ has a Fréchet differential of order s at x_0, i.e., $F^{(s+1)}(x_0)$ exists.

Now we prove the necessity. Assume that the conditions (i), (ii), and (iii) are necessary for $n = s$ and assume that $F^{(s+1)}(x_0)$ exists. Then $F, F', \ldots, F^{(s)}$ exist and are continuous in a neighborhood of x_0. We let $L_j = F^{(j)}(x_0)$ and

$$(7.4) \quad \alpha_{s+1}(h) = F(x_0+h) - F(x_0) - F'(x_0)h - \ldots - \frac{1}{(s+1)!}F^{(s+1)}(x_0)h^{s+1}$$

Clearly $\alpha_{s+1}(h)$ is well defined and $d^s\alpha_{s+1}(h;\cdot)$ exists in a deleted neighborhood of 0, since we have assumed that $F^{(s+1)}(x_0)$ exists. Thus (ii) is satisfied. Condition (i) is the statement of Taylor's theorem in W. H. Young's form (Theorem 6.1). To prove (iii), differentiate (7.4) to obtain

$$d\alpha_{s+1}(h;k) = dF(x_0+h;k) - F'(x_0)k - \frac{1}{s!} F^{(s+1)}(x_0) h^s k$$

in view of the symmetry of the j-linear forms on the right side, for $j = 2, \ldots, s+1$. Applying the induction hypothesis to the sth derivative of $F'(x_0;k)$, we conclude that

$$\lim_{h \to 0} \frac{\|d^{j+1}\alpha_{s+1}(h;k,\cdot)\|}{\|h\|^{s-j}} = \lim_{h \to 0} \frac{\|d^i\alpha_{s+1}(h;k,\cdot)\|}{\|h\|^{s+1-i}} = 0,$$

$$i = j+1 = 1, 2, \ldots, s.$$

Remark. In view of Proposition 1.2, conditions (i) and (iii) of Theorem 7.3 can be replaced by the equivalent conditions

(i)' $\quad \lim_{\tau \to 0} \tau^{-n} \|\alpha_n(\tau h)\| = 0$, and

(iii)' $\quad \lim_{\tau \to 0} \tau^{j-n} \|\alpha_n^{(j)}(\tau h)\| = 0, \quad j = 1, \ldots, n-1$,

uniformly with respect to h on the set $\{h: \|h\| = 1\}$.

The above theorems which are special cases of Theorem 1 in Nashed [188], are converses to W. H. Young's form of Taylor's Theorem. They provide characterizations of higher order differentiability <u>at a point</u>. Different characterizations can be given for higher order differentiability <u>over a region.</u> The next theorem asserts that if a mapping has at each point of an open convex set a Taylor's formula in Young's form with continuous coefficients up to the nth term, then the mapping is continuously n times differentiable.

Theorem 7.4. Let X and Y be normed linear spaces, Ω an open convex subset of X. A mapping $F: \Omega \to Y$ has continuous n^{th} Fréchet derivative on Ω if and only if there exist mappings $L_i: \Omega \to \mathcal{L}_i^s(X,Y)$ which are continuous, $i = 1, \ldots, n$, and such that the mapping r defined for $x \in \Omega$ and all $h \in X$ with $x + h \in \Omega$ by

$$r(x;h) = F(x+h) - F(x) - \sum_{i=1}^{n} \frac{1}{i!} L_i(x) h^i$$

satisfies the following condition for each $x_0 \in \Omega$,

$$\frac{\|r(x;h)\|}{\|h\|^n} \to 0 \quad \text{as} \quad (x,h) \to (x_0, 0) .$$

This theorem was given in the one dimensional case by Marcinkiewicz and Zygmund [187]. Glaeser [156] proved the theorem for the case of a mapping on $\mathbb{R}^n$ to $\mathbb{R}$. The proof of the general case was reduced to this special case by Abraham and Robbin [230].

Another characterization of higher order differentiability can be given in terms of a uniform remainder. Let $F: U \to Y$, where U, X, Y are as in Theorem 7.2. Suppose that for each $x \in U$, there exist $L_k \in \mathcal{L}_k^s(X,Y)$, $k = 1, \ldots, n$, such that

$$F(x+h) = F(x) + L_1 h + \ldots + \frac{1}{n!} L_n h^n + r(x;h) ,$$

where

$$\frac{\|r(x;h)\|}{\|h\|^n} \to 0 \quad \text{as} \quad h \to 0$$

<u>uniformly with respect to</u> $x \in U$. Then F is n times differentiable in U, and conversely.

Converses of Taylor's Theorem in the setting of (H, β)-differentiability in locally convex topological vector spaces are given in [188], [161].

8. Direct Higher Order Differentials and Difference Differentials

In Sections 3 and 4 we considered higher order variations and differentials which were defined inductively. Some problems of approximation theory, trigometric series, and constructive theory of functions, have led to the consideration of several notions of nth order differentials which are defined <u>directly</u> without reference to differentials of order less than n. In this section, we introduce and discuss briefly several notions of direct differentials on normed and linear topological spaces.

Let X be a linear space, Y a linear topological space.

<u>Riemann Differentials</u>. For $F: X \to Y$, define the nth central difference by

$$\Delta_h^n F(x_0) = \sum_{k=1}^{n} (-1)^{n-k} \binom{n}{k} F(x_0 + (k - \frac{n}{2})h), \quad h \in X.$$

If $\lim_{t \to 0} \frac{1}{t^n} \Delta_{th}^n F(x_0)$ exists, we say that F has n^{th} order

<u>Riemann variation</u>. Now let H be a linear topological subspace of X, and take $h \in H$. We say that F has (H, β)-Riemann differential of order n, denoted by $R^n F(x_0; h)$, if

$$\lim_{t \to 0} \frac{1}{t^n} \Delta_{th}^n F(x_0) = R^n F(x_0; h)$$

uniformly with respect to $h \in S$ for each $S \in \beta$, where β is a system of bounded subsets of H. In particular when H is normed, and $\beta = \beta_b$ is the system of <u>all</u> bounded subsets of H, then F has n^{th} order Riemann (H, β_b)-differential at x_0 if and only if

$$\frac{\|\Delta_h^n F(x_0) - R^n F(x_0;h)\|}{\|h\|^n} \to 0 \text{ as } h \to 0.$$

Note that the existence of an nth order Riemann differential at x_0 does not necessarily imply the existence of all Riemann differential of order less n at x_0.

Peano Differentials. We say that F has an nth order (H, β)-Peano differential at x_0 if there exist $P_0 \in Y$, $P_1 \in \mathcal{L}_1(H,Y)$, $P_k \in \mathcal{L}'_k(H,Y)$, $2 \leq k \leq n$ (see Section 4 for notation), such that for all $h \in H$,

(8.1)
$$\begin{cases} F(x_0 + h) = P_0 + P_1 h + \ldots + P_n h^n + r_n(h), \text{ where} \\ \dfrac{r_n(th)}{t^n} \to 0 \text{ as } t \to 0 \end{cases}$$

uniformly with respect to $h \in S$ for each $S \in \beta$. $\dfrac{1}{n!} P_n h^n$ is then called the nth (H, β)-Peano differential.

Similarly we say that F has an nth order <u>Peano variation</u> if there exist functions $Q_i : H \to Y$, $i = 1, \ldots, n$, where Q_i is homogeneous of degree i, such that

(8.2)
$$\begin{cases} F(x_0 + h) = P_0 + Q_1 h + \ldots + Q_n h^n + r_n(h), \text{ where} \\ \dfrac{r_n(th)}{t^n} \to 0 \text{ as } t \to 0. \end{cases}$$

Note that if a representation such as (8.1) or (8.2) exists, it is unique. Variants of the notion of Peano differential in linear topological spaces can be introduced by replacing $P_0 + P_1 h + \ldots + P_n h^n$ in (8.1) by one of the other definitions of a polynomial of degree n which were mentioned at the end of Section 4.

THE ROLE OF DIFFERENTIALS

The existence of a first order Peano variation ((H,β)-Peano differential) is equivalent to the existence of the Gâteaux variation ((H,β)-differential). However, this is not necessarily true for second or higher order Peano variations on differentials. For example, let $f(x) = 0$ if x is rational and $f(x) = x^{n+1}$ if x is irrational. Then f has n^{th} order Peano differential at 0: $P_n(0;h) = 0$ for all h, while $f''(0)$ does not exist.

In the definition of Peano variation of order n, we have not stipulated the existence of any Gâteaux variation of order greater than one. This motivates the following definition.

Taylor Differentials. If the $(n-1)^{st}$ (H,β)-differential of F exists at x_0 and if

$$\lim_{t \to 0} \frac{n!}{t^n} \left\{ F(x_0 + th) - \sum_{k=0}^{n-1} \frac{1}{k!} F^{(k)}(x_0)(th)^k \right\}$$

exists uniformly with respect to $h \in S$ for each $S \in \beta$, then we say that F has n^{th} order (H,β)-Taylor differential. The nth order Taylor variation is defined similarly.

The implications among these notions of higher order differentiability follow easily for a fixed order n and for the same system β. (H,β)-Taylor differentiability implies (H,β)-Peano differential, which in turn implies (H,β)-Riemann differentiability for each fixed order n and a given system of bounded sets in H. Also (H,β)-differentiability of order n implies (H,β)-Taylor differentiability of order n in any locally convex space. The second order Riemann differential of the function $f(x) = x|x|$ exists at $x = 0$ and is equal to zero; however the second order Peano differential of f does not exist at 0. The function

$$f(x) = \begin{cases} e^{-x^{-2}} \sin e^{x^{-2}} & x \neq 0 \\ 0 & x = 0 \end{cases}$$

has Peano differentials of all order at 0, but the third order Taylor differential does not exist at 0 because the second order ordinary differential does not exist at this point.

Direct (H,β)-differentials were introduced in [188] and will be examined further elsewhere. For applications of Peano and Taylor differentials in approximation theory see [181].

As an application of the notion of Taylor differential, we consider the following result on the regularity of a solution of the equation $f(x,y) = 0$. Let X,Y,Z be real Banach spaces, U a neighborhood of $x_0 \in U$, V a neighborhood of $y_0 \in Y$, and let $f: U \times V \to Z$. A continuous mapping $\varphi: W \to Y$ (W is a neighborhood of x_0) is called a p-approximate solution of

(8.3) $\qquad\qquad f(x,y) = 0$

at (x_0, y_0) if $\varphi(x_0) = y_0$ and $f(x, \varphi(x)) = o(\|x - x_0\|^p)$. Let f satisfy the condition $\|f(x,y) - f(x,y')\| \geq A\|y - y'\|$ for all (x,y), (x,y') in a neighborhood $U_0 \times V_0$ of (x_0, y_0). If $\varphi: U_0 \to V_0$ is a p-approximate solution of (8.3) at (x_0, y_0) which has a Taylor differential of order p at x_0, and if the solution u of (8.3) is defined in a neighborhood of x_0, then u has an nth order Taylor differential at x_0. If in addition the solution of (8.3) is defined in U and for each $a \in U$, φ_a is a p-approximate solution at a and $f(x, \varphi_a(x)) = o(\|x - a\|^p)$ uniformly with respect to $a \in U$, then u is p times Fréchet differentiable in U. (See [326]).

Difference Differentials. There are several notions of direct higher order differentials that are based on various higher order differences. These differentials may be called difference differentials; the Riemann differential is an example of such differentials. We consider several other notions of difference differentials in the setting of normed spaces; the extensions to linear topological space require only minor modifications. Let U be an open subset of X and let X and Y be normed spaces. Let $F: U \to Y$ and define

THE ROLE OF DIFFERENTIALS

$$\Delta' \Delta F(x;h,k) = F(x+h+k) - F(x+h) - F(x+k) + F(x) .$$

We say that F has a <u>direct second order variation</u> at x if

$$\lim_{(s,t) \to (0,0)} \frac{\Delta' \Delta F(x;th,sk)}{st} , \quad s,t \in \mathbb{R}$$

exists for any $h,k \in X$. If this limit exists, we denote it by $\delta'\delta F(x;h,k)$ and call it the <u>direct second order variation</u> at x. Clearly $\delta'\delta F(x;h,k)$ is homogeneous in h,k but is not necessarily linear nor continuous in h or k. If $\delta'\delta F(x;h,k)$ is bilinear and continuous in (h,k), we call it the <u>direct second order Gâteaux differential</u> and denote it by $D'DF(x;h,k)$. The mapping F is said to be <u>directly second order differentiable in Fréchet sense</u> at x if there exist a continuous bilinear map $B(x): X \times X \to Y$ such that

$$\lim_{(h,k) \to (0,0)} \frac{\|\Delta'\Delta F(x;h,k) - B(x)hk\|}{\|h\| \|k\|} = 0 .$$

We denote B(x)hk by $d'dF(x;h,k)$ and call it the direct second order differential in Fréchet's sense.

It should be noted that the existence of the <u>direct</u> second order variation at a point does not necessarily imply the existence of the first order variation.

It is easy to show that if F is directly second order differentiable in Fréchet sense, then F has a direct second order variation and $d'dF(x;h,k) = \delta'\delta F(x;h,k)$. If F is directly second order differentiable in Fréchet's sense at x, then F is directly second order continuous at x in the following sense: For every $\varepsilon > 0$ there exists a neighborhood W_ε of the point x such that $x + h \in W_\varepsilon$ and $x + k \in W_\varepsilon$ imply $\|F(x+h+k) - F(x+h) - F(x+k) + F(x)\| < \varepsilon$.

The following mean value theorem for second order direct variations can be easily established: If f is a real valued functional which has direct second order variation in a convex subset Ω, then for x, $x+h+k$, $x+h$, $x+k \in \Omega$, there exist γ, γ' in the interval $(0,1)$ such that

$$f(x+h+k) - f(x+h) - f(x+k) + f(x) = \delta'\delta f(x + \gamma h + \gamma' k; h, k) .$$

Using this mean value theorem it can be shown (see [192]) that if $F: \Omega \subset X \to Y$, where Ω is a convex open set containing x_0 and if $\delta'\delta F(x; h_i, k_j)$ is continuous in x at x_0 for $i = 1, \ldots, n$, $j = 1, \ldots, m$, then

$$\delta'\delta F(x_0; h_1 + \ldots + h_n, k_1 + \ldots + k_m) = \sum_{i=1}^{n} \sum_{j=1}^{m} \delta'\delta F(x_0; h_i, k_j) .$$

If X is a separable space and if the direct second order variation $\delta'\delta F(x; h, k)$ exists for every $x \in \Omega$, then $\delta'\delta F(x; h, k)$ is bilinear and continuous for every x belonging to a residual set in Ω. A similar result for the Gâteaux variation was given by Alexiewicz and Orlicz [122].

We have seen that a continuous Gâteaux differential is a Fréchet differential. A similar result holds for direct second order Gâteaux and Fréchet differentials. Let F have a second order direct Gâteaux variation $\delta'\delta F(x; h, k)$ in a convex neighborhood of x_0. If $\delta'\delta F(x; \cdot, \cdot): X \times X \to Y$ is bilinear and continuous and if $\delta'\delta F(x; \cdot, \cdot)$ is continuous at x_0 in the topology of the norm on $\mathcal{L}(X, Y)$, then the mapping F is directly second order differentiable in Fréchet sense at x_0.

Let $F: [0,1] \to Y$. The second Schwarz derivative at $t_0 \in (0,1)$ is defined by

$$\lim_{h \to 0} \frac{F(t_0 + h) - 2F(t_0) + F(t_0 - h)}{h^2}$$

if this limit exists. A simple extension of Schwarz's theorem [189; p. 37] asserts that if F is continuous on $[0,1]$ and

the second Schwarz derivative vanishes at each point of
(0,1), then $F(x) = ax + b$, where a and b are constant
functions with range in the normed linear space Y. Using
this result, an analogous theorem for direct second order
variations has been given by Lilly Nicolescu [191]: Let
$F: X \to Y$ be a continuous mapping. If F has a direct second
order Gâteaux variation at all points of X, and if $\delta'\delta F(x;
h,k) = 0$ for all x, then for every $x, y \in X$,

$$F(x + y) = F(x) + F(y) - F(0) .$$

9. Remarks on Differentiability in Complex Linear Topological Spaces

There is a fundamental difference between the Gâteaux
variation and differential in real linear topological spaces
and the corresponding notions in complex linear topological
spaces. The crux of the matter is essentially the analogue
of the well known fact that the existence of the first derivative (or the first partial derivatives) of a function of a complex variable (or variables) implies the existence of derivatives of all orders, which is not necessarily true in case of
a function of a real variable (or variables). A different situation prevails in the case of differentials on complex spaces;
we examine briefly some of its aspects.

The development of the theory of differentials in complex spaces relies heavily on several concepts in the theory
of analytic functions on the complex plane to a complex vector space. This latter theory is fully developed as a result
of the work of D. H. Hilbert, F. Riesz, N. Wiener,
N. Dunford, L. Fantappi, I. Gelfand, E. R. Lorch, M. H.
Stone, A. E. Taylor and others (see Dunford and Schwartz
[2], Hille and Phillips [6] for details and references).

Let W be a domain in the complex t-plane, $t = \xi + i\eta$,
and let x be a function on W to a linear topological space
X. The function x is said to be holomorphic on W if
$x^*(x)$ is analytic in Cauchy's sense for each choice of the
continuous linear functional $x^* \in X^*$. A function $x: W \to X$

is called differentiable (weakly differentiable) at $t = t_0$ if there exists an element $x'(t_0) \in X$ such that $\dfrac{x(t) - x(t_0)}{t - t_0}$ tends strongly (weakly) to $x'(t_0)$ as $t \to t_0$. $x'(t_0)$ is called the derivative (weak derivative) of x at t_0. The function x is called strongly (weakly) continuous at t_0 if $\lim_{t \to t_0} [x(t) - x(t_0)] = 0$ in the topology of X (respectively, $\lim_{t \to t_0} x^*[x(t) - x(t_0)] = 0$ for all $x^* \in X^*$). If x is holomorphic on W, then x is strongly continuous and strongly differentiable on W, uniformly with respect to t on any bounded domain W_0 strictly interior to W (see [6]).

Let X and Y be linear topological spaces over the field of <u>complex</u> numbers and let F be an operator on an open set $U \subset X$ to Y. F is said to have a Gâteaux variation at $x_0 \in U$ if for each $h \in X$, $t^{-1}\{F(x_0 + th) - F(x_0)\}$ tends to an element of Y as $t \to 0$ in the complex plane $\mathbb{C}$. We say that F is Fréchet differentiable at $x_0 \in U$ if there exists a continuous linear operator $L: X \to Y$ such that

$$F(x_0 + h) - F(x_0) = Lh + r(x_0; h)$$

for all $h \in X$ with $x_0 + h \in U$, where $\lim_{t \to 0} \dfrac{r(x_0; th)}{t} = 0$ for $t \in \mathbb{C}$, uniformly with respect to h on each bounded subset of X. Clearly the Gâteaux variation $\delta F(x_0; h)$ is again homogeneous in h of degree one: $\delta F(x_0; \lambda h) = \lambda \delta F(x_0; h)$, $\lambda \in \mathbb{C}$, and the existence of a Fréchet differential at x_0 implies the existence of the Gâteaux variation. F has a Gâteaux variation on U if and only if for every $x \in U$ and $h_1, \ldots, h_n \in X$, the function defined by $F(x + \sum_{i=1}^{n} t_i h_i)$ is partially differentiable with respect to t_i, $(i = 1, \ldots, n)$ in the open set $\{(t_1, \ldots, t_n): x + \sum_{i=1}^{n} t_i h_i \in U\}$ of the n-dimensional complex space $\mathbb{C}^n$.

Higher order Gâteaux variations are defined inductively. For complex spaces, if F has a Gâteaux variation $\delta F(x;h)$ for $x \in U$, $h \in X$, then $\delta F(x; \cdot)$ has a Gâteaux variation, i.e.

$$\delta^2 F(x; h_1, h_2) = \frac{\partial^2}{\partial t_2 \partial t_1} F(x + t_1 h_1 + t_2 h_2) \Big|_{t_1 = t_2 = 0}$$

exists. Indeed, since $\delta F(x; \cdot)$ exists, the function $\varphi(t_1, t_2) = F(x + t_1 h_1 + t_2 h_2)$ is partially differentiable, and hence φ has partial derivatives of all orders. Here we are using a well known theorem that if a function from the n-dimensional complex space $\mathbb{C}^n$ to a linear topological space Y has first order partial derivatives with respect to all the variables, then it has partial derivatives of all orders. Furthermore the mixed partial derivatives are independent of the order of differentiation. Hence

$$\frac{\partial^2}{\partial t_1 \partial t_2} F(x + t_1 h_1 + t_2 h_2) \Big|_{t_1 = t_2 = 0}$$

exists and is symmetric in h_1 and h_2. Thus if in a complex linear topological space, an operator F has a Gâteaux variation, it has Gâteaux variations of all orders and the higher order variations are symmetric.

In the case of complex spaces, $\delta F(x; h)$ is automatically linear in h. Consider the function φ on a neighborhood of $(0,0)$ in $\mathbb{C}^2$ into Y defined by $\varphi(t_1, t_2) = F(x + t_1 h_1 + t_2 h_2)$. Since this function is partially differentiable, it follows from a fundamental theorem in the theory of two complex variables that φ has a total differential, i.e.

$$F(x+t_1h_1+t_2h_2) = F(x) + t_1 \frac{\partial \varphi}{\partial t_1}(0,0) + t_2 \frac{\partial \varphi}{\partial t_2}(0,0) + \varepsilon(\|t\|) \|t\|,$$

where $\varepsilon(\|t\|) \to 0$ as $\|t\| \to 0$. Hence,

$$F(x+t_1h_1+t_2h_2) = F(x) + t_1 \delta F(x;h_1) + t_2 \delta F(x;h_2) + \varepsilon(\|t\|) \|t\|.$$

In particular for $t_1 = t_2 = s$, we have

$$F(x+s(h_1+h_2)) = F(x) + s[\delta F(x;h_1) + \delta F(x;h_2)] + \psi(|s|)|s|,$$

where $\psi(|s|) \to 0$ as $s \to 0$. This implies that for $s \in \mathbb{C}$ $\lim_{s \to 0} s^{-1}[F(x+s(h_1+h_2)) - F(x)]$ exists and is equal to $\delta F(x;h_1) + \delta F(x;h_2)$. This shows that $\delta F(x; \cdot)$ is additive and completes the proof since $\delta F(x; \cdot)$ is always homogeneous.

Conditions under which a Gâteaux variation on a complex space becomes a Fréchet differential are considerably weaker than conditions in the case of a real space. For instance the mere existence of the Gâteaux variation on an open set implies the existence of the Fréchet differential on that set. Other conditions involving only the Gâteaux variation at a point, such as continuity, local boundedness and essential boundedness were considered by several authors (E. Hille, A. E. Taylor, M. A. Zorn and others). See Zorn [70, 71] and Hille and Phillips [6] for references and discussion of these conditions.

In particular, if Y is a Banach space and $F: X \to Y$, then the existence of the Gâteaux variation on an open set U of X implies that $\delta F(x; h)$ is linear and continuous in h, $t^{-1}\|F(x+h) - F(x) - t\delta F(x;h)\| \to 0$ as $t \to 0$, uniformly with respect to h on each bounded subset of X, and that $\delta F(x;h)$ is continuous with respect to x uniformly with respect to h. The Gâteaux variation of $\delta F(x, \cdot)$ exists for all $x \in U$ and has similar properties.

Some aspects of differentiability in complex linear topological spaces are also given in the papers of Silva [116].

We now consider some connections between the direct second order differentials and the Gâteaux variation in complex spaces. A mapping $F: X \to Y$ is said to be <u>directly second order differentiable</u> in Gâteaux's sense at a point $x_0 \in X$ if the quotient

$$(\lambda\mu)^{-1} \{F(x + \lambda h + \mu h') - F(x + \lambda h) - F(x + \mu h') + F(x)\}$$

has a limit as $\lambda, \mu \to 0$, $\lambda, \mu \in \mathbb{C}$, for any $h, h' \in X$. This limit is denoted by $\delta'\delta F(x; h, h')$.

If $F: X \to Y$ has a Gâteaux variation at x_0, then it follows easily that f is directly second order Gâteaux differentiable at x_0, and $\delta'\delta F(x_0; h_1, h_2) = \delta^2 F(x_0; h_1, h_2)$. Therefore, if F is Gâteaux differentiable, then the direct second order differential is Gâteaux differentiable. The converse question was considered by Nicolescu [194, 195]; the completeness of the space Y is essential in this case. To this end, we consider some definitions and results which are also of intrinsic interest.

Let f be a function of a complex variable on a domain $W \subset \mathbb{C}$ with values in a complex Banach space B. The function f is said to be weakly (strongly) directly second order differentiable at the point $t \in \mathbb{C}$ if there exists an element $\varphi(t) \in B$ such that $\lambda^{-2}\{f(t+2\lambda) - 2f(t+\lambda) + f(t)\}$ tends weakly (strongly) to $\varphi(t)$ as $\lambda \to 0$, $\lambda \in \mathbb{C}$. It can be shown that a function $f: W \to B$ which is directly second order weakly differentiable in W and which has partial second order derivatives with respect to ξ and η ($t = \xi + i\eta$) is directly second order strongly differentiable in W, uniformly with respect to t in any bounded domain W_0 strictly interior to W.

A mapping $f: \mathbb{C} \times \mathbb{C} \to B$ is called <u>weakly (strongly) directly second order differentiable</u> at the point $(s, t) \in \mathbb{C} \times \mathbb{C}$ if there exists an element $\psi(s, t) \in B$ such that

$$(\lambda\mu)^{-1} \{f(s + \lambda, t + \mu) - f(s + \lambda, t) - f(s, t + \mu) + f(s, t)\}$$

tends weakly (strongly) to $\psi(s,t)$ as $\lambda, \mu \to 0$ in $\mathbb{C}$; denote $\psi(s,t)$ by $\delta^2 f(s,t)$. Let W be a domain in $\mathbb{C} \times \mathbb{C}$. If $f: W \to B$ is directly second order weakly differentiable on W and has partial second order partial derivatives with respect to ξ, η, ξ', η' ($s = \xi + i\eta$, $t = \xi' + i\eta'$), then f is directly second order strongly differentiable, uniformly with respect to (s,t) in any domain W_0 that is bounded and strictly interior to W.

For $F: X \to Y$ we denote $F(x + sh + th') = f(s,t)$. It is obvious that F is directly second order differentiable in Gâteaux sense at the point $x + s_0 h + t_0 h'$ if and only if $f: \mathbb{C} \times \mathbb{C} \to Y$ is directly second order differentiable at (s_0, t_0), and $\delta'\delta F(x + s_0 h + t_0 h') = \delta^2 f(s_0, t_0)$. Assume that f has partial second order weak derivatives with respect to s and t. Then F is directly second order differentiable in Gâteaux's sense if and only if f is directly second order weakly differentiable. If these conditions are satisfied, $\delta'\delta F$ is Gâteaux differentiable. Also if f is directly second order weakly differentiable with partial second order weak derivatives with respect to ξ, η, ξ', η', and if F is Baire continuous in a domain $D \subset X$, then $\delta'\delta F(x; h, h')$ is Fréchet differentiable and hence an analytic function on D. (See [194]).

10. Remarks on Weak Differentials

We first consider a mapping φ defined on $I = [0,1]$ with values in a real or complex normed linear space Y. We say φ is <u>weakly differentiable</u> at $t \in (0,1)$ if there is an element $w\varphi'(t)$, called the <u>weak derivative</u> of φ at t, such that

$$\lim_{h \to 0} \frac{1}{h} \{y^*(\varphi(t+h) - \varphi(t))\} = y^*(w\varphi'(t))$$

for all $y^* \in Y^*$, where Y^* is the space of continuous linear functionals on Y. The map φ is said to have a <u>pseudo-derivative</u> if there exists a function $p\varphi'$ defined on a measurable $E \subseteq (0,1)$ such that for every $y^* \in Y^*$,

$$\lim_{h \to 0} \frac{1}{h} \{y^*(\varphi(t+h)) - \varphi(t))\} = y^*(p\varphi'(t))$$

for almost every $t \in E$. Note that a pseudoderivative, unlike the weak derivative, is not necessarily unique. Two pseudoderivatives of φ need not be almost everywhere equal (see, for instance [138]). However, if there exists a countable set $\Lambda \subset Y^*$ such that $\|y\| = \sup\{|y^*(y)|; y^* \in \Lambda\}$ for all $y \in Y$, then any two pseudoderivatives of φ must be almost everywhere equal. In particular, this is true if Y is separable.

An operator $F: X \to Y$ is said to have at $x_0 \in X$ a <u>weak Gâteaux variation</u> $w\delta F(x_0; h)$ if the map $t \to F(x_0 + th)$ on $(-\tau, \tau) \to Y$, for $\tau > 0$, is weakly differentiable at 0, i.e. if

$$\lim_{t \to 0} \frac{1}{t} \{y^*[F(x_0 + th) - F(x_0)]\} = y^*(w\, \delta F(x_0; h))$$

exists for every $h \in X$ and $y^* \in Y^*$.

Several mean value theorems analogous to the theorems of Section 5 can be easily established using weak differentials. We illustrate this with two mean value theorems. Let $F: \Omega \to Y$, where Ω is a convex subset of a normed linear space X and let F have a weak Gâteaux variation on Ω. If $x_0, x_0 + h \in \Omega$, then for every $y^* \in Y^*$ there exists $\tau = \tau(y^*)$, $0 < \tau < 1$, such that

(10.1) $\quad y^*[F(x_0 + h) - F(x_0)] = y^*[w\delta F(x_0 + \tau h; h)]$.

It may appear at a first glance, by applying carelessly a corollary to the Hahn-Banach theorem to (10.1), that

$$F(x_0 + h) - F(x_0) = w\delta F(x_0 + \tau h; h) \text{ for } 0 < \tau < 1.$$

However this is not true since τ depends on y^*.

We now consider another mean value theorem. Let $\varphi: [a,b] \to Y$, where $a < b$, both finite. If $\lim_{t \to a^+} \varphi(t) = \varphi(a)$ and $\lim_{t \to b^-} \varphi(t) = \varphi(b)$, both weakly, and if the weak derivative $w\varphi'(t)$ of φ exists on (a,b), then there is a number τ, $a < \tau < b$, such that

(10.2) $$\|\varphi(b) - \varphi(a)\| \leq \|\varphi'(\tau)\| (b-a) .$$

To prove this assertion, let y^* be any continuous linear functional on Y. Then the mean value for a real function of a real variable applies to $y^*(\varphi)$ and yields

(10.3) $$y^*[\varphi(b) - \varphi(a)] = y^*(\varphi'(\tau)) (b-a)$$

for some $\tau \in (0,1)$. Note again that τ depends on y^*. Suppose $\varphi(b) \neq \varphi(a)$ (otherwise (10.2) is trivial) and let x be the functional defined on the one-dimensional subspace of Y spanned by $\varphi(b) - \varphi(a)$, as follows:

$$x(r[\varphi(b) - \varphi(a)]) = r\|\varphi(b) - \varphi(a)\|, \quad r \in \mathbb{R} .$$

Then $\|x\| = 1$ and the functional x may thus be extended by the Hahn-Banach theorem [6; pp. 28-30] to a continuous linear functional y^* defined on Y, with $\|y^*\| = 1$, and

$$y^*(\varphi(b) - \varphi(a)) = \|\varphi(b) - \varphi(a)\| .$$

Using this particular functional y^* in (10.3), the preceding relation yields

$$\|\varphi(b) - \varphi(a)\| = |y^*(\varphi(b) - \varphi(a))| = |y^*(\varphi'(\tau))| (b-a)$$
$$\leq \|y^*\| \|\varphi'(\tau)\| (b-a) = \|\varphi'(\tau)\| (b-a) .$$

Similarly, Theorem 5.8 holds if only the weak bidimensional derivative exists.

Various notions of weak differentials (Gâteaux, Fréchet, Hadamard, (H,β), etc.) on normed or linear topological spaces can be formulated and investigated. For example, we say that a mapping $F: X \to Y$ between two normed linear spaces has a <u>weak Fréchet differential</u> $wdF(x_0;h)$ at $x_0 \in X$ if for any $y^* \in Y^*$

$$y^*(F(x_0+h) - F(x_0)) = y^*(wdF(x_0;h)) + y^*(r(x_0;h)) ,$$

where $wdF(x_0;h)$ is linear in h and

$$\lim_{\|h\| \to 0} \frac{|y^*(r(x_0;h))|}{\|h\|} = 0 .$$

If $wdF(x_0;h)$ is bounded on some sphere with center at 0, we shall say that F has a <u>weak Fréchet derivative</u> $wF'(x_0)$ at x_0.

It follows immediately that if F has a weak Gâteaux variation at $x_0 \in X$, then F is <u>directionally demicontinuous</u> at x_0, i.e. $\lim_{t \to 0} y^*(F(x_0+th) - F(x_0)) = 0$ for any $y^* \in Y^*$ and each $h \in X$; however this limit is not uniform in h. On the other hand if F has a weak Fréchet derivative at x_0, then F is <u>demicontinuous</u> at x_0, i.e. $x_n \to x_0$ implies $y^*(F(x_n)) \to y^*(F(x_0))$ for each $y^* \in Y^*$.

It is easy to show that if F has a weak Gâteaux variation in some neighborhood of $x_0 \in X$ and if $w\delta F(x;h)$ is demicontinuous at x_0 for every fixed $h \in X$, then $w\delta F(x;h)$ is a weak Fréchet differential. The interested reader should have no difficulty formulating results for weak differentials analogous to various theorems stated in the preceding sections. Weak differentials are useful in the study of differential operators in Sobolev spaces, for instance in the space W_2^1, and in some problems in function spaces.

II. Some Aspects of the Role of Differentials in Nonlinear Functional Analysis

11. The Role of Differentials in Extrema of Functionals: Infinitesimal Conditions for Continuity and Lower Semicontinuity in the Weak Topology

If f is a continuous real function on a closed bounded set V in a finite dimensional space, then f is uniformly continuous and attains its infimum and supremum in V. If f is lower (upper) semicontinuous, then f attains its infimum (supremum) in V. These statements are not necessarily true in infinite dimensional spaces, since a closed bounded set S may fail to be compact or sequentially compact. For example in ℓ_2, the Hilbert space of square summable sequences $x = (x_1, x_2, \ldots)$, the unit sphere $S = \{x: \sum_{i=1}^{\infty} x_i^2 = 1\}$ is closed and bounded but not compact, and the functional $f(x) = \sum_{i=1}^{\infty} 2^{-i} x^i$ is continuous on S but does not attain its greatest lower bound in S. The proof of the Weierstrass theorem is based on a compactness argument which can be extended to continuous functionals defined on a compact set of an arbitrary metric space. That is, if a set V in a metric space X is compact, then every continuous functional is bounded on V and attains its greatest upper and least lower bounds (see, for instance, [9]). The usefulness of this result is severely limited by the fact that many important sets (spheres, convex sets, etc.) are not compact in infinite dimensional normed spaces. However, this situation is salvaged by the fact that the above-mentioned result extends to arbitrary topological spaces. In this section we discuss the role of the weak topology and differentials in the theory of extrema.

11.1. Lower Semicontinuity

Let X be a topological space and f a functional on X

into $\mathbb{R}$. Recall that by $\lim\inf_{x \to x_0} f(x)$ we mean the number s uniquely determined by the properties (i) for each $\delta > 0$ there exists a neighborhood $U_\delta(x_0)$ of x_0 such that $f(x) > s - \delta$ for $x \in U_\delta(x_0) - x_0$; and (ii) for each neighborhood V of x_0 and $\delta > 0$, there exists some $x_1 \in V - x_0$ such that $f(x_1) < s + \delta$.

<u>Definition 11.1.</u> A functional $f: X \to \mathbb{R}$ is <u>lower semicontinuous</u> at x_0 if

$$\lim_{x \to x_0} \inf f(x) \geq f(x_0),$$

or, equivalently, if for every $\varepsilon > 0$ there exists a neighborhood $U(x_0)$ of x_0 such that

$$f(x) \geq f(x_0) - \varepsilon \text{ for all } x \in U(x_0).$$

It is easy to show that a functional f is lower semicontinuous relative to a given topology if and only if the set $\{x: f(x) > r\}$ is open (in the same topology) for each real number r. Also, $\lim\inf_{x \to x_0} f(x) \leq b$ if and only if $x_0 \in \bigcap_{a > b} Cl\{x: f(x) \leq a\}$, where Cl denotes closure.

We recall that a topological space X is called compact if every open covering of X contains a finite subcovering. A subset S of a topological space X is called compact if S as a topological space with the relative topology induced by that of X is compact. ([2; 4.12] or [8, p. 37]).

<u>Theorem 11.1.</u> If f is a lower semicontinuous functional on a compact set S in a topological space X, then f is bounded below and attains its greatest lower bound in S.

<u>Proof:</u> Suppose f is not bounded below on S. Then each of the sets $A_n = \{x \in S: f(x) \leq -n\}$ is nonempty. Since f

is lower semicontinuous, each of the sets A_n is closed; also $A_{n+1} \subset A_n$ for all n. Hence $A = \bigcap_{n=1}^{\infty} A_n$ is nonempty. But this is a contradiction since $f(x) = -\infty$ for $x \in A$. Thus f is bounded below on S.

Let γ be the greatest lower bound of f on S, and let $B_n = \{x \in S: f(x) \leq \gamma + \frac{1}{n}\}$. A similar argument shows that $B = \bigcap_{n=1}^{\infty} B_n$ is nonempty. Clearly $f(x) = \gamma$ for $x \in B$, and thus the greatest lower bound of f is attained.

<u>Corollary 11.1.</u> If f is upper semicontinuous functional on a compact set S in a topological space X, then f is bounded above and attains its least upper bound in S.

Theorem 11.1 provides the flexibility that one needs to develop a theory of extrema. It motivates introducing (this turns out to be possible) on a Banach space a topology relative to which a ball or a closed convex bounded set is compact. Certainly the normed topology will not work since the ball $\{x: \|x\| \leq 1\}$ in a normed space E is compact in the normed topology if and only if the space is finite dimensional. On the other hand, the weak topology does the job in reflexive Banach spaces.

11.2. <u>The Weak Topology in Banach Spaces and Existence of Extrema</u>

Let E be a Banach space. We introduce on E several topologies using appropriate definitions of neighborhoods, equivalently of open sets. A strong ε-neighborhood of a point $x_0 \in E$ is defined by $N(x_0; \varepsilon) = \{x: \|x-x_0\| < \varepsilon\}$. The <u>strong topology</u> on E is induced by taking as "open sets" the empty set and all unions of strong neighborhoods.

Let $\varepsilon > 0$ and let $\ell_i \in E^*$, $i = 1, \ldots, n$. A <u>weak</u> neighborhood of θ is defined by

$$N(\ell_1, \ldots, \ell_n; \varepsilon) = \{x \in E: |\ell_i(x)| < \varepsilon, \ i = 1, \ldots, n\}.$$

Let $\eta(\theta)$ be the family of all weak neighborhoods of θ. Let $\mathfrak{F}(\theta)$ be the family of all sets in E which contain a finite intersection of elements of $\eta(\theta)$, and let $\mathfrak{F}(x)$ be the family of all sets of the form $x + U(\theta)$, where $U(\theta) \in \mathfrak{F}(\theta)$. We take the elements of $\mathfrak{F}(x)$ as neighborhoods of x and call the induced topology the <u>weak topology</u> on E, or the E^* topology of E. It can be shown that with this topology the space E becomes a linear topological Hausdorff locally convex space [2; V.3.3]. A set which is closed in the weak topology of E is called weakly closed; similarly for other topological notions.

The weak topology of the dual space E^* may be defined as in the case of the space E by using continuous linear functionals on E^*, i.e. using elements in the second dual space E^{**}. However, there is a more important topology on E^* defined by restricting the functionals to E. We define a weak* neighborhood of $\theta \in E^*$ to be a set of the form

$$N(x_1, \ldots, x_n; \varepsilon) = \{\ell \in E^* : |\ell(x_i)| < \varepsilon, \quad i = 1, \ldots, n\}$$

where ε is a positive number and $x_1, \ldots, x_n \in E$. The topology induced using this definition of neighborhood is called the <u>weak* topology</u> of E^* or the E topology of E^*.

The role of the weak topology and reflexive Banach spaces in the theory of extrema is derived from two important theorems. The first theorem, which was proved independently by Alaoglu, Bourbaki, and Kakutani, states that the ball $\{\ell \in E^* : \|\ell\| \leq 1\}$ is compact in the weak* topology of E^*. For reflexive spaces, the weak topology of E coincides with the weak* topology of $E^{**} = E$ and thus the unit ball is weakly compact. Eberlein has shown that this characterizes reflexive spaces.

<u>Theorem 11.2.</u> A Banach space E is reflexive if and only if every bounded and weakly closed subset of E is weakly compact.

For a proof, see [2; V.4.8].

A weakly closed set is strongly closed but the converse is not necessarily true (consider a complete orthonormal set in an infinite dimensional space). However, the converse is true for convex sets in Banach spaces. This is the essence of the second theorem alluded to earlier.

<u>Theorem 11.3</u>. Every strongly closed convex set in a Banach space is weakly closed.

This theorem was first proved by Mazur. See [2; V.3.13].

<u>Corollary 11.3</u>. In a reflexive Banach space every bounded closed convex set is weakly compact.

Combining Theorems 11.1 and 11.3 we obtain

<u>Theorem 11.4</u>. Let f be a weakly lower semicontinuous real valued function on a bounded weakly closed subset Ω of a reflexive Banach space. Then f attains a minimum at some point of Ω. In particular, the assertion is true if Ω is bounded, closed and convex.

If f is upper semicontinuous, then it attains a maximum at some point of Ω.

Note that a weakly continuous real valued functional defined on a reflexive Banach space E need not attain its infimum on <u>all</u> of E; for example consider $f(x) = e^x$ on $(-\infty, \infty)$. The following simple observation is useful in minimizing a functional on the whole space.

<u>Theorem 11.5</u>. Let E be a reflexive Banach space and let f be a functional which satisfies the following properties:

(i) f is weakly lower semicontinuous on E;

(ii) there exist $x_0 \in E$ and a $K > 0$ such that for $r > K$, $\inf\{f(x): \|x\| = r\} \geq f(x_0)$.

Then f is bounded below and attains its infimum in E.
<u>Proof</u>: Let $a = \max(\|x_0\|, K)$, $S_r = \{x: \|x\| \leq r\}$.

By Theorem 11.4, $\min\{f(x): x \in S_r\}$ exists. On the other hand, we obviously have

$$\inf\{f(x): x \in E\} = \inf\{f(x): x \in S_a\}$$

and the theorem is proved.

Remark. Property (ii) in Theorem 11.5 is guaranteed by the growth condition

$$\lim_{\|x\| \to \infty} f(x) = +\infty ,$$

which is easier to verify than (ii).

The utility of Theorem 11.4 in extremal problems depends on establishing sufficient conditions (a) for a set Ω to be a weakly closed and bounded subset of a reflexive Banach space, and (b) for a functional f to be weakly lower semicontinuous (or upper semicontinuous for maximum). A sufficient condition for (a) is that Ω be convex and closed in the norm topology. The rest of this section will be devoted to sufficient conditions for (b). This will show the relevance and importance of differentials for providing sufficient conditions in extremal problems. We shall see also how these conditions are intimately connected with various notions of convexity for functionals.

11.3. Weak and Sequential Weak Continuity

A partially ordered set $\{\mathcal{J}, >\}$ is called directed if for every pair of elements α, β of $\mathcal{J}$ there exists another element γ such that $\gamma > \alpha$ and $\gamma > \beta$. A <u>generalized sequence</u> in a topological space X is a map $\alpha \to x_\alpha$ of $\mathcal{J}$ into X. An ordinary sequence is obtained by taking $\mathcal{J}$ to be the set of natural numbers in their natural order. A generalized sequence $\{x_\alpha : \alpha \in \mathcal{J}\}$ is said to converge to x_0, $x_\alpha \to x_0$ if and only if for any neighborhood U of x_0, there exists an α_0 such that $x_\alpha \in U$ for $\alpha > \alpha_0$. If X is a Banach space, we write $x_\alpha \xrightarrow{w} x_0$ to indicate weak convergence of a generalized sequence. It follows readily that $x_\alpha \xrightarrow{w} x_0$ if and only if $\ell(x_\alpha) \to \ell(x_0)$ for every $\ell \in X^*$. It is easy to show that a real valued function f defined on a topological space X is <u>continuous</u> at x_0 if and only if

(11.1) $$\lim f(x_\alpha) = f(x_0)$$

for any generalized sequence $\{x_\alpha\}$ in X which converge to x_0. (See, for instance, [2; I.7.4] or [8]). In particular, if X is a Banach space then f is <u>weakly continuous</u> at x_0 if and only if (11.1) holds for any generalized sequence satisfying $\ell(x_\alpha) \to \ell(x_0)$ for every $\ell \in X^*$.

A real valued function f on a subset of Banach space E is called <u>sequentially weakly</u> continuous at x_0 if (11.1) holds for each <u>ordinary</u> sequence for which $\ell(x_\alpha) \to \ell(x_0)$ for every $\ell \in E^*$. If f is weakly continuous at x_0, then it is also sequentially weakly continuous, but not conversely. It can be shown that on <u>bounded</u> sets in a Banach space E whose dual space E^* is separable, the notions of weak and sequential weak continuity coincide. Now if the conjugate space X^* of a Banach space is separable, then X is separable but not conversely [2; II3.16]. On the other hand, if E is reflexive and separable, then by the preceding statement E^* is separable. Thus we have the important result: <u>On bounded sets</u> of a <u>separable</u> and <u>reflexive</u> Banach space, the notions of weak and sequential weak lower semicontinuity coincide.† This statement is false in general if the word "bounded" is dropped. In contrast, continuity and sequential continuity are equivalent in the strong topology in any normed space.

---

† Some analysts (for instance [11]) use the phrase, "weakly lower semicontinuous" for what is called here "sequentially lower semicontinuous". The context usually makes it clear which notion is being used.

Finally we mention that in the case of a metric space, the notion of compactness of a set S (as defined earlier in terms of the finite covering property) is equivalent to the property that every sequence in S has a subsequence which converges to a point in S. In the rest of this section, only Banach spaces will be used and thus the two notions may be used interchangeably.

11.4. "Infinitesimal" Criterion for Weak Semicontinuity

We are now in a position to state the relationship (which was first noted by Rothe [342], [343]) between the weak continuity of a functional on a Banach space and the compactness of its gradient, and thereby establish conditions under which a differentiable functional is weakly continuous.

Theorem 11.6 [343; Theorem 3.3]. Let f be a real valued function on a bounded convex subset Ω of a Banach space E. Suppose that the Gâteaux differential of f exists for each $x \in \Omega$ and that the gradient of f considered as a map on Ω into E^* is a compact map. Then for every $\varepsilon > 0$ there exist $\delta > 0$ and $\ell_1, \ldots, \ell_r \in E^*$ such that

(11.2) $\qquad |f(x+h) - f(x)| < \varepsilon \|h\| \quad x, x+h \in \Omega$

for all h satisfying

(11.3) $\qquad |\ell_i(h)| < \delta \|h\|, \qquad i = 1, \ldots, r .$

Before we prove this theorem we define the Leray-Schauder approximation for compact sets and compact mappings.

Definition 11.2. Let X be a Banach space and B a subset of X with compact closure $\bar{B}$. Let η be a positive number and let $e_1, \ldots, e_r$ be points of B such that for every $y \in B$,

$$\|y - e_i\| < \eta \quad \text{for at least one } i ;$$

the existence of such e_i is guaranteed by the compactness of $\bar{B}$ (see, for instance I.6.15 in [2]). For $j=1,\ldots,r$, let μ_j be the continuous function defined by

$$\mu_j(y) = \begin{cases} \eta - \|y - e_j\| & \text{if } \|y - e_j\| < \eta \\ 0 & \text{if } \|y - e_j\| \geq \eta \end{cases}.$$

Then

(11.4) $$a(y) = \frac{\sum_{j=1}^{r} \mu_j(y) e_j}{\sum_{j=1}^{r} \mu_j(y)}, \quad y \in B$$

is called the Leray-Schauder η approximation of the set B. If G is a compact operator and $B = G(A)$, where A is bounded, then

(11.5) $$G_\eta(x) = a(G(x)) \quad x \in A$$

is called the <u>Leray-Schauder approximation</u> of the map G.

It follows from Definition 11.2 that $a(y)$ is defined and continuous for all $y \in B$; $G_\eta(x)$ is defined for all $x \in A$ and is continuous if G is completely continuous. Moreover,

(11.6) $$\|y - a(y)\| < \eta, \quad y \in B$$

(11.7) $$\|G(x) - G_\eta(x)\| < \eta, \quad x \in A.$$

<u>Proof of Theorem 11.6</u>. We apply the Leray-Schauder approximation taking in Definition 11.2, $X = E^*$, $G = F(x) = \operatorname{grad} f(x)$, $\eta = \varepsilon/2$. Using (11.7) we get

$$|F(x)h| \leq |(F(x) - F_\eta(x))h| + |F_\eta(x)h|$$

$$< |F_\eta(x)h| + \eta\|h\| \, .$$

On the other hand, from the definition of the Leray-Schauder approximation, (11.4) and (11.5),

$$|F_\eta(x)h| = \frac{|\sum_1^r \mu_j(F(x))\ell_j(h)|}{\sum_1^r \mu_j(F(x))} \, , \quad \ell_j \in E^*$$

$$\leq \max_{1 \leq j \leq r} |\ell_j(h)| \, .$$

Using this estimate in the preceding inequality we get

$$|F(x)h| \leq \eta\|h\| + \max_{1 \leq j \leq r} |\ell_j(h)| = \frac{\varepsilon}{2}\|h\| + \max_{1 \leq j \leq r} |\ell_j(h)| \, .$$

Thus for all h satisfying (11.3) with $\delta = \frac{\varepsilon}{2}$,

$$|F(x)h| < \varepsilon\|h\| \text{ for all } x \in \Omega \, .$$

But by the mean-value theorem (see (5.4))

$$f(x+h) - f(x) = \delta f(x + \tau h; h) \text{ for } 0 < \tau < 1$$

or

$$|f(x+h) - f(x)| = \|F(x+\tau h)h\| < \varepsilon\|h\| \quad x \in \Omega \, .$$

This completes the proof.

Corollary 11.6. Under the assumptions of Theorem 11.6, f is weakly continuous on Ω.

Proof: We have to show that to every positive number ε' there exists a positive number δ and $\ell_1, \ldots, \ell_r \in E^*$ such that $|\ell_j(h)| < \delta$, $j = 1, \ldots, r$ implies $|f(x+h) - f(x)| < \varepsilon'$. This follows from Theorem 11.6 by taking $\varepsilon = \frac{\varepsilon'}{2R}$ where $\|x\| \leq R$ for $x \in \Omega$.

Rothe [343], [64] has shown that the converse of Corollary 11.6 holds in Hilbert spaces. In particular, a necessary and sufficient condition for a Fréchet differentiable functional on a Hilbert space to be weakly continuous is that grad f be completely continuous. Rothe also showed that the converse holds for Banach spaces of a special type.

Theorems establishing converses to Theorem 11.6 under further restrictions were given by many authors. Citlanadze [254] (see also Theorem 7.2 in [11]) proved the following theorem: Let E be a reflexive Banach space, $D_r = \{x \in E: \|x\| < r\}$, and let f be a continuous functional on E. If (a) E has a base, (b) f is weakly continuous on $D_{r+\alpha}$ for some $\alpha > 0$, and (c) f is Fréchet differentiable with uniform remainder on D_r, then the gradient of f is compact on D_r. Vainberg replaced (c) by the condition (c)' df(x;h) is uniform on D_r. Kadec [285] showed that for separable spaces the condition (a) may be dropped using the result that in any separable space E there is an equivalent norm relative to which E is uniformly convex. Citlanadze's results were extended by Ansoov to nonreflexive spaces which satisfy certain conditions. Another result in this direction was obtained by Rothe [64], [346], [349] for a Banach space E with property (P). This means that there exist a linearly independent sequence $\{e_i\}$ in E^* and a number $M > 0$ with the following properties: the span of $\{e_i\}$ is dense in E^* and for each positive integer there exists a linear projection of norm at most M on the intersection $\bigcap_{i=1}^{n} N_i$, where $N_i = \{x \in E: e_i(x) = 0\}$. The main result of [346] is as follows.

Let E be a Banach space with property (P), f a functional which has a continuous Fréchet derivative on a convex subset Ω of E. Then the following condition is sufficient for the gradient of f to be completely continuous on Ω: For each $\eta > 0$ there exist $\ell_i \in E^*$, $i = 1, \ldots, N$ such that

$$|f(x+h) - f(x)| < \eta \|h\| \qquad x, x+h \in \Omega$$

for all h for which

$$|\ell_i(h)| < \frac{\eta}{2} \|h\| \qquad (i = 1, \ldots, N) .$$

Note that from Theorem 11.6 this condition is also necessary. Every reflexive Banach space with a base has property (P) but not conversely. The proof of the converse of Theorem 11.6 given in [343] depends heavily on the existence of a projection of norm one on every closed linear subspace. It is well known that Hilbert spaces are the only Banach spaces (of dimension at least 3) which have this property. Rothe was led to impose property (P) by the observation that for the proof of the converse to Theorem 11.6, it is not necessary to have projections of norm one (or even of uniformly bounded norm) on all closed linear subspace; rather, it suffices to have such projections exist on a sufficiently large collection of subspaces. Ando [233] established sufficient conditions for the compactness of grad f without requiring the space to have property (P). With the slight modification of using the Gâteaux derivative instead of the Fréchet derivative, Ando's theorem is as follows.

Theorem 11.7. Let f be a real valued function on a bounded convex subset Ω of a Banach space E. Suppose that the Gâteaux differential of f exists for each $x \in \Omega$ and that $F(x) = \text{grad } f(x)$ is bounded on Ω, i.e., $\|F(x)\| < m$, $x \in \Omega$. Assume that for each $\varepsilon > 0$ there exist p elements $\ell_j \in E^*$ such that (11.2) holds for all h satisfying

$$\ell_j(h) = 0 \text{ and } \|h\| < \varepsilon, \quad j = 1, \ldots, p .$$

Then F is a compact operator on Ω into E^*.

Fichera [270] also considered theorems of this type in the context of the calculus of variations.

Corollary 11.6 leads to a geometric interpretation which gives some insight into the role that the weak topology plays in connection with extrema of functionals. First we recall that an operator $F: X \to Y$ is completely continuous on a bounded set $\Omega \subset X$ (X and Y are Banach spaces) if and only if for each $\varepsilon > 0$, there exists a continuous finite dimensional mapping $F_\varepsilon : \Omega \to Y_k$, where Y_k is a finite dimensional subspace of Y and $\|F(x) - F_\varepsilon(x)\| < \varepsilon$ for every x in Ω. (See, for instance, [342], Lemma 5.2; [11], Theorem 1.1; [278], Chapter II). In the norm topology, the Weierstrass theorem holds if the continuous functional is restricted to compact sets. The restriction to compact sets is quite severe in the case of infinite dimensional spaces and we resort instead to the <u>weak topology</u>, relative to which a bounded closed convex subset of a reflexive Banach space is compact. The requirement of continuity of f in the weak topology is tantamount to the complete continuity of grad f. Thus in the case of Theorem 11.1 in the strong topology, the domain is approximated by finite sets (the finite covering theorem), whereas using the weak topology grad f is approximated by a finite dimensional mapping. Roughly speaking, we are trading compactness of the domain in the norm topology with weak continuity of the functional, or with complete continuity of its gradient. In the latter case, the range of grad f is approximated by a set contained in a finite dimensional space. Thus in both cases some notion of finite dimensional approximation prevails.

Finally we remark that if $F(x) = \text{grad } f(x)$ is strongly continuous (i.e., maps weakly convergent sequences into strongly convergent sequences), on a closed ball S in a reflexive Banach space X, then $F(x)$ is compact and uniformly continuous on S (see Theorem 1.3 in Vainberg [11]), and thus f is weakly continuous on S. Various sufficient conditions can be given for grad f(x) to be strongly continuous. For example (see [292], Theorem 1), suppose that f is a convex subadditive functional on an open convex subset

G (containing a closed ball S) of a reflexive Banach space, $f(0) = 0$, and that f is bounded above on some convex open subset of G . If f has a Fréchet differential with uniform remainder on S, then the grad f is strongly continuous on S . Other related results are given in Kolomý [292].

11.5. "Infinitesimal" Conditions for Weak Lower Semicontinuity

Theorem 11.8. Let Ω be a convex closed subset of a Banach space E . A sufficient condition for a functional f defined on Ω to be weakly lower semicontinuous is the existence of an element $\ell \in E^*$ such that

$$f(x) - f(x_0) \geq \ell(x - x_0) \text{ for all } x, x_0 \in \Omega .$$

This proposition is an immediate consequence of the definition of weak neighborhoods and lower semicontinuity ([344], Theorem 4.1). As an application of it, we state

Theorem 11.9. Suppose that f has a second order Gâteaux differential $D^2 f(x; h, k)$ for every $x \in \Omega$, $h, k \in E$ and that

(11.8) $\quad D^2 f(x; h, h) \geq 0$ for every $x \in \Omega$, $h \in E$.

Then f is weakly lower semicontinuous on Ω .

Proof: From the Taylor formula with remainder (see Theorem 6.2)

$$f(x_0 + h) - f(x_0) = Df(x_0; h) + \int_0^1 D^2 f(x_0 + th; h, h) (1-t) dt .$$

Thus, using (11.8),

$$f(x_0 + h) - f(x_0) \geq Df(x_0; h) .$$

Hence the assumptions of Theorem 11.8 are satisfied with $\ell(x - x_0) = Df(x_0; x - x_0)$.

As an application of Theorems 11.5 and 11.9 we give the following theorem.

Theorem 11.10. Let E be a reflexive Banach space and let f be a functional on E which has second order Gâteaux differential on E which satisfies the inequality

(11.9) $\quad D^2 f(x; h, h) \geq \gamma(\|h\|) \|h\|$ for all $h \in E$,

where γ is a nonnegative real valued function on $[0, \infty)$ such that

(11.10) $\qquad \lim_{R \to \infty} \int_0^1 \gamma(tR) dt = \infty$.

Assume also that $D^2 f(tx; h, h)$ is continuous for $t \in [0, 1]$. Then f is bounded below on E and attains its minimum at some point in E. Moreover, if $\gamma(t) > 0$ for $t > 0$, then there exists a unique point at which the minimum is attained.

Proof: From (11.9) we have $D^2 f(x; h, h) \geq 0$, and hence by Theorem 11.9, f is weakly lower semicontinuous on each sphere in E. Now we show that the assumptions of Theorem 11.5 are satisfied. Let $F(x) = \text{grad } f(x)$. Then (see (5.2))

$$Df(x; h) - Df(0; h) = \int_0^1 D^2 f(tx; x, h) dt, \quad \text{or}$$

$$F(x) h = F(0) h + \int_0^1 DF(tx; x) h \, dt .$$

In particular, we have for $h = x$, using (11.9)

(11.11) $F(x)x = F(0)x + \int_0^1 DF(tx, x) x \, dt \geq F(0)x + \|x\| \gamma(\|x\|)$.

On the other hand,

$$f(x) = f(0) + \int_0^1 F(tx)x\,dt ,$$

which using (11.11) implies

$$f(x) \geq f(0) + \int_0^1 \{F(0)tx + \|tx\|\gamma(\|tx\|)\}\frac{dt}{t}$$

$$= f(0) + F(0)x + R\int_0^1 \gamma(tR)dt$$

on the sphere $\|x\| = R$. Therefore,

(11.12) $\quad f(x) \geq f(0) + R(-\|F(0)\| + \int_0^1 \gamma(tR)dt)$.

Thus for $\varepsilon > 0$, there exists, on account of (11.10) and (11.12), $R_0 > 0$ such that for $R > R_0$, $f(x) \geq f(0) + \varepsilon$ on the sphere $\|x\| = R$, i.e., $\inf\{f(x) : \|x\| = R\} \geq f(0)$. Thus Theorem 11.5 applies and f attains its infimum on E. The second part of the theorem is trivial.

One of the early general criteria for existence of minima was given by Krasnoselskii [295]: Let f be a functional on a Hilbert space H and let f have Fréchet differential with a bounded remainder on every sphere. Assume that the gradient of f can be written in the form $F(x) = \operatorname{grad} f(x) = F_1(x) + F_2(x)$, where F_1 is a linear continuous operator with continuous inverse and F_2 is completely continuous. Finally, assume that $\lim_{\|x\| \to \infty} f(x) = \infty$. Then f attains its infimum on H at some point $x^* \in H$, and $F(x^*) = 0$. This criterion, which is subsumed in the setting of the more general theorems of this section, includes almost all existence theorems proved for nonlinear integral equations by variational methods until the late 1950's.

11.6. Weakly Semicontinuous Quadratic Functionals

A positive self-adjoint operator T defined on a real Hilbert space H determines a weakly lower semicontinuous quadratic functional $Q(x) = \langle Tx, x \rangle$. Weakly lower semicontinuous quadratic functionals can be completely characterized. Let T be a self-adjoint (continuous) linear operator on a real Hilbert space H. It follows from the spectral theorem that T determines an orthogonal decomposition $H = H_+ \oplus H_-$ such that $Q(x) = \langle Tx, x \rangle \geq 0$ on H_+ and $Q(x) \leq 0$ on H_-. Furthermore $T = T_+ + T_-$ where $\langle T_+ x, x \rangle \geq 0$ and $\langle T_- x, x \rangle \leq 0$ for $x \in H$, and $T_+ H_- = T_- H_+ = \{0\}$. Hestenes [280] obtained an interesting characterization of weakly lower semicontinuous quadratic forms: A necessary and sufficient condition for $Q(x)$ to be weakly lower semicontinuous is that the negative part T_- of T be a compact operator; equivalently the form $Q_-(x) = \langle T_- x, x \rangle$ be weakly continuous.

An important class of weakly lower semicontinuous functionals are the Legendre forms that arise in the calculus of variations. A self-adjoint operator T is called a <u>Legendre operator</u> if $T = P + K$, where P is positive definite: $\langle Px, x \rangle \geq m \|x\|^2$, $m > 0$, and K is compact. A quadratic form induced by a Legendre operator is called a Legendre form. A necessary and sufficient condition for T to be a Legendre operator on a Hilbert space H is that there exists a finite dimensional subspace H_n such that the restriction of T to H_n is positive definite. An excellent exposition on semicontinuous quadratic functional and applications to quadratic variational problems is given in the recent paper of Hestenes [281].

11.7. Convexity and Weak Lower Semicontinuity

We recall that the positiveness of the Hessian of a twice differentiable functional $f: \Omega \to \mathbb{R}$ implies convexity of f. Thus in view of Theorem 11.10 and Section 11.6, one might suspect that "convexity" of f would be a natural assumption for extremal problems, just as convexity of the set on which f is to be minimized was a useful assumption for the setting of the weak topology. We conclude this section by reviewing briefly some notions of convexity which play a decisive role in extremal problems in reflexive Banach spaces.

A functional f defined on a convex set Ω is said to be <u>quasiconvex</u> if for each real number c, the set

$$S_c = \{x: f(x) \le c, \ x \in \Omega\}$$

is convex. Note that if f is convex, then the set S_c is convex, but not conversely. It is easy to show that f is quasiconvex if and only if for all $\lambda \in (0,1)$, and $x_0, x \in \Omega$,

(11.13) $\qquad f(\lambda x + (1-\lambda)x_0) \le \max\{f(x), f(x_0)\}$.

Also if f has a linear Gâteaux variation on Ω, then f is quasiconvex if and only if for every $x, x_0 \in \Omega$,

$$f(x) \le f(x_0) \quad \text{implies} \quad \delta f(x_0; x-x_0) \le 0 \ .$$

This follows easily by considering the functional $\varphi(t) = f(tx + (1-t)x_0)$ and noting that $\varphi'(0) \le 0$. If a quasiconvex functional defined on a convex set Ω is lower semicontinuous in the norm topology, then it is weakly lower semicontinuous. The result is generalized in the following proposition ([319], Theorem 2.4) which applies to functionals that arise in the calculus of variations (see also [348], [249], [336]).

Let E be a Banach space, ϕ a functional on $E \times E$ that satisfies the following properties: (i) for each fixed $y \in E$, $\phi(x,y)$ is lower semicontinuous (i.e., in the strong topology) and quasiconvex in x; (ii) $\phi(x, \cdot)$ is weakly lower semicontinuous on bounded sets in E, uniformly in x on each bounded subset of E. Then the functional f defined by $f(x) = \phi(x,x)$ is weakly lower semicontinuous.

A functional f defined on a convex set Ω is called strongly quasiconvex if the strict inequality holds in (11.13) for all $x \neq x_0$. It follows from this definition that a strongly quasiconvex functional can not attain its infimum on Ω at more than one point. Thus combining this with the weak lower semicontinuity property stated earlier we have the following result: A strongly quasiconvex and lower semi continuous functional f on a closed bounded convex set Ω in a reflexive Banach space E attains its minimum at a unique point in Ω.

A partial converse of this result underlines the relevance of strongly quasiconvex functionals to minimization problems. It asserts that if a lower semicontinuous functional f attains a unique minimum point on each bounded closed convex set in a reflexive Banach space, then f must be strictly quasiconvex.

For applications of minimization of functionals see the companion paper of Daniel in this volume (also [261]). Various interesting applications are given in Levitin and Poljak [300], Poljak [335], Cheney and Goldstein [245], Demayanov and Rubinov [263], and others. Poljak [336] provides various sufficient (and in some cases also necessary) conditions for lower semicontinuity and continuity of integral functionals arising in extremal problems, with respect to various types of convergence in the spaces L_p, $1 \leq p \leq \infty$, C, and the Sobolev spaces W_p^1 ($1 \leq p \leq \infty$).

Finally we remark that some nonstandard problems (such as certain optimization problems arising in control theory) require considerations which are not covered in the setting of this section, either because the underlying space is not reflexive or the set on which the functional is defined is not convex. Considerable attention has been given in the recent literature to such extremal problems, as well as to constrained problems. We refer the interested reader to [231], [232], [239], [241], [242], [243], [252], [263], [266], [279], [300], [324], [325] for perspectives in various settings of extremal problems, in addition to references cited earlier.

12. Tangents and Cones of Tangents

If $f: J \to \mathbb{R}$ where J is a subset of $\mathbb{R}$, then differentiability of f can be interpreted in terms of tangency, i.e., if f is differentiable at x_0, then the graph of the function $\phi: \mathbb{R} \to \mathbb{R}$ defined by

$$\phi(x) = f(x_0) + f'(x_0)(x - x_0), \quad x \in \mathbb{R}$$

is the tangent line to the graph of f at the point $(x_0, f(x_0))$. In the case of a mapping $F: X \to Y$, where X, Y are normed linear space over the same (real or complex) scalars, one can envisage not only tangents to the graph of F, but also tangents to the range of F and to the level surfaces of F. More generally, one may seek a notion of tangent to an arbitrary set in a normed space.

Let x_0, x be two distinct points in X, and let $h = x - x_0$. The set $\{x_0 + \lambda h: \lambda \in \mathbb{R}, \lambda \geq 0\}$ is called a ray with origin x_0 and direction $\frac{h}{\|h\|}$ or equivalently a ray from x_0 through x. Let A_0 be a subset of X, let x_0 be an accumulation point in A_0, and $A = A_0 \setminus \{x_0\}$. If

$$\lim_{\substack{x \to x_0 \\ x \in A}} \frac{x - x_0}{\|x - x_0\|}$$

exists (say equal to $u \in X$), then the ray in X with origin x_0 and direction u is called <u>the tangent ray</u> to A_0 at x_0. For a given set $S \subset X$ and a point $x_0 \in \bar{S}$, consider all subsets A_0 of S having a tangent ray at x_0. The union of all these tangent rays is called the <u>tangent cone</u> to S at x_0. For example, if in $\mathbb{R}^n$, with the Euclidean norm, we take $S = \{x: \|x\| \leq 1\}$, then the tangent cone to S at a boundary point of S is a half-space in $\mathbb{R}^n$, whereas the tangent cone to S at an interior point of S is all of $\mathbb{R}^n$.

Let $F: \mathcal{D} \subset X \to Y$ and denote the range and the graph of F by $\mathcal{R}(F)$ and $\mathcal{G}(F)$ respectively. The level surface of F through the point $x_0 \in \mathcal{D}$ is defined as the set $\{x: F(x) = F(x_0)\}$.

<u>Theorem 12.1.</u> Let F be Fréchet differentiable at $x_0 \in \mathcal{D}$, let $y_0 = F(x_0)$, $L = F'(x_0)$ and $\phi: X \to Y$ be defined by

$$\phi(x) = y_0 + L(x - x_0) \quad \text{for } x \in X.$$

Then

(i) the tangent cone to $\mathcal{G}(f)$ at (x_0, y_0) is $\mathcal{G}(\phi)$,

(ii) for each neighborhood U of x_0 contained in $\mathcal{D}$ the tangent cone to $F(U)$ at y_0 contains $\mathcal{R}(\phi)$,

(iii) for any closed vector subspace Z of Y the tangent cone to $F^{-1}(y_0 + Z)$ at x_0 is contained in $\phi^{-1}(y_0 + Z)$.

In particular, the tangent cone at x_0 to the level surface of F through x_0 is contained in the level surface of ϕ through x_0.

For a proof of Theorem 12.1, as well as Theorems 12.2–12.5 below, see Flett [271]. A more leisurely presentation of some of these results for finite dimensional space is given in Flett's book [272].

Note that the assertions of Theorem 12.1 are based solely on the assumption of Fréchet differentiability of F. One may ask whether Fréchet differentiability can be <u>characterized</u> by the existence of an appropriate tangent cone to the graph of F. The following theorem settles the question positively in the finite dimensional case; the question in the infinite dimensional case is still open.

<u>Theorem 12.2.</u> Let X, Y be finite dimensional normed vector spaces, let $\mathcal{D} \subset X$ and $F: \mathcal{D} \to Y$. Let x_0 be an interior

point at which F is continuous, and let $y_0 = F(x_0)$. If there exists a linear map $L: X \to Y$ such that the tangent cone to $\mathcal{G}(F)$ at (x_0, y_0) is contained in $(x_0, y_0) + \mathcal{G}(F)$, then F is Fréchet differentiable at x_0, and $L = F'(x_0)$.

Tangent lines and tangent planes are also carefully defined in a pithy note by Roetman [341] and used to characterize differentiability. The earliest definition of a <u>tangent cone</u> seems to be given by Bouligand [245] who used the term <u>contingent</u>.

The inclusion relations in (ii) and (iii) of Theorem 12.1 cannot be replaced in general by equality. The following two theorems give sufficient conditions under which equality holds in (ii) and (iii) respectively.

Theorem 12.3. Under the assumptions of Theorem 12.1 suppose that L is a homeomorphism of X onto $\Re(L)$ and that $\Re(L)$ is closed in Y. Then for every sufficiently small neighborhood U of x_0 contained in $\emptyset$, the tangent cone to F(U) at x_0 is $\Re(\phi)$.

Theorem 12.4. Let X and Y be Banach spaces, let $F: \emptyset \subset X \to Y$ and let F have a continuous Fréchet derivative at x_0. Let $L = F'(x_0)$, $y = F(x_0)$ and $\phi: X \to Y$ as in Theorem 12.1. If $\Re(L) = Y$, then for any closed subspace Z of Y the tangent cone to $F^{-1}(y_0 + Z)$ at x_0 is $\phi^{-1}(y_0 + Z)$. In particular, the tangent cone at x_0 to the level surface of F through x_0 is the level surface of ϕ through x_0.

As an application of Theorem 12.4, we mention the following theorem which unifies the results on Lagrange's multipliers for the finite dimensional case as well as for some classes of variational problems.

Theorem 12.5. Let X, Y be Banach space, and let W be an open set in X. Suppose that $F: W \to Y$ has a continuous Fréchet derivative on W, and that

(12.1) $\qquad \Re[F'(x)] = Y$.

Let g be a real valued function on W which is Fréchet differentiable at each point of W. If g has a relative minimum or maximum at a point $x_0 \in W$ subject to the constraint $F(x) = 0$, then

(i) $N[F'(x_0)] \subset N[g'(x_0)]$, where N denotes null space;

(ii) there exists a unique function $\Lambda: Y \to \mathbf{R}$ such that $g'(x_0) = \Lambda \circ F'(x_0)$, and, moreover, Λ is linear and continuous.

Note that the assumption (12.1) is a <u>constraint qualification</u>. Even in the finite dimensional case, (12.1) provides more insight than the usual assumptions involving various Jacobians. Conclusion (i) of Theorem 12.5 expresses the intuitively obvious fact that if $x_0 \in S = \{x: F(x) = 0\}$ gives a relative extremum of g subject to the constraint $x \in S$, then every ray tangent to S at x_0 is also tangent at x_0 to the level surface Γ of g through x_0, i.e., the tangent cone to S at x_0 is contained in the tangent cone to Γ at x_0.

This notion of tangent cone as considered above is too strong for most nonlinear programming problems in Banach spaces. We shall describe briefly another notion of tangent cone which is more appropriate for such problems. Recall that a subset C in X is called a cone if $\alpha C \subset C$ for all $\alpha \geq 0$, and that the polar cone of C, denoted by C^*, is defined as the set $\{y \in X^*: <x,y> \leq 0 \ x \in C\}$, where $<x,y>$ denotes the value of the continuous linear functional y at x. Let S be a nonempty subset of a normed space X. We define the <u>cone of weak tangents</u> to S at x_0 to be the set of all $x \in X$ with the property that there is a sequence $\{x_n\}$ in S, converging <u>strongly</u> to x_0, and a sequence of nonnegative numbers λ_n such that $\lambda_n(x_n - x_0)$ converges <u>weakly</u> to x. We denote this cone by $T(S, x_0)$. This definition, which was introduced in [239], is similar to the cone of tangents introduced by Abadie [229] and Varaiya [355] who required strong convergence of both sequences. Let g be a functional on X, and let $T(g, x_0)$ be an abbreviation for the cone of weak tangents $T(\{x: g(x) \leq 0\}, x_0)$. Let $T^*(g, x_0)$

denote the polar cone of $T(g, x_0)$. We note that if $g(x) < 0$ for each S in some neighborhood of x_0, then $T^*(g, x_0) = \{0\}$. If g is Fréchet differentiable at x_0, $g(x_0) = 0$ and grad $g(x_0) \neq 0$, then it is not hard to show that

$$T^*(g, x_0) = \{\lambda \text{ grad } g(x_0): \lambda \geq 0\} .$$

Bazaraa et al. [239] used the polar cone of weak tangents to derive a necessary and sufficient constraint qualification for a generalization of the Kuhn-Tucker conditions in nonlinear programming in Banach spaces, without requiring the functional or the constraints to be differentiable.

13. Gradient Mappings. Integrability and Hyperbolic Integrability

The most direct reformulation of linear problems of mathematical physics (boundary value problems, integral equations, etc.) as variational problems is based on the symmetry or self-adjointness of the linear operator in question. This principle was well recognized by Courant and Hilbert [257] and is thoroughly discussed with various applications in the monographs of Mikhlin [308], [309].

There is a direct analogy between conservative force fields and variational methods for solving linear and nonlinear operator equations. To understand this analogy and to motivate some results on gradient mappings, we consider a vector field $\vec{F}$ on a region $\Omega \subset \mathbb{R}^3$. For convenience we represent $\vec{F}$ in rectangular coordinates $\vec{F} = (F_1, F_2, F_3)$ where $F_i: \Omega \to \mathbb{R}$, $i = 1, 2, 3$, and assume that Ω is a simply connected region. There are two well-known criteria for $\vec{F}$ to be a conservative vector field, i.e., that there exists a Fréchet differentiable function $f: \Omega \to \mathbb{R}$ such that grad $f = \vec{F}$ on Ω. The first criterion states that if $F \in C(\Omega)$, then a necessary and sufficient condition for F to be the gradient of a scalar function is that the line integral $\int_{P_1}^{P_2} \vec{F} \cdot d\vec{r}$

between any two points P_1 and P_2 in Ω be independent of the path joining them, i.e., $\oint_\Gamma \vec{F} \cdot d\vec{r} = 0$ for any closed path Γ in Ω. The second criterion states that if $F \in C'(\Omega)$, then a necessary and sufficient condition for $F = \text{grad } f$ is that curl $\vec{F} = 0$ in Ω. It appears therefore that the formulation of analogous criteria in infinite dimensional spaces entails a definition of line integral, which was given in Section 5, and the formulation of notion of curl or, more precisely, "the vanishing of the curl". Let us examine the condition curl $\vec{F} = 0$ from a different light. In rectangular coordinates we have

$$\text{curl } \vec{F} = \begin{vmatrix} \vec{i} & \vec{j} & \vec{k} \\ \dfrac{\partial}{\partial x^1} & \dfrac{\partial}{\partial x^2} & \dfrac{\partial}{\partial x^3} \\ F_1 & F_2 & F_3 \end{vmatrix} = 0 ,$$

i.e., $\dfrac{\partial F_i}{\partial x_j} = \dfrac{\partial F_j}{\partial x_i}$, $i, j = 1, 2, 3$. The vanishing of the curl of $\vec{F}$ is thus equivalent to the <u>symmetry</u> of the Jacobian matrix $\left[\dfrac{\partial F_i}{\partial x_j} \right]$; the latter is just the Fréchet derivative of the operator F induced by the vector field. (See Example 1.8). Now we are on familiar grounds for infinite dimensional generalizations: one would expect that the analogous condition in infinite dimensional spaces would be the symmetry of the Fréchet derivative of the nonlinear operator.

Let f be a functional on a subset Ω of a Banach space X. If f is Fréchet differentiable on Ω, then the differential $df(x;h)$ induces a mapping $df: \Omega \times X \to \mathbb{R}$. In the opposite direction, we consider the following general integrability problem which contains the case of gradient mappings as a special case.

Let X and Y be Banach spaces, Ω a convex subset

of X. Given a mapping $\varphi: \Omega \times X \to Y$ which is linear and continuous in the second argument, under what conditions is such a mapping a Fréchet differential?

To answer this question, we first define for any triangle Δ determined by $u, y, z \in \Omega$, the line integral

$$\int_\Delta \varphi(x; \cdot) dx = \int_0^1 \varphi[(1-t)u + ty; y-u] dt$$

$$+ \int_0^1 \varphi[(1-t)y + tz; z-y] dt + \int_0^1 \varphi[(1-t)z + tu; u-z] dt .$$

Theorem 13.1. The mapping $\varphi: \Omega \times X \to Y$ continuous with respect to the first argument, linear and continuous with respect to the second argument is a Fréchet differential if and only if

(13.1) $$\int_\Delta \varphi(x; \cdot) dx = 0$$

for any triangle in Ω.

If φ is a Fréchet differential and f is an operator whose Fréchet differential is φ, then for any $x, y \in \Omega$,

$$f(y) - f(x) = \int_{xy} \varphi(x; y-x) dx = \int_0^1 \varphi[(1-t)x + ty; y-x] dt .$$

In particular, assuming $0 \in \Omega$ and $f(0) = 0$, we have $f(x) = \int_0^1 \varphi(tx; x) dt$. Thus $\varphi(x; h)$ is a Fréchet differential if and only if for all $x, y \in \Omega$,

$$\int_0^1 \varphi(ty; y) dt - \int_0^1 \varphi(tx; x) dt = \int_0^1 \varphi[(1-t)x + ty; y-x] dt .$$

As an application of Theorem 13.1 we give the following simple result which is of interest in minimization problems in which the functional is replaced by a sequence of functionals.

<u>Theorem-13.2.</u> Let $\{\varphi_n(x;h)\}$, $x \in \Omega$, $h \in X$, be a sequence of continuous Fréchet differentials uniformly convergent on the convex set Ω to $\varphi(x;h)$. Then $\varphi(x;h)$ is a Fréchet differential. In particular, if $\{F_n(x)\}$ is a sequence of continuous gradient mappings which converges uniformly to $F(x)$ in Ω, then $F(x)$ is a gradient mapping.

<u>Proof:</u> By hypothesis $\int_\Delta \varphi_n(x; \cdot)dx = 0$ for any triangle Δ in Ω. But

$$\lim_{n \to \infty} \int_\Delta \varphi_n(x; \cdot)dx = \int_\Delta \lim_n \varphi_n(x; \cdot)dx = \int_\Delta \varphi(x; \cdot)dx .$$

Therefore $\int_\Delta \varphi(x; \cdot)dx = 0$ and $\varphi(x;h)$ is a Fréchet differential.

<u>Remark 13.1.</u> In Theorem 13.2 instead of continuity it suffices to require that the functions $t \to \varphi_n[(1-t)x + tz; z-x]$ are Riemann-Graves integrable on $[0,1]$.

The integrability condition (13.1), or the condition for a mapping to be a potential operator, takes a simple form if the mapping is continuously differentiable. This form has a local character, in contrast with the integrability condition (13.1), and may be stated without the notion of the integral.

<u>Theorem 13.3.</u> Let Ω be an open convex subset of a Banach space E, $F: \Omega \to E^*$ which has a Gâteaux differential $DF(x;h)$ on Ω. Suppose that $DF(x;h)$ is continuous with respect to x in every two dimensional manifold, i.e., for each fixed $x \in \Omega$, $h \in E$, $DF(x+th_1+sh_2;h) \to DF(x;h)$ as $(t,s) \to (0,0)$ for each fixed $h_1, h_2 \in E$. Then a necessary and sufficient

condition for F to be a potential operator (gradient mapping) on Ω is that the bilinear functional $<k, DF(x;h)>$ be symmetric for $x \in \Omega$, i.e.,

$$<h, DF(x;k)> = <k, DF(x;h)> \text{ for } x \in \Omega, h,k \in E .$$

The earliest version of this theorem was given by Kerner [52]. For a proof, see Vainberg [11], where numerous applications of this principle are given. The main results in [11] are summarized by Rall in [337]. See also Saaty [350], Mikhlin and Smolitskiy [310], and Krasnoselskii [296].

We remark that Theorem 13.3 can be modified so as to apply to operators which are densely defined, and provides a basis for the solution by variational methods of equations with unbounded operators.

Theorem 13.4. Let M be a dense linear manifold in a real Hilbert space H and let $F: M \to H$. Assume that the Gâteaux differential of F exists for $x, h \in M$, and is continuous with respect to x in every two dimensional manifold. Then a necessary and sufficient condition for F to be a gradient mapping is that $F'(x)$ be symmetric for every $x \in M$, i.e., for every $x, h, k \in M$,

$$<h, F'(x)k> = <k, F'(x)h> .$$

Assume further that $<F'(x)h, h> > 0$ for $h \neq 0$. If the equation $F(x) = y$ has a solution for $y \in H$, then this solution is unique and minimizes the functional

$$f(x) = \int_0^1 <F(tx), x> dt .$$

Note that in the case of a linear operator, the above theorem reduces to the minimum principle for quadratic functionals which is extensively treated in [308]. The symmetry of the operator in this case plays an important role.

The two criteria for an operator to be a potential operator remain valid in a locally convex linear topological space. (See also [269]).

The principal results in variational methods in the existence proofs for solutions of nonlinear operator equations, eigenvalue problems and bifurcation theory, known until 1956, are given in the monograph of Vainberg [11]. A synthesis of some of the results in this direction which have been developed in the last few years is given in Nečas [322], with particular emphasis on the variational method for the generalized divergence equation and elliptic nonlinear partial differential equations.

Remark 13.2. In case an operator F is not a potential operator, there arises the possibility of replacing the equation $F(x) = 0$ by an equivalent equation $G(x) = 0$, where G is a potential operator. This problem has been solved in some spaces (in real Hilbert spaces and in L^p, with $p > 2$) for operators of the form $F = LA$ where L is a linear operator satisfying certain conditions and A is a potential operator. For example if L is an operator decomposable in the form TT^*, where T is a bounded linear operator on a Hilbert space and A is a potential operator, then the operator T^*AT is a potential operator and the equations $LAx = 0$ and $T^*ATx = 0$ are equivalent.

Another extension of the variational method can be made by applying the theory of symmetrizable linear operators to $F'(x)$ and by renorming the space. Applications of these and related ideas to solution of operator equations by the steepest descent method are given in Section III of [321].

Remark 13.3. The theory of integrability conditions for abstract vector fields provides another extension of the variational method to nonpotential operators. We recall that if $\vec{F}$ is a vector field in $\mathbb{R}^3$, $\vec{F} = \{F_i(x^1, x^2, x^3)\}$, and if F_i, $i = 1, 2, 3$, have continuous partial derivatives in a simply connected region, then a necessary and sufficient condition for the existence of a scalar function $\varphi(x^1, x^2, x^3)$ such that curl $(\varphi \vec{F}) = 0$ is that $\vec{F} \cdot$ curl $\vec{F} = 0$, i.e., in rectangular coordinates,

$$F_1(\frac{\partial F_3}{\partial x^2} - \frac{\partial F_2}{\partial x^3}) + F_2(\frac{\partial F_1}{\partial x^3} - \frac{\partial F_3}{\partial x^1}) + F_3(\frac{\partial F_2}{\partial x^1} - \frac{\partial F_1}{\partial x^2}) = 0 .$$

An appropriate generalization of this result for abstract vector fields has been given by Minty [313]. Let F be an operator defined on a real Banach space X into its dual space X^*, and assume that the Fréchet derivative of F exists and is continuous in a neighborhood U of a point $x_0 \in X$ with $F(x_0) \neq 0$. Define the trilinear form

$$I(x; h,j,k) = <F(x), h> \{<F'(x)j,k> - <F'(x)k,j>\}$$

$$+ <F(x),j> \{<F'(x)k,h> - <F'(x)h,k>\} + <F(x),k> \{<F'(x)h,j>$$

$$- <F'(x)j,h>\}$$

for $x \in U$, $h,j,k \in X$. If $I(x;h,j,k) = 0$ for all $x \in U$, $h,j,k \in X$, then there exists a positive functional φ and a functional f, both defined on a neighborhood of x_0, such that

$$\varphi(x) F(x) = \text{grad } f(x)$$

for all x in this neighborhood.

Remark 13.4. The theory of potential operators motivates the introduction of a notion of adjoint of a differentiable nonlinear mapping on a Hilbert space by analogy with the case of a linear operator. Let $\mathcal{D}$ denote the set of all continuously Fréchet differentiable mappings F on H into H such that $F(0) = 0$. A mapping $F \in \mathcal{D}$ is said to have an <u>adjoint</u> if there exists a mapping $G \in \mathcal{D}$ such that $G'(x) = [F'(x)]^*$, where $[F'(x)]^*$ is the adjoint of the linear operator $F'(x)$. If there exists such a G, then it is easy to show that it is unique and is given by

$$G(x) = \int_0^1 [F'(tx)]^* x \, dt .$$

We denote G by F^* and call it the adjoint of F. A map
$F \in \mathcal{S}$, which has an adjoint, is said to be <u>symmetric</u> if
$F = F^*$; it is said to be <u>skew-symmetric</u> if $F^* = -F$. Thus
a mapping $F \in \mathcal{S}$ is symmetric if and only if it is a potential
operator. A characterization of skew-symmetric mappings
can also be easily given: A mapping $F \in \mathcal{S}$ is skew-symmetric if and only if it is linear and $<F(x), x> = 0$ for all
$x \in H$. These facts lead to several characterizations of non-linear mappings which have adjoints (see Yamamuro [356]):
(a) $F \in \mathcal{S}$ has an adjoint if and only if there exists a skew-symmetric continuous linear operator L such that $F + L$ is
a potential operator; (b) $F \in \mathcal{S}$ has an adjoint if and only if
$F'(x) - [F'(x)]^*$ is independent of x; (c) if F is twice differentiable, then F has an adjoint if and only if the linear
operator $F''(x)x$ is symmetric for every $x \in H$. It follows
easily from Theorem 14.3 and (b) that a mapping $F \in \mathcal{S}$ which
has an adjoint is completely continuous if and only if its
adjoint F^* is completely continuous.

The problem of integrability can also be posed in terms
of the so-called hyperbolic differentials as in [326]. Let
X, Y, Z be Banach spaces. A mapping $\varphi(x, y; h, k)$ which is
bilinear and continuous in $h \in X$, $k \in Y$ is said to be a
<u>hyperbolic differential</u> in Fréchet's sense if there exists a
function $f: X \times Y \to Z$ such that

$$\lim_{h,k \to 0} \frac{\|f(x+h, y+k) - f(x+h, y) - f(x, y+k) + f(x, y) - \varphi(x, y; h, k)\|}{\|h\| \|k\|} = 0$$

<u>Theorem 13.5.</u> A mapping $\varphi(x, y; h, k), x, h \in X, y, k \in Y$, which
is bilinear and continuous in h, k and continuous in x, y,
is a hyperbolic differential in the sense of Fréchet if and only
if

$$\int_{\overline{xz}} \int_{\Delta'} \varphi(y', z'; \cdot, \cdot) dy' dz' = 0$$

and

$$\int_{\Delta} \int_{\overline{yv}} \varphi(y', z'; \cdot, \cdot) dy' dz' = 0$$

for any triangle Δ determined by the points x,z,u of the space X and for any triangle Δ' determined by the points y,v,w of the space Y.

Remark 13.5. Let $x,z \in X$, $y,v \in Y$. Define $\psi: [0,1] \times [0,1] \to Z$ by

$$\psi(t,\tau) = \varphi[(1-t)x + tz, (1-\tau)y + \tau v; z-x, v-y].$$

Then under the assumption of Theorem 13.5, ψ is continuous and we write

$$\int_0^1 d\tau \int_0^1 \psi(t,\tau)dt = \int_{\overline{xz}} \int_{\overline{yv}} \varphi(x,z; \cdot, \cdot)dx\,dz.$$

The integrals in Theorem 13.5 then have the usual meaning.

In his thesis, P. Montel has studied the following equation

$$\frac{\partial^2 z}{\partial x \partial y} = f(x,y,z)$$

where the derivative is a hyperbolic derivative. As an application of Theorem 13.5 we consider the corresponding equation in Banach spaces. Let X,Y,Z be real Banach spaces and $W = X \times Y$ with norm $\|w\| = \|x\| + \|y\|$. Let $d'd\varphi(x,y; h,k)$ denote the hyperbolic Fréchet differential of a function φ. We seek a solution of the following "hyperbolic" differential equation in Banach space:

(13.2) $d'd\varphi(x,y; h,k) = f(x,y,\varphi(x,y);h,k)$

where $f: W \times Z \times W \to Z$ is a given continuous mapping, which is bilinear in $(h,k) \in X \times Y$. That is, does there exist a mapping $\varphi: X \times Y \to Z$, which has a hyperbolic Fréchet differential such that (13.2) holds for any (x,y), $(h,k) \in W$?

We seek a solution φ which satisfies the conditions

(13.3) $\varphi(x_0, y_0) = z_0$, $\varphi(x, y_0) = \alpha(x)$, $\varphi(x_0, y) = \beta(y)$

where z_0 is given, $\alpha(x)$ and $\beta(y)$ are prescribed mappings of $X \times y_0$, respectively $x_0 \times Y$, into Z. Define the sequence $\{\varphi_n\}$ as follows

$$\varphi_0(x, y) = \alpha(x) + \beta(y) - z_0$$

$$\varphi_n(x, y) = \varphi_0(x, y) + \int_{\overline{x_0 x}} \int_{\overline{y_0 y}} f(u, v, \varphi_{n-1}(u, v); h, k) du\, dv.$$

Theorem 13.6. With the above notation, suppose that a constant $K > 0$ exists such that

$$\|f(x, y, z_1, h, k) - f(x, y, z_2, h, k)\| \le K \|z_1 - z_2\| \|h\| \|k\|$$

for all (x, y), $(h, k) \in W$. Then the system (13.2)-(13.3) has a unique solution in the closed sphere $S(x_0, y_0; r)$ with $r < \frac{1}{K}$, if and only if

$$\lim_{n \to \infty} \int\int_{\Delta} \int_{\overline{y_0 y}} f(u, v, \varphi_n(u, v); \cdot, \cdot) du\, dv = 0$$

$$\lim_{n \to \infty} \int_{\overline{x_0 x}} \int_{\Delta'} f(u, v, \varphi_n(u, v); \cdot, \cdot) du\, dv = 0$$

for any triangles Δ, Δ' such that $\Delta \times \Delta' \subset S(x_0, y_0; r)$ with one of the vertices at x_0, respectively y_0.

Note that if (13.2) has a solution, then $f(x, y, \varphi(x, y); h, k)$ is a hyperbolic Fréchet differential; thus by Theorem 13.5

$$\int_{\overline{xz}} \int_{\Delta'} f(u, v, \varphi(u, v); \cdot, \cdot) du\, dv = 0$$

$$\int\int_{\Delta \overline{yv}} f(u,v,\varphi(u,v);\cdot,\cdot)\,du\,dv = 0$$

for any triangles Δ in X and Δ' in Y. Thus the conditions of Theorem 13.6 are quite natural. For proofs, see [326],[327].

Finally we remark that Theorem 13.6 is an analogue of a theorem of Michal and Elconin [59] for ordinary differential equations, using integrability conditions of Theorem 13.3. Existence and uniqueness theorems for differential equations in Montel spaces (using integrability conditions analogous to Theorems 13.1 or 13.3) were given by Dubinsky [85].

14. Compact, Collectively Compact Operators and Differentiability

The purpose of this section is to show that differentiability and certain properties of an operator F on a Banach space are implied by related properties of a sequence $\{F_n\}$ of operators converging to F either in the operator norm or pointwise. In most of these implications compactness or collective compactness of the sequence of operators plays an important role. We remark that theorems of this type are useful in approximate minimization problems and in approximate methods for the solution of integral and operator equations. We begin with a simple result.

Theorem 14.1. Let X and Y be Banach spaces, Ω be an open convex subset of X, and $F_n: \Omega \to Y$ be a sequence of Fréchet differentiable operators such that $F_n': \Omega \to \mathfrak{L}(X,Y)$ converges uniformly on Ω to $G: \Omega \to \mathfrak{L}(X,Y)$. Assume also that for some point $x_0 \in \Omega$, $F_n(x_0)$ converges to some element in Y. Then for each $x \in \Omega$, the sequence $\{F_n(x)\}$ has a limit in Y (denoted by $F(x)$). Furthermore, the convergence of $\{F_n\}$ to F is uniform on each bounded set of Ω and F is Fréchet differentiable with $F' = G$.

Proof: By the mean value theorem (Theorem 5.4)

$$\|F_p(x) - F_p(x_0) - F_q(x) + F_q(x_0)\|$$

(14.1)
$$\leq \|x - x_0\| \sup_{u \in \Omega} \|F_p'(u) - F_q'(u)\|,$$

which on account of uniform convergence of $\{F_n'\}$ on Ω implies that the first member of (14.1) converges to zero as $p, q \to \infty$, uniformly with respect to x on each bounded set in Ω. Since by assumption $\{F_p(x_0)\}$ is convergent, $F_p(x_0) - F_q(x_0) \to 0$ as $p, q \to \infty$. Thus $F_p(x) - F_q(x) \to 0$ uniformly in x on each bounded set in Ω, as $p, q \to \infty$. Since the space is complete $\{F_n(x)\}$ converges to some element $F(x) \in Y$. Clearly convergence is uniform on bounded sets of Ω. Now we show that F is Fréchet differentiable at each $x \in \Omega$, and that $F' = G$. Consider

(14.2) $\|F(x+h) - F(x) - G(x)h\| \leq \|F(x+h) - F(x) - [F_n(x+h) - F_n(x)]\|$

$+ \|F_n(x+h) - F_n(x) - F_n'(x)h\| + \|F_n'(x)h - G(x)h\|.$

Let $\varepsilon > 0$. There exists, on account of (14.1) and the definition of $G(x)$, a positive integer N such that for $p, n \geq N$,

$$\|F_p(x+h) - F_p(x) - [F_n(x+h) - F_n(x)]\| \leq \varepsilon \|h\|,$$

and
$$\|F_n'(x) - G(x)\| \leq \varepsilon.$$

Passing to the limit as $p \to \infty$, we get

$$\|F(x+h) - F(x) - F_n(x+h) + F_n(x)\| \leq \varepsilon \|h\| \quad \text{for } n > N.$$

Also,

$$\|F_n'(x)h - G(x)h\| \leq \varepsilon \|h\|.$$

Now let N_0 be a fixed integer $N_0 \geq N$. Then by definition of the derivative $F'_n(x)$, the second term on the right side of the inequality (14.2) can be majorized for sufficiently small h, i.e., for $n = N_0$,

$$\|F_n(x+h) - F_n(x) - F'_n(x)h\| \leq \varepsilon \|h\| \quad \text{for} \quad \|h\| \leq \delta .$$

Thus combining all these estimates in (14.2) we have

$$\|F(x+h) - F(x) - G(x)h\| \leq 3\varepsilon \|h\| \quad \text{for} \quad \|h\| \leq \delta ,$$

which shows that F is Fréchet differentiable and $F' = G$.
An easy consequence of this theorem is the following

Theorem 14.2. Let $F_n : \Omega \subset X \to Y$, where Ω, X, Y are as in Theorem 14.1, be Fréchet differentiable mappings on Ω. Assume that for each $x \in \Omega$, there exists a sphere of center x in which the sequence $\{F'_n\}$ converges uniformly, and that for some $x_0 \in \Omega$, the sequence $\{F_n(x_0)\}$ is convergent. Then for each $x \in \Omega$, the sequence $\{F_n(x)\}$ has a limit $F(x)$ and each point of Ω has a neighborhood in which the convergence of the sequence $\{F_n\}$ to F is uniform. Moreover F is differentiable and $\{F'_n\}$ converges to F'.

For the rest of this section we consider compact and collectively compact operators.

Theorem 14.3. Let $F : X \to Y$ be Fréchet differentiable at x_0. If F is a completely continuous operator, then the Fréchet derivative $F'(x_0)$ at x_0 is also a completely continuous operator on X into Y.

<u>Proof</u>: Suppose $F'(x_0)$ is not completely continuous. Then there exists an $\varepsilon > 0$ and a bounded sequence $\{h_n\}$, $\|h_n\| < r$, such that

$$\|F'(x_0)h_i - F'(x_0)h_j\| > 3\varepsilon, \, i \neq j, \quad i,j = 1, 2, \ldots .$$

Since F is Fréchet differentiable at x_0, given $\varepsilon > 0$ there exists $\eta > 0$ such that

$$\|r(x_0;h)\| < \varepsilon \|h\| \text{ if } \|h\| < \eta, \text{ where}$$

$$r(x_0;h) = F(x_0+h) - F(x_0) - F'(x_0)h \ .$$

Thus we have

$$\|F(x_0 + \eta h_i) - F(x_0 + \eta h_j)\| =$$

$$\|\eta F'(x_0)h_i - \eta F'(x_0)h_j + r(x_0;\eta h_i) - r(x_0;\eta h_j)\|$$

$$\geq \eta \|F'(x_0)h_i - F'(x_0)h_j\| - \|r(x_0;\eta h_i)\| - \|r(x_0;\eta h_j)\| > \eta \varepsilon \ .$$

But this inequality contradicts the assertion that F is a completely continuous operator, which proves the theorem.

Bonic [363] has shown that complete continuity of $F'(x)$ does not necessarily imply complete continuity of F. However we have the following partial converse to Theorem 14.3. (See Theorem 4.8 in [11]).

Theorem 14.4. Let $F: X \to Y$ be Fréchet differentiable. If for each fixed $x \in X$, $F'(x)$ is a completely continuous operator from X to Y and the Fréchet derivative F' is a compact operator on X into $\mathcal{L}(X,Y)$, then F is a completely continuous operator

Recently, Palmer [330] gave conditions on the derivative F' which imply strong continuity of F (i.e., weak convergence of $\{x_n\}$ to x implies strong convergence of $\{F(x_n)$ to $F(x))$.

Theorem 14.4. Let X be a reflexive Banach space and let $F: X \to Y$ be uniformly differentiable on every open sphere S, i.e., given $\varepsilon > 0$ there exists a $\delta > 0$ such that $\|h\| < \delta$ and $x, x+h \in S$ imply $\|F(x+h) - F(x) - F'(x)h\| < \varepsilon \|h\|$. The

F is strongly continuous if and only if $F'(x)$ is completely continuous and $F': X \to \mathcal{L}(X,Y)$ is compact. Strong continuity cannot be replaced by compactness.

Theorem 14.3 on the complete continuity of the Fréchet derivative of a completely continuous operator has been generalized in several directions. One such generalization is given in the next theorem due to Melamed and Perov [307]; another generalization using collectively compact operators was given by Moore [314] and is stated in Theorem 14.6.

Theorem 14.5. Let F be a nonlinear operator mapping a sphere S (with center at the origin) of a Banach space E into E. Suppose F satisfies the following conditions

$$F(x) = \sum_{i=1}^{n} C_{k_i} x + Bx$$

where

$$C_{k_i}(tx) = t^{k_i} C_{k_i} x, \quad x \in E, \quad t \geq 0,$$

$$\|C_{k_i} x\| \leq M \|x\|^{k_i}, \quad 0 < k_1 < k_2 < \ldots < k_n$$

and

$$\lim_{\|x\| \to 0} \frac{\|Bx\|}{\|x\|^{k_n}} = 0.$$

If F is completely continuous, then the operator C_{k_i} (i = 1, ..., n) and B are completely continuous. The same assertion also holds if the operators are defined in the <u>exterior</u> of some sphere, $k_1 > k_2 > \ldots > k_n > 0$ and one takes the limit as $\|x\| \to \infty$. (The second part of this theorem generalizes the result that the <u>asymptotic</u> Fréchet derivative of a completely continuous operator is completely continuous; see Section 17).

Remark 14.1. A compact operator is of course not necessarily differentiable. In view of the fact that a compact operator can be approximated by a finite dimensional mapping, i.e., a mapping with a finite dimensional range (see p. 226 for a more precise statement of the sense of this approximation), it is of interest to characterize the set of all everywhere Fréchet differentiable finite dimensional mappings. It is not hard to show (see [331]) that a mapping F on a Banach space E is a Fréchet differentiable finite dimensional mapping if and only if there exists a finite dimensional subspace E_0 of E such that $\cup \{\Re(F'(x)): x \in E\} \subset E_0$, where $\Re(F'(x))$ denotes the range of $F'(x)$, considered as a linear operator on E.

The set of all compact operators on a Banach space is closed in the topology of uniform convergence, but is not closed in the topology of pointwise convergence, i.e., if $\{F_n\}$ is a sequence of compact operators such that $\|F_n x - Fx\| \to 0$ for each x, then F is not necessarily compact. This motivates in part the introduction of the notion of a collectively compact family of operators, which was studied by Anselone, Moore, and others (see [235], [236]; the second reference is a lucid exposition on the theory and applications of collectively compact operators and contains an extensive bibliography).

A family $\mathcal{F}$ of operators on a Banach space X is said to be <u>collectively compact</u> if for every bounded set $B \subset X$, the set $\cup \{F(B): F \in \mathcal{F}\}$ has compact closure. Note that if $\mathcal{F}$ is a collectively compact family of operators on X, and T is is in the pointwise closure of $\mathcal{F}$, then T is compact. For if B is any bounded subset of X, then $T(B) \subset \overline{\cup \{F(B): F \in \mathcal{F}\}}$ and the last set is compact.

Remark 14.2. Recall that a subset S of a metric space X is called <u>totally bounded</u> if for each $\varepsilon > 0$, S can be covered by a finite number of spheres with radius ε, i.e., there exists a finite set (an ε-net), $x_1, \ldots, x_m$, such that for any

$x \in S$, $\min_{1 \le i \le m} \|x - x_i\| < \varepsilon$. A totally bounded subset of a metric space is clearly bounded and separable. A bounded set is not necessarily totally bounded. A necessary condition for a subset S of a metric space X to be sequentially compact (equivalently, S has a compact closure) is that it is totally bounded; the condition is also sufficient if X is complete. S is totally bounded if and only if there is a totally bounded ε-net (not necessarily finite) for S.

Theorem 14.6. Let $\mathfrak{F}$ be a collectively compact family of nonlinear operators on a Banach space X, and let $\mathfrak{F}$ be Fréchet equidifferentiable (see Section 1.6) at $x_0 \in X$. Then the family of operators $\{F'(x_0): F \in \mathfrak{F}\}$ is collectively compact.

Proof: Let $\varepsilon > 0$ be given and let B be the unit ball. Then by equidifferentiability of $\mathfrak{F}$ at x_0, there exists $\delta > 0$ such that for all $F \in \mathfrak{F}$, $h \in B$ implies

$$\|\delta^{-1}\{F(x_0 + \delta h) - F(x_0)\} - F'(x_0)h\| < \varepsilon \|h\| \le \varepsilon .$$

Since $\mathfrak{F}$ is collectively compact, the set

$$\bigcup_{h \in B} \bigcup_{F \in \mathfrak{F}} \delta^{-1}\{F(x_0 + \delta h) - F(x_0)\}$$

is totally bounded, and the last inequality shows that this set is an ε-net for $\bigcup_{F \in \mathfrak{F}} \{F'(x_0)h: \|h\| \le 1\}$. This completes the proof in view of the last part of Remark 14.2.

Differentiability of an operator T is implied by the existence of an equidifferentiable collectively compact sequence which converges pointwise to T. More precisely, we have

Theorem 14.7. Let $\{T_n\}$ be a sequence of operators on a Banach space X. Suppose that for some $x_0 \in X$, $\{T_n\}$ is equidifferentiable and the set of linear operators $\{T'_n(x_0)\}$ is collectively compact. If $\|T_n x - Tx\| \to 0$ for all x in a neighborhood of x_0, then T is Fréchet differentiable at x_0 and $\|T'_n(x_0)h - T'(x_0)h\| \to 0$ for each $h \in X$.

For a proof and a number of related results, see [314]. Applications of collectively compact operators to approximate solutions of nonlinear operator equations and Newton's method are given in [315].

Finally we state a result which generalizes Corollary 11.6 to collectively compact sets of gradient mappings (see [262] for other related results and applications to approximate minimization of functionals).

<u>Theorem 14.8</u>. Let $\mathcal{F}$ be a family of real-valued functions f defined on a bounded convex subset Ω of a Banach space E. Suppose that grad f exists for all $f \in \mathcal{F}$ and $x \in \Omega$ and that the family $\{\text{grad } f : f \in \mathcal{F}\}$ of maps on Ω into E^* is collectively compact. Then $\mathcal{F}$ is weakly equicontinuous on Ω.

15. Monotone Operators and Convex Functionals: Differential Characterizations

An operator $F: D \subset H \to H$, where H is a (complex) Hilbert space is said to be (i) <u>monotone</u> on D if

(15.1) $\text{Re} \langle Fu - Fv, u - v \rangle \geq 0$ for all $u, v \in D$;

(ii) <u>strictly monotone</u> if the strict inequality in (15.1) holds for $u \neq v$;

(iii) <u>strongly monotone</u> if for some $m > 0$,

$$\text{Re} \langle Fu - Fv, u - v \rangle \geq m \|u - v\|^2 .$$

The notion of a monotone operator was introduced by Zarantonello [357], Minty [311] and Kačurovskii [283]. Monotonicity conditions in the context of variational methods for nonlinear operator equations were also used by Kachurovskii and Vainberg [354]. The notion has been extended to Banach spaces by several authors. The theory of monotone operators and its applications to nonlinear partial differential equations, evolution equations, variational inequalities, etc. has evolved into a substantial chapter in nonlinear functional analysis. We do not wish to dwell on this theory here; rather we shall limit our discussion to some remarks on the connections between monotonicity, convexity, and related properties of differentials. The reader interested in the theory of monotone operators may refer to several papers in [250] (see the introduction and references in [334]), the survey papers [284], [247], [248] and to the more comprehensive account given in [250] Part II. More recent aspects are examined by several authors in [360], along with applications.

Let $F: \Omega \subset E \to E^*$, where E is a Banach space. The definition of a monotone operator carries over to this setting by taking $<e, x>$ to mean the value of $e \in E^*$ at x.

For each $x \in \Omega$, let $T(x) \in \mathfrak{L}(E, E^*)$. T is said to be positive on Ω if for all $x \in \Omega$, $h \in E$, $<T(x)h, h> \geq 0$; T is strictly positive if for $h \neq 0$ the strict inequality holds; T is uniformly positive definite if $<T(x)h, h> \geq m \|h\|^2$ for some $m > 0$ and all $x \in \Omega$, $h \in E$. Note that if Ω is compact and $T(x)$ is continuous, then strict positiveness of $T(x)$ implies uniform positive definiteness, so that the two notions are equivalent for continuous $T(x)$ on compact sets.

Theorem 15.1. Let Ω be a convex subset of a Banach space E and let $F: \Omega \subset E \to E^*$ have a Gâteaux derivative which is continuous on line segments in Ω, i.e., for any fixed x_1, $x_2 \in \Omega$, $\|F'(x_1 + t(x_2 - x_1)) - F'(x_1)\| \to 0$ as $t \to 0$. Then:

(a) F is monotone if and only if $F'(x)$ is positive.

(b) F is strongly monotone if and only if $F'(x)$ is uniformly positive definite on Ω.

(c) If $F'(x)$ is positive and strictly positive except possibly on a set which contains no line segment, then F is strictly monotone; the converse is not necessarily true.

The theorem follows using simple manipulations based on the mean value theorem

$$<F(u) - F(v), u-v> = \int_0^1 <F'(tv + (1-t)u)(u-v), u-v> dt .$$

In view of Theorem 15.1 monotone operators may be considered as a natural generalization of positive linear operators (in the inner product sense) and of differentiable nonlinear operators with a positive Fréchet derivative. Using monotone operators one can develop existence and uniqueness theorems similar to those obtained by variational methods (potential operators) when such methods do not apply. The reader may refer to Dolph and Minty [264] for motivation and a retrospective analysis with reference to integral equations and Hammerstein operators, where monotonicity and mild continuity and boundedness hypotheses replace the usual hypotheses of self-adjointness and complete continuity required in the variational method. The theory of Hammerstein integral equations has been considerably advanced in the last few years using monotonicity and related methods.

Monotonicity methods are useful also in minimization problems. They enable one to dispense with second order differentiability of the functional to be minimized, and to replace the positive definiteness of the second order derivative by some monotonicity conditions on the gradient of the functional.

In addition to the notions of convex and strictly convex functional, we define the notion of a <u>strongly convex</u> functional. A real-valued functional f on a convex subset Ω of E is said to be strongly convex if there exists a function $\alpha: [0, \infty) \to [0, \infty)$ having the property that $\alpha(x_n) \to 0$ implies $x_n \to 0$, such that for all $t \in [0,1]$, α is nondecreasing and

$$f(tu + (1-t)v) \leq tf(u) + (1-t) f(v) - \|u-v\| \max\{t\alpha[(1-t)\|u-v\|],$$
$$(1-t) \alpha[t\|u-v\|]\} .$$

An operator $F: \Omega \to E^*$, where E is a normed real space, is said to be <u>α-monotone</u> if

$$<Fu - Fv, u-v> \geq \|u-v\| \alpha(\|u-v\|), \quad u,v \in \Omega,$$

for some α with the property mentioned above.

We are now in a position to state relations between these notions of convexity of a functional and monotonicity of the gradient of the functional.

<u>Theorem 15.2.</u> Let f be a real-valued functional on an open convex subset Ω of a normed real space E. Let f have a Gâteaux derivative which is continuous on line segments in Ω. Then:

(a) f is convex on Ω if and only if grad f is a monotone operator on Ω.

(b) f is strictly convex if and only if grad f is strictly monotone on Ω

(c) f is strongly convex if and only if grad f is α-monotone on Ω.

In particular, taking $\alpha(t) = mt$ with $m > 0$, part (c) implies that the gradient of f is strongly monotone if and only if

(13.1) $f(tu + (1-t)v) \leq tf(u) + (1-t)f(v) - mt(1-t)\|u-v\|^2$.

This special case of a strongly convex functional was introduced by Poljak [335] who used it to establish convergence of minimizing sequences in extremal problems. An application of (13.1) to the so-called complementarity problem in nonlinear programming was given by Karamedian [286]. The general notion of a strongly convex functional was introduced by Elkin [268] in his study of minimization algorithms.

A functional f which has a Gâteaux derivative on an open convex set Ω is convex on Ω if and only if for all $u,v \in \Omega$, $f'(v)(u-v) \leq f(u) - f(v) \leq f'(u)(u-v)$.

A similar characterization can be given when f is not

differentiable [319]: f is convex if and only if for each $u \in \Omega$ there exists a gauge functional g_u (i.e., a convex and positively homogeneous functional) such that

(13.2) $\quad f(v) - f(u) \geq g_u(v-u)$ for all $u, v \in \Omega$.

Furthermore if for each $u \in \Omega$, we let G_u denote the set of all gauge functionals such that (13.2) holds, then

$$\sup\{g_u(h) : g_u \in G_u\} = \lim_{t \to 0^+} \frac{f(u+th) - f(u)}{t}.$$

Similar characterizations can be given for strictly or strongly convex functionals. For example, a functional f which has a Gâteaux derivative on an open convex set is strongly convex if and only if

$$f(v) - f(u) \geq f'(u)(v-u) + \|u-v\| \alpha(\|u-v\|).$$

Part (a) of Theorem 15.2 asserts that the gradient of a convex functional is a monotone operator. Similarly in the case of not-necessarily-differentiable convex functional, the graph of the generalized gradient is a monotone relation (see [312]). A <u>subgradient</u> ([312], [316], [318], [246]) of a convex functional f at a point u in its domain $\mathcal{D} \subset E$ is a vector $u^* \in E^*$ such that

$$f(v) \geq f(u) + <u^*, v-u> \text{ for all } v \in \mathcal{D}.$$

Let $\partial f(u)$ denote the set of all subgradients of f at u. If $\partial f(u)$ is nonempty, f is said to be subdifferentiable at u. The subgradient mapping ∂f is a (multi-valued) mapping assigning to each $u \in \mathcal{D}$ all its subgradients. For example if $f(x) = |x|$, then

$$\partial f(x) = \begin{cases} \dfrac{|x|}{x} & \text{if } x \neq 0 \\ [-1, 1] & \text{if } x = 0 \end{cases}$$

For a comprehensive study of subgradients and related results, see [317], [318], [338], [339], [340]. Various aspects of differentiability of convex functionals are also studied in [237], [238], [293], [294], [260], [332] among others.

16. Numerical Range and Spectrum of Some Classes of Nonlinear Operators. Spectral Radii and Points of Attraction

16.1. Numerical Range and Related Spectral Theorems

The numerical range of a bounded linear operator T on a Hilbert space H into H is defined as the set $\{<Tx,x>: x \in H, \|x\| = 1\}$. It is well known that <u>the closure of the numerical range contains the spectrum</u> of T. Zarantonello [358], [359] gave the first extension of this result to nonlinear operators. Let H be a real or complex Hilbert space, and let T be an operator defined on a domain $\mathcal{D}(T) \subset H$ into H. Zarantonello defined the numerical range of T to be the set

$$W(T) = \left\{ \frac{<Tx_1 - Tx_2, x_1 - x_2>}{\|x_1 - x_2\|^2} : x_1 \neq x_2, x_1, x_2 \in \mathcal{D}(T) \right\}.$$

For simplicity, we only state here a special case of the "existence part" of Zarantonello's theorem under stronger conditions than used in [358]

<u>Theorem 16.1.</u> Let $T: H \to H$ be continuous and assume that it maps bounded sets into bounded sets. For any complex number λ at a positive distance $d(\lambda, T)$ from the numerical range W(T), the equation $Tx - \lambda x = y$ has a unique solution for each $y \in H$ and the operator $(T - \lambda I)^{-1}$ is Lipschitzian with Lipschitz constant $1/d(\lambda, T)$.

In [358] the assumptions of continuity and boundedness are relaxed, the numerical range is localized, and the conclusion is sharpened. Zarantonello's remarkable paper [359] sets a tone for research in three aspects of nonlinear operators which he has beautifully interwoven: existence theory for monotone operator equations, constructive methods for solutions, and spectral theory for nonlinear operators. Recently Zarantonello has been successful in developing a framework for spectral theory of nonlinear operators where nonlinear projections play a significant role.†

There are several areas of analysis which have interface with spectral theory of nonlinear operators. These include nonlinear eigenvalue problems and bifurcation theory, Fredholm alternative theorems for nonlinear operators, abstract variational problems with constraints, etc. Recent advances in some of these areas are examined by several authors in [360]. Here we shall mention only a few simple results which seem to be quite motivated by the linear theory and in which the differential serves as guiding light for the generalizations.

Let E be a Banach space and T be an operator on E into E. We recall that a number λ is said to be a <u>regular value</u> of T if the operator $R_\lambda = (T - \lambda I)^{-1}$ exists and satisfies the Lipschitz condition $\|R_\lambda(x) - R_\lambda(y)\| \leq K\|x-y\|$ for some constant K, and for all $x, y \in E$. The complement of the set of all regular values in the complex plane is called the <u>spectrum</u> of T. If $T(0) = 0$ and $T(x_\lambda) = \lambda x_\lambda$, $x_\lambda \neq 0$, then the vector x_λ is called an eigenvector of T corresponding to the eigenvalue λ. Clearly the eigenvalues are points of the spectrum.

In [357], the following quantities were introduced for any operator T on a subset $\mathcal{D}$ of a Hilbert space H into itself:

$$(16.1) \qquad \mu^+(T) = \sup_{\substack{x_1, x_2 \in \mathcal{D} \\ x_1 \neq x_2}} \frac{\langle Tx_1 - Tx_2, x_1 - x_2 \rangle}{\langle x_1 - x_2, x_1 - x_2 \rangle}$$

---

† Some aspects of this theory will be given by Zarantonello in the Proceedings of the Symposium on Nonlinear Functional Analysis [360].

$$(16.2) \qquad \mu^-(T) = \inf_{\substack{x_1, x_2 \in \mathcal{D} \\ x_1 \neq x_2}} \frac{<Tx_1 - Tx_2, x_1 - x_2>}{<x_1 - x_2, x_1 - x_2>}$$

$$(16.3) \qquad |\!|\!|T|\!|\!|^2 = \sup_{\substack{x_1, x_2 \in \mathcal{D} \\ x_1 \neq x_2}} \frac{<Tx_1 - Tx_2, Tx_1 - Tx_2>}{<x_1 - x_2, x_1 - x_2>} .$$

Obviously $\mu^-(T) = -\mu^+(-T)$, $\mu^-(T) \leq \mu^+(T)$, $|\mu^+(T)| \leq |\!|\!|T|\!|\!|$.
If T is Fréchet differentiable on a convex set $\mathcal{D}$, then it follows easily from the mean value theorem (Theorem 5.4),

$$(16.4) \qquad |\!|\!|T|\!|\!| = \sup_{x \in \mathcal{D}} \|T'(x)\| ,$$

$$(16.5) \qquad \mu^+(T) = \sup_{x \in \mathcal{D}} \sup_{\|h\|=1} <T'(x)h, h> ,$$

and

$$(16.6) \qquad \mu^-(T) = \inf_{x \in \mathcal{D}} \inf_{\|h\|=1} <T'(x)h, h> .$$

Theorem 16.2. Let H be a Hilbert space and T be a continuous operator on H into H such that $<Tx - Ty, x-y>$ is real for all $x, y \in H$. Then:

    (a) Each number $\lambda = \alpha + \beta i$, $\beta \neq 0$ is a regular value of T;

    (b) the spectrum of T lies in $[\mu^-(T), \mu^+(T)]$;

and
    (c) if the numbers μ^+ and μ^- are finite and T

satisfies the Lipschitz condition $\|T(x) - T(y)\| \le$ max$\{|\mu^-|, |\mu^+|\} \|x-y\|$ for all $x, y \in H$, then at least one of the numbers μ^-, μ^+ is a point of the spectrum of T.

<u>Theorem 16.3.</u> Let T be a completely continuous gradient mapping in H. Suppose that $T(\alpha x) = \alpha^\nu T(x)$ for some $\nu > 0$ and all $x \in H$ and $\alpha \ge 0$, and that $<Tx, x>$ is real for all $x \in H$.

(a) If $\nu \ne 1$ and there exists an element $u_1 \in H$ such that $<Tu_1, u_1> > 0$, then every positive number is an eigenvalue of T.

(b) If $\nu \ne 1$ and there exists an element $u_2 \in H$ such that $<Tu_2, u_2> < 0$, then every negative number is an eigenvalue of T.

(c) If T is an even operator (i.e., $T(-u) = Tu$ for all $u \in H$), $Tx \ne 0$ in H and $\nu \ne 1$, then every real number $r \ne 0$ is an eigenvalue of T.

(d) If $\nu = 1$, then the numbers $m = \inf_{\|x\|=1} <Tx, x>$, $M = \sup_{\|x\|=1} <Tx, x>$ are finite and at least one of them is an eigenvalue of T.

Theorems 16.2 and 16.3 are due to Kačurovskii [282] who also showed that if zero is a point of the spectrum of a completely continuous homogeneous operator T on a Banach space E, then the set of regular values of T is an open set and the spectrum is a closed set.

Generalizations of the result on the numerical range stated at the beginning of this section to Banach spaces can be made by endowing the space with a semi-inner product as was done by Lumer and Phillips [304] in their work on dissipative operators. Let E be a complex reflexive Banach space and assume that the norm in E has a Gâteaux variation for each $x \ne 0$ and $y \in E$; for $t \in \mathbb{R}$ let $G(x, y) = \lim_{t \to 0} \frac{1}{t} \{\|x + ty\| - \|x\|\}$. The vector x is called orthogonal to y ($x \perp y$) if for all complex numbers k, $\|x+ky\| \geq \|x\|$. Then [274] given any $x, y \in E$, $x \neq 0$, there is a unique complex number $a(x,y)$ such that $x \perp ax + y$; $a(x,y)$ is given by

$$a(x,y) = \{iG(x, iy) - G(x,y)\} \|x\|^{-1}.$$

If a is extended to a functional on all of $E \times E$ by setting $a(0,y) = 0$ for $y \in E$, then it is easy to verify that $[x,y] = -\|y\|^2 a(y,x)$ is a semi-inner product on E. That is, it is linear in x, $[x, ty] = t[x,y]$ for all real t, $[x,x] = \|x\|^2$, and $|[x,y]| \leq \|x\| \|y\|$. Define the <u>numerical range</u> of a bounded linear transformation T on X into X to be the set $W(T) = \{-a(x, Tx) : x \in X, \|x\| = 1\}$. George [274] has shown that the spectrum of T is contained in the closure of $W(T)$ for any reflexive Banach space. Since $a(y,x) = -\dfrac{[x,y]}{\|y\|^2}$, the natural definition for the numerical range of a nonlinear operator on a Banach space would be the set

$$W(T) = \left\{ \frac{[Tx - Tx, x-y]}{\|x-y\|^2} : x, y \in E, x \neq y \right\}.$$

If E is a Hilbert space this agrees with Zarantonello's definition, while for linear operators on Banach space it coincides with the definition of Lumer and Phillips. Edmunds [267] has shown that if E is a reflexive complex Banach space with a norm which has a Gâteaux variation at each $x \neq 0$, and if $T : E \to E$ has a continuous Fréchet derivative at each point of E, then for any complex number λ at a positive distance $d(\lambda, T)$ from $W(T)$, the equation $Tx - \lambda x = y$ has a unique solution for each $y \in E$ and the solution depends continuously on y. Note that this result is only a weak analogue of Zarantonello's Hilbert space result, the latter does not require T to be Fréchet differentiable.

16.2. Spectral Radii and Points of Attractions

Let F be a mapping on a subset S of a Banach space X into itself, and suppose that F has a fixed point u in the interior of S. If there exists a neighborhood of u such that for any x_0 in this neighborhood the sequence of successive approximations $x_{n+1} = F(x_n)$ converges to u, then u is called a <u>point of attraction</u>. Ostrowski [329; Theorem 22.1] has shown (in the case $X = R^n$) that if F is Fréchet differentiable at a fixed point u of F and if the spectral radius of $F'(u)$ is less than one, then u is a point of attraction. The proof given in [329] makes use of the Jordan canonical form. Using the spectral radius formula $\rho(T) = \lim_{n \to \infty} \|T^n\|^{\frac{1}{n}}$, an easy proof can be given which avoids matrices; see [289], [328]. The proof can be further simplified using the following formula for the spectral radius:

(16.7) $\qquad \rho(T) = \inf\{ |T| : |\cdot| \in N \}$

where the infimum is taken over the set N of all norms which are equivalent to the given norm $\|\cdot\|$ on the space X.

<u>Theorem 16.4</u>. Suppose $F : S \subset X \to X$ has a fixed point u at an interior point of S and that F is Fréchet differentiable at u. If $\rho[F'(u)] < 1$, then u is a point of attraction.

<u>Proof</u>: Without loss of generality we may suppose that $u = 0$ and $F(0) = 0$. Then by differentiability of F at 0,

$$F(x) = Lx + r(x) \quad \text{where} \quad L = F'(0) \quad \text{and} \quad \lim_{x \to 0} \frac{\|r(x)\|}{\|x\|} = 0.$$

Since $\rho(L) < 1$, there exists on account of (16.7) a norm $|\ |$ which is equivalent to the given norm on X and such that $|L| < 1$. Pick a number δ such that $|L| < \delta < 1$. Then there exists an $\varepsilon > 0$ such that $\|r(x)\| < (1-\delta)\|x\|$ whenever $\|x\| < \varepsilon$. Let $V = \{x \in X : \|x\| < \varepsilon\}$. Then for $x \in V$, $\|F(x)\| < (|L| + 1 - \delta)\|x\| = \beta\|x\|$ where $\beta = |L| + 1 - \delta < 1$, and by induction $\|F^{(n)}(x)\| \leq \beta^n \|x\|$. This implies that $F^{(n)}(x)$ converges to the fixed point $u = 0$ since $\beta < 1$.

17. Asymptotic Derivatives, Quasibounded Operators and Related Applications

Fréchet differentiability of an operator F at x_0 means that F is "close" to a continuous linear operator in a sufficiently small neighborhood of x_0. In particular if $F: X \to Y$, where X and Y are normed spaces and $F(0) = 0$, then Fréchet differentiability of F at 0 is equivalent to the existence of a continuous linear operator $L: X \to Y$ such that

$$(17.1) \qquad \lim_{\|x\| \to 0} \frac{\|F(x) - Lx\|}{\|x\|} = 0.$$

By analogy with (17.1) it is natural to consider <u>asymptotic</u> approximation of a nonlinear operator by a linear one.

<u>Definition 17.1.</u> Let X and Y be normed spaces. An operator $F: X \to Y$ is said to asymptotic to a continuous linear operator L if

$$(17.2) \qquad \lim_{\|x\| \to \infty} \frac{\|F(x) - Lx\|}{\|x\|} = 0.$$

Note that there exists at most one linear operator which satisfies (17.2). For if F is asymptotic to L and $\tilde{L}$, then $\lim_{\|x\| \to \infty} \frac{\|Lx - \tilde{L}x\|}{\|x\|} = 0$; in particular, taking $x = th$ with $\|h\| = 1$, we have $\|Lh - \tilde{L}h\| = 0$ and hence $L = \tilde{L}$. We call L the <u>asymptotic Fréchet derivative</u> (or the Fréchet derivative at ∞) of F, and denote it also by F'_∞ .

This notion was first introduced by Dubrovskii [265] in the context of integral equations and used later by Krasnoselskii [295], who also gave a variety of sufficient conditions under which nonlinear integral operators acting in various function spaces are asymptotic to a Fredholm integral operator [295], [296].

Asymptotic Gâteaux variation and differential as well as <u>asymptotic (H-σ)-derivatives</u> in topological spaces may be defined in a similar manner. For example, F has an asymptotic (H-σ)-derivative (see Section 2 for notation) if there exists a continuous linear operator L such that

$$\lim_{t \to 0} \frac{\|F(th) - tLh\|}{t} = 0$$

uniformly with respect to $h \in S$ for each $S \in \sigma$.

We remark that if F is completely continuous and asymptotically Fréchet differentiable, then F'_∞ is also completely continuous.

Another related notion was introduced by Granas [278] who called an operator F <u>quasibounded</u> if the number defined by

$$|F| = \inf_{0 \leq \rho \leq \infty} \left\{ \sup_{\|x\| \geq \rho} \frac{\|F(x)\|}{\|x\|} \right\} = \lim_{\|x\| \to \infty} \sup \frac{\|F(x)\|}{\|x\|}$$

is finite. The number $|F|$ is called the <u>quasinorm</u> of F. It is easy to see that F is quasibounded if and only if there exist $\alpha, \beta > 0$ such that

$$\|F(x)\| \leq \beta \|x\| \quad \text{for} \quad \|x\| \geq \alpha .$$

It readily follows that if F is <u>asymptotically</u> Fréchet differentiable, then F is quasibounded and

(17.3) $\|F(x)\| \le \|F'_\infty\| \|x\|$ for $\|x\| \ge \alpha$ for some $\alpha > 0$.

In contrast, we remark that if F is Fréchet differentiable <u>at 0</u> and $F(0) = 0$, then

(17.4) $\|F(x)\| \le \|F'(0)\| \|x\|$ for $\|x\| \le \delta$ for some $\delta > 0$.

A comparison of (17.3), (17.2) with (17.4) and (17.1) respectively shows why the notion of an operator asymptotic to a continuous linear operator may be considered as Fréchet differentiability at ∞. As an example of the usefulness of these notions, we mention the following two theorems ([320], Theorems 3 and 4).

<u>Theorem 17.1.</u> Let A,B be two operators on a Banach space X into itself. Let A be a strict contraction, i.e., $\|Ax - Ay\| < \gamma \|x-y\|$, $\gamma < 1$, for all $x,y \in X$, and B be completely continuous and quasibounded on X, with quasinorm $|B|$. If $|B| < 1 - \gamma$, then the equation $x = Ax + Bx + y$ has a solution for each $y \in X$.

Theorem 17.1 includes as special cases a result of Granas (when $A = 0$), and an earlier result of Dubrovskii (when $A = 0$ and $|B| = 0$).

<u>Theorem 17.2.</u> Let A be a bounded linear operator on X such that A^p is a strict contraction (with constant γ) for some $p > 1$, and let B be quasibounded and completely continuous on X. If $|B| < 1 - \gamma$, then the equation $x = Ax + Bx + y$ has a solution for each $y \in X$.

These theorems may be considered as variants of a fixed point theorem of Krasnoselskii [298] (see also Bonsall [244]) which was generalized by Zabreiko, Kachurouskii and Krasnoselskii [299] to the following version:

<u>Theorem 17.3.</u> Let Ω be a bounded, closed, and convex set in a real Hilbert space H. Let A satisfy the condition

$$\|Ax - Ay\| \le \gamma \|x-y\| \text{ for } x,y \in \Omega.$$

Let B be an operator on Ω such that $T = A + B$ maps Ω into itself and assume that one of the following conditions holds: (a) $\gamma < 1$ and B is continuous and compact, (b) $\gamma = 1$ and B maps weakly convergent sequences into strongly convergent sequences. Then T has at least one fixed point in Ω.

The utility of Theorems 17.1 and 17.2 results from the fact that, unlike the standard form of the Schauder-Tychonoff theorem or Theorem 17.3, they do not require <u>a priori</u> that a certain closed bounded convex set is mapped into itself by the completely continuous operator. The hypothesis of Theorem 17.1 or 17.2 guarantee the existence of <u>some</u> closed ball which is mapped into itself by a certain completely continuous operator L whose fixed points coincide with the fixed point of the operator $Fx = Ax + Bx + y$. Applications of Theorems 17.1 and 17.2 to existence of solutions of nonlinear integral equations are given in [320]. Applications of Krasnoselskii's theorem to integral equations are also given in the companion paper of Collatz [255] in this volume. Generalizations of the notion of quasiboundedness and fixed point theorems for sum of two operators in locally convex spaces along the lines of Theorems 17.1 and 17.3, as well as consideration of stability of solutions as in [287], were recently given by Cain and Nashed [251].

Another application of the notion of asymptotic derivative was given by George [273]: Let A be a linear operator, not necessarily bounded, with domain $D(A) \subset X$, and range $\Re(A) = X$, where X is a Banach space, and suppose that A has a completely continuous inverse. Let P be a continuous operator which is asymptotic to the zero operator, and which maps bounded sets into bounded sets. Then $\Re(A + P) = X$. Other applications of asymptotic derivatives in nonlinear problems may be found in [333], [290], [296], [352], [251].

18. One-Sided and Logarithmic Derivatives and Norms

Many extremal problems in analysis involve functionals which have only one-sided Gâteaux variations (directional derivatives). For such problems as well as in other problems of convex analysis, differential inequalities, nonlinear operator equations, etc., one-sided derivatives provide useful tools.

Let S be a subset of a vector space E. A function $f: S \to \mathbb{R}$ is said to have a one-sided Gâteaux variation (directional derivative) at $x_0 \in S$ if for each fixed $h \in E$ for which $x_0 + th \in S$ for sufficiently small positive t,

$$V^+ f(x_0; h) = \lim_{t \to 0^+} \frac{f(x_0 + th) - f(x_0)}{t}$$

exists in the extended real number. Suppose f has a one-sided Gâteaux variation at a point $x_0 \in S$. Then a necessary condition for f to have a relative minimum at x_0 is that

(18.1) $$V^+ f(x_0; x - x_0) \geq 0$$

for all $x \in S$ for which $V^+ f(x_0; h)$ exists. For convex functionals, condition (18.1) is also sufficient (see [126]).

Now let E be a real or complex Banach space and let $\mathcal{L}$ denote the space of all bounded linear operators on E into E. The <u>logarithmic norm</u> of $A \in \mathcal{L}$ is defined by

(18.2) $$\mu(A) = \lim_{\lambda \to 0^+} \frac{\|I + \lambda A\| - 1}{\lambda}.$$

The limit in (18.2) always exists since the function $t \to \|I + tA\|$ is convex. The following properties of $\mu(A)$ follow easily from the definition.

(18.3) $$\begin{cases} \mu(rA) = r\mu(A) \text{ for positive number } r, \\ \mu(A + B) \leq \mu(A) + \mu(B) \\ |\mu(A)| \leq \|A\|, \quad |\mu(A) - \mu(B)| \leq \|A - B\| \\ \mu(A + \alpha I) = \mu(A) + \operatorname{Re} \alpha \text{ for every complex number } \alpha \end{cases}$$

and

$$\mu(A) = \lim_{t \to 0^+} \frac{\|\exp(tA)\| - 1}{t}, \quad \text{where} \quad \exp A = \sum_{n=0}^{\infty} \frac{A^n}{n!}.$$

Logarithmic norms of matrices were introduced by Lozinskii [301] and Dahlquist [259] in the study of error bounds in the numerical integration of ordinary differential equations and have been extended in [302] to matrix-valued norms.

Let Lip E denote the space of all Lipschitzian mappings on E into E and let $\|\|T\|\|$ denote the Lipschitzian norm of T (see (16.3)). Obviously, $\|\|\cdot\|\|$ is a seminorm on E, with $\|\|T\|\| = 0$ if and only if T is a constant mapping. Thus for each fixed $x_0 \in E$, $N(T) = \|\|T\|\| + \|Tx_0\|$ is a norm on Lip E, and the resulting normed space is complete.

A logarithmic norm of a Lipschitzian mapping, generalizing the logarithmic norm of a bounded linear operator, may be defined by

$$(18.4) \qquad M(T) = \lim_{\lambda \to 0^+} \frac{\|\|I + \lambda T\|\| - 1}{\lambda}$$

for each $T \in $ Lip E . Properties (18.3) hold for M(T) with μ replaced by M and $\|\cdot\|$ by the Lipshitzian norm $\|\|\cdot\|\|$. The relationship $\|\|T\|\| = \sup \{\|T'(x)\| : x \in \Omega\}$ that holds for any Fréchet differentiable Lipschitzian mapping on a convex subset Ω, remains valid when logarithmic norms are used, i.e.,

$$M(T) = \sup\{\mu[T'(x)] : x \in \Omega\} .$$

The <u>logarithmic norm</u> of a square matrix A is a standard tool in the theory of linear ordinary differential equations and is used for instance to obtain bounds for the solutions of u' = Au . Martin [306] introduced the notion of <u>logarithmic derivative</u> which plays a similar role for nonlinear differential equations and used it in [305], [306] and other related papers

THE ROLE OF DIFFERENTIALS

to obtain results on the existence, stability, and bounds of solutions of nonlinear differential equations in a Banach space. Let A be a mapping on E into E. For any $x, y \in E$, define

$$(18.5) \quad D_+[x,y;A] = \lim_{\lambda \to 0^+} \frac{1}{\lambda} \{\|x-y+\lambda(Ax-Ay)\| - \|x-y\|\},$$

noting that the limit in (18.5) always exists. If there exists a number K such that for all $x, y \in \aleph \subset E$

$$(18.6) \quad D_+[x,y;A] \le K \|x-y\|,$$

then the infimum of all numbers K for which the inequality in (18.6) holds is called the <u>logarithmic derivative</u> of A on $\aleph$ and is denoted by L[A]. The class of all mappings on $\aleph \subset E$ to E which have logarithmic derivative is denoted by Ln($\aleph$, E). Note that if A is a bounded linear operator, then $\mu(A) = \sup\{D_+(x;0,A): \|x\|=1\}$ and that if $A \in \text{Lip}(\aleph, E)$, then $A \in \text{Ln}(\aleph, E)$ but not conversely. If $A \in \text{Lip}(\aleph, E)$ then $L[A] \le M(A)$.

There is a close and easily deducible connection between mappings which have logarithmic derivative and accretive operators (considered by Browder, Kato and others; see, e.g., [250], [258]). Let $A: \aleph \to E$ where $\aleph$ is a subset of a Banach space E, and let c be a real number. Then these are equivalent: (i) $A \in \text{Ln}(\aleph, E)$ with $L[A] \le c$; (ii) $cI - A$ is accretive on $\aleph$. In particular $-A$ is accretive on $\aleph$ if and only if $A \in \text{Ln}(\aleph, E)$ and $L[A] \le 0$. If E is a Hilbert space then $A \in \text{Ln}(E)$ if and only if $-A$ is strongly monotone, i.e. $\text{Re} \langle Ax - Ay, x-y \rangle \le m\|x-y\|^2$ and L[A] is the smallest number m for which this inequality holds.

REFERENCES

I. DIFFERENTIAL CALCULUS IN NORMED AND TOPOLOGICAL SPACES

   A. BOOKS

General Background in Linear Topological Spaces

[1]   N. Bourbaki, Espaces vectoriels topologiques, V. Herman, Paris, 1955.

[2]   N. Dunford and J. Schwartz, Linear Operators. I: General Theory, Interscience, New York, 1958.

[3]   H. H. Schaefer, Topological Vector Spaces, Macmillan, New York 1966.

   See [2], [5], [6], [8], [9] for background in functional analysis, and [8; Chapter XI] for background in linear topological spaces, which is adequate for this paper.

Calculus in Normed Spaces

[4]   H. Cartan, Calcul différentiel, Herman, Paris, 1967.

[5]   J. Dieudonné, Foundations of Modern Analysis, Academic Press, New York, 1960.

[6]   E. Hille and R. S. Phillips, Functional Analysis and Semigroups, Amer. Math. Soc. Colloq. Publ., Vol. 31, Providence, R. I., 1957.

[7]   A. D. Michal, Le calcul différentiel daus les espaces de Banach, Vol. 1, 2, Gauthier-Villars, Paris, 1958, 1964.

[8]   L. V. Kantorovich and G. P. Akilov, Functional Analysi in Normed Spaces, Pergamon Press, New York, 1964.

[9] L. A. Liusternik and V. J. Sobolev, <u>Elements of Functional Analysis</u>, Ungar, New York, 1961.

[10] L. B. Rall, <u>Computational Solution of Nonlinear Operator Equations</u>, Wiley, New York, 1969.

[11] M. M. Vainberg, <u>Variational Methods for the Study of Nonlinear Operators</u>, Holden-Day, San Francisco, 1964.

Some aspects of differential calculus are also given in the books of Krasnoselskii [296], [297], Lang [103], Ortega and Rheinboldt [328], Schwartz [351], and others.

Calculus in Finite Dimensional Spaces

[12] W. Fleming, <u>Functions of Several Variables</u>, Addison-Wesley, 1965.

[13] L. H. Loomis and S. Sternberg, <u>Advanced Calculus</u>, Addison-Wesley, 1968.

Calculus of Variations

[14] I. M. Gelfand and S. V. Fomin, <u>Calculus of Variations</u>, Prentice-Hall, Englewood Cliffs, N.J., 1963.

[15] M. R. Hestenes, <u>Calculus of Variations and Optimal Control</u>, Wiley, 1966.

[16] H. Sagan, <u>Introduction to the Calculus of Variations</u>, McGraw-Hill, New York, 1969.

[17] L. C. Young, <u>Lectures on the Calculus of Variations and Optimal Control Theory</u>, Saunders, Philadelphia, 1969.

B. DIFFERENTIAL CALCULUS IN NORMED SPACES: EARLY CONTRIBUTIONS

[18] C. Arzelá, Funzioni di linei, Atti Accad. Lincei Rend. 5, 1 (1889).

[19] G. A. Bliss, A note on functions of lines, Proc. Nat. Acad. Sci. USA 1 (1915), 173-177.

[20] O. Bolza, An application of the notion of "general analysis" to a problem of the calculus of variations, Bull. Amer. Math. Soc. 16 (1910), 402-407.

[21] P. J. Daniell, The derivative of a functional, Bull. Amer. Math. Soc. 25 (1919), 414-416.

[22] Ky Fan, Sur quelques notions fondamentals de l'analyse générale, J. Math. Pures Appl. 21 (1942), 289-368.

[23] C. Fischer, A generalization of Volterra's derivative of a function of a curve, Amer. J. Math. 35 (1913), 369-395.

[24] M. Fréchet, La notion de différentielle totale, Nouv. Ann. Math. 12 (1912), 385-403, 433-449.

[25] M. Fréchet, La notion de différentielle dans l'analyse générale, C. R. Acad. Sci. (Paris), 180 (1925), 806-809.

[26] M. Fréchet, La notion de différentielle dans l'analyse générale, Ann. École Norm. Sup. (3) 42 (1925), 293-323.

[27] M. Fréchet, Sur la notion de différentielle dans l'analyse générale, J. Math. Pures Appl. 16 (1937), 233-250.

[28] M. Fréchet, Sur quelques points du calcul fonctionnel (Thése), Rend. Circolo Matem. Palermo, 22 (1906), 1-74.

[29] M. Fréchet, Sur les fonctionelles continues, Ann. École Norm. Sup. (3) 27 (1910), 93-216.

[30] M. Fréchet. Les polynomes abstraits, Journal de mathématique pures et appliquées (9) 8 (1929), 71-92.

[31] M. Fréchet, Une définition fonctionelle des polynômes, Nouv. Ann. Math. (4) 9 (1909), 145-162.

[32] M. Fréchet, Pages choiseés d'Analyse générale, Gauthier-Villars, Paris, 1952.

[33] M. Fréchet, Les espaces abstraits et leur théorie considéreé comme introduction à l'analyse générale, Gauthier-Villars, Paris, 1928.

[34] M. Fréchet, Sur la notion de differéntielle, C. R. Acad. Sci. (Paris), 152 (1911), 845-847, 1050-1051.

[35] G. Fubini, Al cumi nuovi problemi di calcolo delle variazioni, Ann. Mat. Pura Appl. Ser. 3, No. 20 (1913), 217-244.

[36] R. Gâteaux, Sur les fonctionelles continues et les fonctionelles analytiques, C. R. Acad. Sci. Paris 157 (1913), 325-327 (its expanded version was published posthumously in Bull. Soc. Math. France, 50 (1922), 1-21).

[37] R. Gâteaux, Fonction d'une infinité des variables independantes, Bull. Soc. Math. France 47 (1919), 70-96.

[38] M. K. Gavurin, On k-ple linear operations in Banch spaces, Dokl. Akad. Nauk SSSR 22 (1939), 547-551.

[39] M. K. Gavurin, Über die stieltjessche Integration abstrakten Funkionen, Fund. Math. 27 (1936), 254-268.

[40] M. K. Gavurin, Toward the construction of a differential and integral calculus in Banach spaces, Dokl. Akad. Nauk SSSR 22 (1939), 552-556.

[41] M. Golomb, Zur Theorie der nicht-linearen Integralgleichungen, Integralgleichungs-systeme und allgemeinen Functionalgleichungen, Math. Zeit. 39 (1934), 47-75.

[42] L. M. Graves, Riemann integration and Taylor's theorem in general analysis, Trans. Amer. Math. Soc. 29 (1927), 163-177.

[43] L. M. Graves, Topics in functional calculus, Bull. Amer. Math. Soc. 41 (1935), 641-662.

[44] T. H. Hildebrandt and L. M. Graves, Implicit functions and their differentials in general analyses, Trans. Amer. Math. Soc. 29 (1927), 127-153.

[45] I. E. Highberg, A note on abstract polynomials in complex spaces, J. de Math. Pures et Appl., (9) 16 (1937), 307-314.

[46] J. Hadamard, La notion de différentielle dans l'enseignement, Scripta Univ. Ab. Hierosolymitanarum, Jerusalem 1 (1923), 3.

[47] J. Hadamard, Sur un problème mixte aux dérivées partielles, Bull. Soc. Math. France, 31 (1903), 208-224.

[48] J. Hadamard, Sur les transformations ponctuelles, Bull. Soc. Math. France 34 (1906), 71-84.

[49] M. Kerner, Zur Theorie der Impliziten Funktional Operationen, Studia Math. 3 (1931), 156-173.

[50] M. Kerner, Gewöhnliche Differentialgleichungen der allgemienen Analysis, Prace Matematyczno-Fizyczne 40 (1932), 47-67.

[51] M. Kerner, Sur les variations faibles et fortes d'une fonctionelle, Annali di Matematica Pura ed Applicata (4) 10 (1932), 145-164.

[52] M. Kerner, Die Differentiale in der allgemeinen Analysis, Ann. of Math. (2) 34 (1933), 546-572.

[53] M. Kerner, L'extremum dans l'espace hilbertien, Annali di Mat. Pura ed Appl. (4) 10 (1933), 183-202.

[54] K. W. Lamson, A general implicit function theorem with an application to problems of relative minima, Amer. J. Math. 42 (1920), 243-256.

[55] P. Lévy, Sur les derivees des fonctions des lignes planes, C. R. Acad. Sci. Paris 152 (1911), 178-180.

[56] P. Lévy, Sur problèmes concrets d'analyse fonctionelle, Gauthier-Villars, Paris, 1921; 2nd rev. ed. with supplement, 1951.

[57] P. Lévy, Sur les fonctions de lignes implicites, Bull. Soc. Math. France 48 (1920), 13-27.

[58] R. S. Martin, Contribution to the Theory of Functionals, Thesis, California Institute of Technology, 1932.

[59] A. D. Michal and V. Elconin, Completely integrable differential equations in abstract spaces, Acta Math. 68 (1936), 71-107.

[60] A. D. Michal and R. S. Martin, Some expansions in vector space, J. Mathematiques Pures et Appliques (9) 13 (1934), 69.

[61]   A. D. Michal, The Fréchet differential of regular power series in normed linear spaces, Duke Math. J. 13 (1946), 57-69.

[62]   A. D. Michal, Bounds of polynomials in hyperspheres and Fréchet-Michal derivatives on complex normed linear spaces, Math. Mag. 27 (1954), 119-126.

[63]   J. Pierpont, Theory of functions of real variables, Vol. 1, Boston, 1905.

[64]   E. H. Rothe, Gradient mappings, Bull. Amer. Math. Soc., 59 (1953), 5-19.

[65]   Selected Papers on Calculus, Reprinted from the Amer. Math. Monthly, Volumes 1-75, and the Math. Mag., Vol. 1-40; edited by T. M. Apostol et al., Math. Assoc. of America, 1969.

[65]   O. Stolz, Grundzü ge der Differential-und Integralrechnung, Bd. 1, Leipzig, 1893.

[66a] G. A. Suchomlinov, Analytic functionals, Bulletin of the Moscow University, Ser. A, 1 (1937), 1-19.

[66b] A. E. Taylor, Analytic functions in general analysis, Annali della R. Scuola Normale Superiore di Pisa, (2), 6 (1937), 277-292.

[67]   V. Volterra, Sopra le funzioni che dipendeno da altre funzioni, Atti Reale Accad. Lincei Rend. 3: 2 (1887), 97-105, 141-146, 153-158.

[68]   W. H. Young, On differentials, Proc. London Math. Soc. (2) 7 (1909), 157-180.

[69]   W. H. Young, The Fundamental Theorems of Differential Calculus, Univ. Press, Cambridge, 1910.

[70] M. Zorn, Gâteaux differentiability and essential boundedness, Duke Math. J. 12 (1945), 579-583.

[71] M. Zorn, Derivatives and Fréchet differentials, Bull. Amer. Math. Soc. 52 (1946), 133-137.

C. DIFFERENTIAL CALCULUS IN LINEAR TOPOLOGICAL SPACES

[72] V. I. Averbukh and O. G. Smolyanov, The theory of differentiation in linear topological spaces, Uspekhi Mat. Nauk 22:6(1967), 201-260.
= Russian Math. Surveys 22: 6(1967), 201-258.

[73] V. I. Averbukh and O. G. Smolyanov, The various definitions of the derivative in linear topological spaces, Uspekhi Mat. Nauk 23: 4 (1968), 67-114.

[74] V. I. Averbukh, On higher order derivatives in linear topological spaces, Vestnik Moskovoskogo Universiteta Serija I(1970), 32-37.

[75] V. I. Averbukh and O. G. Smolyanov, Differentiation in linear topological spaces, Dokl. Akad. Nauk SSSR, 173 (1967), 735-738. = Soviet Math. Dokl. 8 (1967).

[76] M. Balanzat, La différentielle d'Hadamard-Fréchet dans les espaces vectoriels topologiques, C. R. Acad. Sci. Paris 251 (1960), 2459-2461.

[77] M. Balanzat, Théorème des accroissements finis pour les applications différentiables, au sens d'Hadamard-Fréchet, dans les espaces vectoriels topologiques, C. R. Acad. Sci. Paris 253(1961), 1240-1242.

[78] M. Balanzat, La différential en les espacios métricas affines, Matem. Notae 9 (1949), 29-51; 19 (1964), 43-62.

[79] A. Bastiani, Différentiabilité dans les espaces localement conveyes, Distructures, These Doct. Sci. Math. Fac. Scie., Univ. Paris (1962).

[80] A. Bastiani, Applications différentiables et variétés differentiables de dimension infinie, J. Anal. Math. 13 (1964), 1-114.

[81] E. Binz, Ein Differenzierbarkeitsbegriff in limitierten Vektorräumen, Comm. Math. Helv. 41 (1966), 137-156.

[82] E. Binz and H. H. Keller, Funktionenräume in der Kategorie der Limesräume, Ann. Acad. Sci. Fenn. A, 383 (1966), 1-21.

[83] E. Binz and W. Meier-Solfrian, Zur Differentialrechnung in limitierten Vektorräumen, Comm. Math. Helv. 42 (1967), 285-296.

[84] J. Dieudonné, Recent developments in the theory of locally convex spaces, Bull. Amer. Math. Soc. 59 (1953), 495-512.

[85] E. Dubinsky, Differential equations and differential calculus in Montel spaces, Trans. Amer. Math. Soc. 110 (1964), 1-21.

[86] P. L. Falb and M. Q. Jacobs, On differentials in locally convex spaces, J. Differential Equations 4 (1968), 444-459. A correction, ibid, 6 (1969), 395-396.

[87] S. Fernandez Long de Foglio, Extension de la differentielle d' Hadamard-Fréchet aux applications entre deux espaces vectoriels L, C. R. Acad. Sci. Paris 248 (1959), 1108-1110.

[88] Susana Fernandez Long de Foglio, La différentielle au sens d' Hadamard-Fréchet dans les espaces L vectoriels, Portugal. Math. 19 (1960), 165-184.

[89] H. R. Fischer, Limesräume, Math. Annalen 137 (1959), 269-303.

[90] H. R. Fischer, Differentialkalkül für nicht-metrische Strukturen, Ann. Acad. Sci. Fenn. 247 (1957), 1-14.

[91] M. Fréchet, La notion de différentielle sur un groupe abélian, Portugal. Math. 7 (1948), 59-72.

[92] M. Fréchet, La différentielle sur deux semi-espaces de Banach, C. R. Acad. Sci. Paris 252 (1961), 481-483.

[93] A. Frölicher and W. Bucher, Calculus in Vector Spaces without Norms, Lecture Notes in Mathematics No. 30, Springer-Verlag, Berlin-New York, 1966.

[94] J. Gil de Lamadrid, Topology of mappings in locally convex topological vector spaces, their differentiation and integration and application to gradient mappings, Thesis, Univ. of Michigan, 1955.

[95] J. Gil de Lamadrid, Topology of mappings and differentiation processes, Illinois J. Math. 3 (1959), 408-420.

[96] D. H. Hyers, A generalization of Fréchet differential, Proc. Nat. Acad. Sci. USA 27 (1941), 315-316.

[97] D. H. Hyers, Linear topological spaces, Bull. Amer. Math. Soc. 51 (1945), 1-21.

[98] D. H. Hyers, A generalization of Fréchet's differential, Rev. Cienc., Lima (Peru) 47 (1945), 645-663.

[99] H. H. Keller, Differenzierbarkeit in topologischen Vektorräumen, Comm. Math. Helv. 38 (1964), 308-320.

[100] H. H. Keller, Räume stetiger multilinearer Abbildungen als Limesräume, Math. Annalen 159 (1965), 259-270.

[101] H. H. Keller, Differentialrechnung in lokalkonvexen Vektorräumen. To appear.

[102] H. H. Keller, Über Probleme die bei einer Differentialrechnung in topologischen Vektorräumen auft reten, H. P. Künzi, A. Pfluger, Festband zum 70 Geburtstag von Rolf Nevanlinna. Springer, 1966.

[103] S. Lang, Introduction to Differentiable Manifolds, Wiley (Interscience), New York, 1962.

[104] J. P. LaSalle, Pseudo normed linear sets over valued rings, California Institute of Technology, Thesis, 1941.

[105] G. Marinescu, Différentielles de Gâteaux et Fréchet dans les espaces localement convexes, Bull. Math. Soc. Sci. Math. Phy. R. P. Roumaine (N.S.) $\underline{1}$ (49) (1957), 77-86.

[106] A. D. Michal, Differential calculus in linear topological spaces, Proc. Nat. Acad. Sci. USA $\underline{24}$ (1938), 340-342.

[107] A. D. Michal, Differentials of functions with argument and values in topological abelian groups, Proc. Nat. Acad. Sci. USA $\underline{26}$ (1940), 356-359 (for its expanded version see Revista de Ciencias $\underline{43}$ (1941), 155-176; ibid $\underline{47}$ (1945), 389-422).

[108] A. D. Michal, Functional analysis in topological group spaces, Math. Mag. $\underline{20}$ (1947), 80-90.

[109] A. D. Michal, General differential geometries and related topics. Bull. Amer. Math. Soc. $\underline{45}$ (1939), 529-563.

[110] A. D. Michal and E. W. Paxson, La différentielle dans les espaces abstraits linéaires avec une topologie, C. R. Acad. Sci. (Paris) $\underline{202}$ (1936), 1741-1743.

[111] J. B. Miller, Generalized Gâteaux and Fréchet derivatives in convolution algebras. Proc. Cambridge Philos. Soc. 59 (1963), 707-718.

[112] K. Millsaps, Differential calculus in topological groups I, II, Revista Ciencias 44 (1942), 485-492; 45 (1943), 45-52.

[113] E. W. Paxson, Analysis in linear topological spaces, California Institute of Technology, Thesis, 1942.

[114] E. W. Paxson, Les équations différentielles dans les espaces linéaires et topologiques, Revista Ciencias 42 (1940), 823-826.

[114a] E. J. Pinney, The calculus of variations in abstract spaces and related topics, California Institute of Technology, Thesis, 1942.

[115] G. A. Reid, Concepts of differentiability and analyticity on certain classes of topological groups, Proc. Cambr. Phil. Soc. 61 (1965), 347-379.

[116] J. Sebastião e Silva, Le calcul différentiel et intégral dans les espaces localement convexes, réels ou complexes, I, II, Atti Accad. Lincei Rend. 20 (1956), 743-750; 21 (1956), 40-46.

[117] J. Sebastião e Silva, Concietos de funcão differenciável em espacos localmente convexos, Publ. Centro Estudes Matem. Lisboa, Inst. Alta Cultura, Lisboa, 1957.

[118] J. Sebastião e Silva, Les espaces à bornés et la notion de fonction différentiable, collogue sur l'analyse fonctionnelle, Louvain, 1960, Paris-Louvain 1961, 57-61.

[119] M. Sova, General theory of differentiability in linear topological spaces, Czech. Math. J. 14 (1964), 485-508 (in Russian).

[120] M. Sova, Conditions of differentiability in linear topological spaces, Czech. Math. J. 16 (1966), 339-362 (in Russian).

[121] M. M. Vainberg and Ya. L. Engel'son, The conditional extremum of functionals in linear topological spaces, Mat. Sb. 45 (1958), 417-422.

D. ADDITIONAL REFERENCES ON DIFFERENTIAL CALCULUS IN NORMED SPACES AND RELATED CITED REFERENCES

[122] A. Alexiewicz and W. Orlicz, On the differentials in Banach spaces, Ann. Soc. Pol. Math. 25 (1952), 95-99.

[123] A. M. Bruckner and J. L. Leonard, Derivatives, Amer. Math. Monthly 73 (1966), No. 4, Part II (Slaught Memorial Papers), 24-56.

[124] M. Esser and O. Shisha, A modified differentiation, Amer. Math. Monthly, 71 (1964), 904-906.

[125] M. Fréchet, Existence de la différentielle d' une integrale du calcul des variations, C. R. Acad. Sci. Paris, 240 (1955), 2038-20

[126] D. M. Friedlen and M. Z. Nashed, A note on one-sided directional derivative, Math. Mag. 41 (1968), 147-150.

[127] M. K. Gavurin, On the fundamental theorems of the differential and integral calculus in linear spaces, Vestnik Leningrad Univ. 13 (1958), 38-48.

[128] M. K. Gavurin, Analytic method for the study of non-linear functional transformations, Leningrad Gas. Univ. Uc. Zap. Ser. Mat. Nauk 19 (1950), 59-154.

[129] J. Kolomý, On the differentiability of mappings in functional spaces, Comment. Math. Univ. Carolinae 8 (1967), 315-329.

[130] J. Kolomý and V. Zizler, Remarks on the differentiability of mappings in linear normed spaces, Comment. Math. Univ. Carolinae 8 (1967), 691-704.

[131] E. B. Leach, A note on inverse function theorem, Proc. Amer. Math. Soc., 12 (1961), 694-697.

[132] G. Marinescu, Asupra diferentialei si derivatei in spatrile normate, Bull. stii nt. Acad. R. P. Romîne, Sect. Mat. si fiz. 6 (1954), 213-219.

[133] L. Lombardi, Sur l'application de la méthode du gradient au calcul des variations. Les mathématiques de l' ingénieur, pp. 216-225. Mém. Publ. Soc. Sci. Arts Lett. Hainaut., 1958.

[134] M. Z. Nashed, Some remarks on variations and differentials, Amer. Math. Monthly 73 (1966), Slaught Memorial Papers, 63-76.

[135] M. Z. Nashed, Higher order differentiability of nonlinear operators on normed spaces I, II, Comment. Math. Univ. Carolinae 10 (1969), 509-533, 535-557.

[136] G. Peano, Sur la definition de la derivee, Mathesis, (2) 2 (1892), 12-14. = Opere Scelte, v. 1, Edizioni Cremonsense, Rome, 1957, 210-212.

[137] G. H. Sindalovskii, On a generalization of derived numbers, Izv. Akad. Nauk SSSR Ser. Mat. 24 (1960), 707-720.

[138] D. W. Solomon, On the differentiability of vector-valued functions of a real variable, Studia Math. 29 (1967), 1-4.

[139] R. Tapia, The Differentiation and Integration of Nonlinear Operators, this volume, pp. 45-101.

[140] E. S. Titlanadze, On the differentiation of functionals, Mat. Sb. 29 (71) (1951), 3-12.

[141] M. M. Vainberg, On the differential and gradient of functionals, Uspehi Mat. Nauk (N.S.) 7 (49) (1952), 139-143.

[142] M. M. Vainberg, Some questions of differential calculus in linear spaces, Uspehi Matem. Nauk (N.S.) 7 (50), 55-102.

[143] V. Zizler, On the differentiability of mappings in Banach spaces, Comment. Math. Univ. Carolinae 8 (1967), 415-430.

[144] A. Alexiewicz, On differentiation of vector-valued functions, Studia Mathematica 11 (1950).

[145] S. Burýsek, Some remarks on polynomial operators, Comment. Math. Univ. Carolinae 10 (1969), 285-306.

[146] N. A. Ivanov, On Gâteaux and Fréchet differentials, Uspehi Mat. Nauk 10 (1955), 161-166.

[147] M. Z. Nashed, On general iterative methods for the solutions of a class of nonlinear operator equations, Math. Comp. 19 (1965), 14-24.

E. MEAN VALUE THEOREMS AND TAYLOR'S THEOREMS

[148] A. K. Aziz and J. B. Diaz, On a mean value theorem of the differential calculus of vector-valued functions, and uniqueness theorems for ordinary differential equations in a linear normed space, Contributions to Differential Equations, 1 (1963), 251-269.

[149] A. K. Aziz and J. B. Diaz, On a mean value theorem of the weak differential calculus of vector vector-valued functions, Contribution to Differential Equations, $\underline{1}$ (1963), 271-273.

[150] A. K. Aziz and J. B. Diaz, On Pompeiu's proof of the mean-value theorem of the differential calculus of real-valued functions, Contributions to Differential Equations, $\underline{1}$ (1963), 467-481.

[151] A. K. Aziz, J. B. Diaz, and W. Mlak, On a mean value theorem for vector-valued functions, with applications to uniqueness theorems for "right-hand-derivative" equations, J. Math. Anal. Appl., $\underline{116}$ (1966), 302-307.

[152] W. Blaschke, Mittelwertsatz der Potentialtheorie, Jber. Deutschen Math. Verein, $\underline{2}$ (1918), 157-160.

[153] J. B. Diaz and R. Výborný, A mean-value theorem for strongly continuous vector valued functions, Czeh. Math. J. $\underline{14}$ (1964), 322.

[154] J. B. Diaz and R. Výborný, On mean value theorems for strongly vector valued functions, Contributions to Differential Equations, $\underline{3}$ (1964), 107-118.

[154a] J. B. Diaz and R. Výborný, A fractional mean value theorem and a Taylor theorem for strongly continuous vector valued functions, Czech. Math. J. $\underline{15}$ (1965), 299-303.

[155] I. S. Gál, On the fundamental theorems of calculus, Trans. Amer. Math. Soc. $\underline{86}$ (1957), 309-320.

[156] G. Glaeser, Étude de quelques algébres Tayloriennes, J. Anal. Math. $\underline{11}$ (1958), 1-118.

[157] P. Hartman and A. Wintner, Mean value theorems and linear operators, Amer. Math. Monthly, 62 (1955), 217-222.

[158] P. Hartman, Remark on Taylor's formula, Bull. Amer. Math. Soc., 51 (1945), 729-732.

[159] R. M. McLeod, Mean value theorems for vector valued functions, Proc. Edinburgh Math. Soc., 14 (1965), 197-209.

[160] W. Mlak, Note on the mean value, Ann. Polon. Math. 3 (1957), 29-31.

[161] M. Z. Nashed, Converses of Taylor's theorems in locally convex spaces, to appear.

[162] L. J. Nicolescu, Asupra unei teoreme de medie, St. Cerc. Mat., Acad. R. P. R. 16 (1964), 987-995.

[163] L. J. Nicolescu, On a weak mean value theorem, Rev. Roum. Math. Pures Appl. 10 (1965), 145-148.

[164] G. Pólya, On the mean-value theorem corresponding to a given linear homogeneous differential equations, Trans. Amer. Math. Soc., 24 (1922), 312-324.

[165] G. Pólya, Quelques théoremès analogues au théorème de Rolle, li és à certaines équations lineaires aux derivées partielles, Comptes Rendus, 199 (1934), 655.

[166] D. Pompieu, Sur le théorème des accroissements finis (Première note), Ann. Sci. Université Jassy, 3 (1906) p. 244). (See, also, pp. 62-66 of D. Pompieu, Opera Mat., Editori Academici Republicii Populare Romine, 1959).

[167] D. Pompieu, Sur le théorème des accroissements finis (Déuxième note), Ann. Sci. Université Jassy, 4, 1906 (also, pp. 67-76 of Dimitrie Pompeiu, Opera Mat., Editore Academici Republicii Populare Romine, 1959).

[168] G. Prasad, Six lectures on the mean-value theorem of the differential calculus, University of Calcutta, 1931.

[169] D. Ugrin-Šparac, A generalization of Taylor's formula and its application in elementary theory of distributions, Društvo Mat. i Fiz. Hrav. Glas. Mat. 5 (1970), 71-80.

[170] T. Wazewski, Une généralisation des théorèmes sur les accroissements finis au cas des espaces de Banach et application à la généralisation du théorème de cas des espaces de Banach et application à la généralisation du théorème de l'Hôpital, Ann. Soc. Polon. Math. 24 (1951), 132-147.

[171] W. H. Young and G. C. Young, On derivatives and the theorem of the mean, Quart. J. Pure Appl. Math., 40 (1909), 1-26.

[172] S. Zaremba, Sur une propriété générale de fonctions harmoniques, Conférence de Réunion Internationale de Mathématiciens, (1937), Paris, 1939, 171-176.

References [173]-[179] are classroom notes on interesting aspects of meanvalue and Taylor's theorem for real functions.

[173] W. R. Ballard, A. E. Livingston, and W. M. Myers, Jr., A variant of Taylor's Theorem, Amer. Math. Monthly, 70 (1963), 865-868.

[174] P. R. Beesack, Taylor's formula and the existence of nth derivatives, Amer. Math. Monthly, 74 (1967), 980-984.

[175] Lipman Bers, On avoiding the mean value theorem, Amer. Math. Monthly 74 (1967), 583.

[176] L. W. Cohen, On being mean to the mean value theorem, Amer. Math. Monthly 74 (1967), 581.

[177] D. B. Goodner, Mean value theorems for functions with finite derivatives, Amer. Math. Monthly, 67 (1960), 852-855.

[178] C. C. Lan, An extension of the mean-value theorem in E_n, Math. Mag. 39 (1966), 91.

[179] O. Shisha, On a Taylor's theorem, J. Res. Nat. Bur. Standards Sect. B, 72 (1968), 5-6.

F. DIRECT HIGHER ORDER DIFFERENTIALS: RIEMANN, PEANO, TAYLOR VARIATIONS AND DIFFERENTIALS, DIFFERENCE DIFFERENTIALS

[180] P. L. Butzer, Beziehungen Zwishen der Riemannschchen, Taylorschen und gewohn lichen Ableitungen reelwertiger Funktionen. Math. Ann. 144 (1961), 275-298.

[181] P. L. Butzer and H. Berens, Semigroups of Operators and Approximation, Springer-Verlag, Berlin-New York, 1967.

[182] P. L. Butzer and W. Kozakiewicz, On the Riemann derivatives for integrable functions, Canadian J. Math. 6 (1954), 572-581.

[183] E. Corominas, Contribution à la theorie de la dérivation d'ordre supérieur, Bull. Soc. Math. France, (Part 3) 81 (1953), 177-222.

[184] A. Denjoy, Sur l'integration des coefficients differentiels d'ordre superieur, Fund. Math. 25 (1935), 273-326.

[185] C. Kassimatis, Functions which have generalized Riemann derivative, Canad. J. Math. 10 (1958), 413-420.

[186] A. Marchaud, Sur les dérivées et sur les différences des fonctions de variable réelles, J. de Math. (9), 6 (1927), 337-425.

[187] J. Marcinkiewicz and A. Zygmund, On the differentiability of functions and summability of trigometrical series, Fund. Math., 26 (1936), 1-43.

[188] M. Z. Nashed, On the representation and differentiability of operators. To appear in Proceedings of the Colloquium on Constructive Theory of Functions (Approximation Theory), Budapest (1969).

[189] I. P. Natanson, Theory of Functions of a Real Variable, Vol. II, Ungar Publishing Co., New York, 1960.

[190] C. J. Neugebauer, Smoothness and differentiability in L_p, Studia Math. 25 (1964), 81-91.

[191] L. J. Nicolescu, On the direct second order differentiability in Fréchet's or Gâteaux's sense, Rev. Roumaine Math. Pures et Appl. 3 (1958), 217-223.

[192] L. J. Nicolescu, An extension of a theorem of Alexiewicz and Orlicz to the direct second order Gâteaux differential, Rev. Roumaine Math. Pures et Appl. 9 (1964), 589-592.

[193] L. J. Nicolescu, On mixed direct differentiation of vector valued functions, Rev. Math. Pures et Appl. 8 (1963), 647-652.

[194] L. J. Nicolescu, On the analyticity of the direct second order differential in Gâteaux's sense, Rev. Math. Pures Appl. 8 (1963), 305-307.

[195] L. J. Nicolescu, On some properties of the direct second order differentials in Gâteaux sense, Rev. Math. Pures Appl. 6 (1959), 271-276.

[196] L. J. Nicolescu, On the direct differential in Gâteaux's sense of the nth order, Rev. Roumaine Math. Pures Appl. 11 (1966), 961-964.

[197] H. Oliver, The exact Peano derivative, Trans. Amer. Math. Soc., 76 (1954), 444-456.

[198] S. Saks, On the generalized derivatives, J. Lond. Math. Soc. 7 (1932), 247-251.

[199] K. Sieklucki, A generalization of a theorem of S. Mazurkiewicz concerning Peano functions, Prace Mat. 12 (1969), 251-254.

[200] V. A. Starcev, Symmetric continuity and symmetric differentiability with respect to sets, Dokl. Akad. Nauk SSSR 185 (1969). = Soviet Math. Dokl. 10 (1969), 517-519.

[201] E. M. Stein and A. Zygmund, Smoothness and differentiability of functions, Ann. Univ. Sci. Budapest, Eötvös Sect. Math., No. 3-4, 1960-61, 295-307.

[202] C. de la Vallee Poussin, Sur l'approximation de fonctions d'une variable reelle et leurs derivees par les polynomes et les suites limitees de Fourier, Bull. Acad. Royale (1908), 193-254.

[203] C. E. Weil, On properties of derivatives, Trans. Amer. Math. Soc. 114 (1965), 363-376.

[204] C. E. Weil, On approximate and Peano derivatives, Proc. Amer. Math. Soc. 20 (1969), 487-490.

[205] A. Zygmund, Trigonometric Series, Vol. 1, Cambridge University Press, 1959.

[206] A. Zygmund, Smooth functions, Duke Math. J. 12 (1945), 47-76.

[206a] J. M. Ash, A characterization of the Peano derivative, Trans. Amer. Math. Soc. 149 (1970), 489-501.

G. SOME APPLICATIONS OF DIFFERENTIAL CALCULUS IN NORMED AND TOPOLOGICAL SPACES TO "HARD" ANALYSIS

[207] A. V. Avidzinskii, A curvilinear integral in a function space, Izv. Vyshch. Uchebn. Zaved. Matematika 1959, No. 1, 199-203.

[208] R. H. Cameron, The first variation of an indefinite Wiener integral, Proc. Amer. Math. Soc. 2 (1951), 914-924.

[209] Yu. L. Daletskii, Elliptic operators with functional derivatives and diffusion equations connected with them, Dokl. Akad. Nauk SSSR 171 (1966), 21-24. = Soviet Math. Doklady 7 (1966), 1399-1402.

[210] Yu. L. Dalteskii, Differential equations with functional derivatives and stochastic equations for generalized stochastic processes. Dokl. Akad. Nauk SSSR 166 (1966), 1035-1038. = Soviet Math. Doklady 7 (1966), 220-223.

[211] M. D. Donsker and J. L. Lions, Fréchet-Volterra variational equations, boundary value problems and function space integrals, Acta Math. 108 (1962), 147-228.

[212] R. V. Gamkrelidze and G. L. Haratišvisli, The theory of the first variation in extremal problems, Sakharth. SSSR Mech. Akad. Moambe 46 (1967), 27-31 (in Russian).

[213] R. V. Gamkrelidze, On the first variation, Kokl. Akad. Nauk, 161 (1965), 23-26. = Soviet Math. Koklady, 6 (1965), 345-348.

[214] E. V. Maïkov, τ-continuity and τ-differentiability of a functional, Dokl. Akad. USSR 155 (1964), 266-269.

[215] E. V. Maikov, τ-smooth functional and integration in functional spaces, Uspekhi Mat. Nauk 18 (1963), 243-244.

[216] P. Murav'ev, A generalized derivative and its application to ordinary differential equations, Izv. Vyss. Ucebn. Zaved. Mathematika 1 (26) (1962), 89-100.

[217] V. I. Tatarskii, The primitive functional and its application to the integration of some equations in function spaces, Uspekhi Mat. Nauk 16 (1961), 179-186.

[218] L. R. Volenvič and S. G. Gindikin, On a class of hypoelliptic polynomials, Math. of the USSR - Sbornik, 4 (1969), 369-384.

H. SOME MEASURE-THEORETIC ASPECTS OF DIFFERENTIABILITY OF FUNCTIONS

An extensive bibliography is given in Brucker and Leonard [123]. We add here a few more references, most of which appeared after [123].

[219] A. M. Bruckner, J. G. Ceder, and M. L. Weiss, On the differentiability structure of real functions, Trans. Amer. Math. Soc. 142 (1969), 1-13.

[220] L. D. Neidleman and E. G. Straus, Functions whose derivatives at one point form a finite set, Trans. Amer. Math. Soc. 140 (1969), 411-422.

[221] H. Rademacher, Partielle und Totale Differenzierbarkeit von Funktionen Mehrerer Variabeln und uber die Transformation der Doppelintegrale, Mathematische Annalen 79 (1919), 340-359.

[222] R. Rado and P. V. Reichelderfer, Continuous Transformations in Analysis, Springer-Verlag, Berlin and New York, 1955.

[223] Ju. G. Rešetnjak, Generalized derivatives and differentiability almost everywhere, Math. of the USSR - Sbornik, 4 (1969), 293-302.

[224] G. H. Sindalovskii, On total differentials, Mat. Sb., 70 (1966), 347-367. = Selected translations in mathematical statistics and probability No. 80 (1969), pp. 153-174.

[225] G. H. Sindalovskii, Derivatives of continuous functions, Math. of the USSR 2 (1968), 943-978.

[226] G. H. Sindalovskii, Differentiability relative to congruent sets. Izv. Akad. Nauk SSSR Ser. Mat. 29 (1965), 11-40.

[227] G. V. Welland, Differentiability almost everywhere of functions of several variables, Proc. Amer. Math. Soc. 19 (1968), 130-134.

[228] H. Wright and W. S. Snyder, On the differentiability of arbitrary real-valued set functions. I, Trans. Amer. Math. Soc. 146 (1969), 439-454.

II. SOME ASPECTS OF THE ROLE OF DIFFERENTIALS IN NONLINEAR FUNCTIONAL ANALYSIS AND APPLICATIONS

[229] J. Abadie, On the Kuhn-Tucker Theorem, in J. Abadie (ed.), Nonlinear Programming, North Holland Publishing Company, Amsterdam, 1967.

[230] R. Abraham and J. Robbin, Transversal Mappings and Flows, W. A. Benjamin, New York, 1967.

[231] M. Altman, A general maximum principle for optimization problems, Studia Math. 31 (1968), 319-329.

[232] M. Altman, A general maximum principle for optimization problems with operator inequalities, Bull. Un. Mat. Ital. 2 (1970), 283-292.

[233] T. Ando, On gradient mappings in Banach spaces, Proc. Amer. Math. Soc. 12 (1961), 297-299.

[234] P. M. Anselone, editor, Nonlinear Integral Equations, University of Wisconsin, Madison, 1964.

[235] P. M. Anselone and R. H. Moore, Approximate solutions of integral and operator equations, J. Math. Anal. Appl. 9 (1964), 268-277.

[236] P. M. Anselone, Collectively Compact Operator Approximations, Tech. Rep. No. 76, Computer Science Department, Stanford University, 1967.

[237] E. A. Asplund, Fréchet differentiability of convex functions, Acta Math. 121 (1968), 30-47.

[238] E. Asplund and R. T. Rockafellar, Gradients of convex functions, Trans. Amer. Math. Soc. 139 (1969), 443-467.

[239] M. S. Bazaraa, J. J. Goode, M. Z. Nashed, and C. M. Shetty, Nonlinear programming without differentiability in Banach spaces: Necessary and sufficient constraint qualifications, J. Applicable Analysis, to appear.

[240] M. S. Berger and M. S. Berger, Perspectives in Nonlinearity, W. A. Benjamin, New York, 1968.

[241] L. D. Berkovitz, Variational methods in problems of control and programming, J. Math. Anal. Appl. 3 (1961), 122-132.

[242] E. K. Blum, Minimization of functionals with equality constraints, J. SIAM Control 3 (1965), 299-316.

[243] E. K. Blum, "The Calculus of Variations, Functional Analysis, and Optimal Control Problems", pp. 417-461 in Topics in Optimization, G. Leitman, ed., Academic Press, New York, 1967.

[244] F. E. Bonsall, Lectures on Some Fixed Point Theorems of Functional Analysis, Tata Institute of Fundamental Research, Bombay, India, 1962.

[245] G. Bouligand, Introduction à la Géométrie Infinitésimale directe, Paris, 1932.

[246] J. Brønsted and R. T. Rockafellar, On the subdifferentiability of convex functions, Proc. Amer. Math. Soc. 16 (1965), 605-611.

[247] F. E. Browder, Les problèmes nonlinéaires, University of Montreal Lecture Notes, 1965.

[248] F. E. Browder, Existence and uniqueness theorems for solutions of nonlinear boundary value problems, Proc. Amer. Math. Soc. Symposium in Applied Math., Vol. 7 (1965), 24-49.

[249] F. E. Browder, Remarks on the direct method of the calculus of variations, Arch. Rational Mech. Anal. 20 (1965), 251-258.

[250] F. E. Browder, editor, Nonlinear Functional Analysis, Proc. Amer. Math. Soc. Symposium in Pure Math., Vol. 18, Part I, Providence, 1970. Part II, to appear.

[251] G. L. Cain, Jr. and M. Z. Nashed, Fixed points and stability for a sum of two operators in locally convex spaces, MRC Report No. 1115, Mathematics Research Center, University of Wisconsin, Madison, 1970.

[252] L. Cesari, Existence theorems for weak and usual optimal solutions in Lagrange problems with unilateral constraints. I, II, Trans. Amer. Math. Soc. 124 (1966), 369-430.

[253] E. W. Cheney and A. A. Goldstein, Tchebycheff approximation and related extremal problems, J. Math. Mech. 14 (1965), 87-98.

[254] (a) E. S. Citlanadze, The variational theory of a class of nonlinear operators in spaces L_p ($p > 1$). Dokl. Akad. Nauk SSSR 71 (1950), 441-444. (b) On the differentiation of functionals, Mat. Sb. 29 (7): 1 (1951), 3-12.

[255] L. Collatz, Some applications of functional analysis to analyses, particularly to nonlinear integral equations, this volume, pp. 1-43.

[256] L. Collatz, Functional Analysis and Numerical Mathematics, Academic Press, New York, 1966.

[257] R. Courant and D. Hilbert, Methods of Mathematical Physics, Vol. I, Interscience, New York, 1953.

[258] D. F. Cudia, The geometry of Banach spaces. Smoothness, Trans. Amer. Math. Soc. 110 (1964), 284-314.

[259] G. Dahlquist, Stability of error bounds in the numerical integration of ordinary differential equations, Kungl. Tekn Högsk Handl. Stockholm, No. 130 (1959), 87 pp.

[260] J. Daneš and J. Kolomý, On the continuity and differentiability properties of convex functionals, Comment. Math. Univ. Carolinae 9 (1968), 329-350.

[261] J. W. Daniel, The Approximate Minimization of Functionals, Prentice-Hall, Englewood Cliffs, 1971. (See also Daniel's paper in this volume).

[262] J. W. Daniel, Collectively compact sets of gradient mappings, Indag. Math. 30 (1968), 270-279.

[263] B. Demayanov and A. M. Rubinov, Approximate Methods of Solving Extremal Problems, Moscow, 1968, (Russian). English translation by American Elsevier, 1970.

[264] C. L. Dolph and G. J. Minty, On nonlinear integral equations of the Hammerstein type, pp. 99-154 in [234].

[265] W. Dubrovskii, Sur certaines èquations integrales non-linéaires, Uc. Zap. Moskov. Gos. Univ. 30 (1939), 49-60.

[266] A. Ya. Dubovitskiy and A. A. Milyutin, Extremum problems in the presence of constraints, Zh. Vychisl. Mat. i. Mat. Fiz. 5 (1965), 395-453.

[267] D. E. Edmunds, Remarks on non-linear functional equations, Math. Annalen 174 (1967), 233-239.

[268] R. M. Elkin, Convergence theorems for Gauss-Seidel and other minimization algorithms, Computer Sci. Report #68-59, University of Maryland, College Park, 1968.

[269] Ya. L. Engel'son, Potential operators in linear topological spaces, Latvijas Valsts Univ. Zinātu Raksti 20 (1958), 27-45.

[270] G. Fichera, Semicontinuity of multiple integrals in ordinary form, Arch. Rational Mech. Anal. 17 (1964), 339-352.

[271] T. M. Flett, On differentiation in normed vector spaces, J. London Math. Soc. 42 (1967), 523-533.

[272] T. M. Flett, <u>Mathematical Analysis</u>, McGraw-Hill, New York and London, 1966.

[273] M. D. George, Completely well-posed problems for nonlinear differential equations, Proc. Amer. Math. Soc. 15 (1964), 96-100.

[274] M. D. George, The spectrum of an operator in Banach space, Proc. Amer. Math. Soc. 16 (1965), 980-982.

[275] H. H. Goldstine, A multiplier rule in abstract spaces, Bull. Amer. Math. Soc. 44 (1938), 388-394.

[276] H. H. Goldstine, The calculus of variations in abstract spaces, Duke Math. J. 9 (1942), 811-822.

[277] H. H. Goldstine, Conditions for a minimum in abstract spaces, Illinois J. Math. 2 (1958), 111-123.

[278] A. Granas, The theory of compact vector fields and some of its applications to topology of functional spaces I, Rozpr. Matematyczne, XXX (1962), 1-93.

[279] H. Halkin and L. W. Neustadt, General necessary conditions for optimization problems, Proc. Nat. Acad. Sci. (U.S.A.) 56 (1966), 1066-1071.

[280] M. R. Hestenes, Applications of the theory of quadratic forms in Hilbert space to the calculus of variations, Pacific J. Math. 1 (1951), 525-582.

[281] M. R. Hestenes, Quadratic variational problems, in Control Theory and the Calculus of Variations, A. V. Balakrishnan, ed., Academic Press, New York, 1969.

[282] R. I. Kačurovskiĭ, The regular points, spectrum and eigenfunctions of nonlinear operators, Soviety Math. Dokl. 10 (1969), 1101-1105.

[283] R. I. Kačurovskiĭ On monotone operators and convex functionals, Uspekhi Mat. 15 (1960), 213-215.

[284] R. I. Kačurovskiĭ, Nonlinear monotone operators in Banach spaces, Uspehi Mat. Nauk 23 (1968), 121-168, (Russian). Also translated in Russian Math. Surveys 23 (1968), 117-165.

[285] M. J. Kadec, Some properties of potential operators in reflexive separable spaces. (Russian). Izv. Vysš. Učebn. Zaved. Matematika 15 (1960), 104-107.

[286] S. Karamedian, The nonlinear complementarity problem with applications, Parts 1,2, J. Optimization Th. Appl. 4 (1969), 87-98, 167-181.

[287] R. H. Kasriel and M. Z. Nashed, Stability of solutions of some classes of nonlinear operator equations, Proc. Amer. Math. Soc. 17 (1966), 1036-1042.

[288] T. Kato, Accetive operators and nonlinear evolution equations in Banach spaces, pp. 138-161 in [250], Part I.

[289] J. W. Kitchen, Concerning the convergence of iterates to fixed points, Studia Math. 27 (1966), 247-249.

[290] J. Kolomý, Some existence theorems for nonlinear problems, Comment. Math. Univ. Carolinae 7 (1966), 207-217.

[291] J. Kolomý, Remarks on nonlinear functionals, Comment. Math. Univ. Carolinae 9 (1968), 145-155.

[292] J. Kolomý, Gradient maps and boundedness of Gâteaux differentials, Comment. Math. Univ. Carolinae 9 (1968), 613-625.

[293] J. Kolomý, Differentiability of convex functionals and boundedness of nonlinear operators and functionals, Comment. Math. Univ. Carolinae 10 (1969), 91-114.

[293a] J. Kolomý, On the differentiability of operators and convex functionals, Comment. Math. Univ. Carolinae 9 (1968), 441-454.

[294] J. Kolomý, On the differentiability of mappings and convex functionals, Comment. Math. Univ. Carolinae 8 (1967), 735-752.

[295] M. A. Krasnoselskii, Some problems of nonlinear analysis, Amer. Math. Soc. Transl. (2) 10 (1958), 345-409.

[296] M. A. Krasnoselskii, Topological Methods in the Theory of Nonlinear Integral Equations, Pergamon, New York, 1964.

[297] M. A. Krasnoselskii, Positive Solutions of Operator Equations, P. Noordhoff, The Netherlands, 1964.

[298] M. A. Krasnoselskii, Two remarks on the method of successive approximations, Uspehi Mat. Nauk 10 (1955), 123-127.

[299] M. A. Krasnoselskii, P. P. Zabreiko, and R. I. Kachurovskii, On a fixed point theorem for operators in Hilbert space, Functional analiz i prilozen 1 (1967), 93-94.

[300] E. S. Levitin and B. T. Poljak, Constrained minimization methods, (Russian), Zh. vych. Mat. mat. Fiz. 6 (1966), 787-823. Translated in USSR Comput. Math. Phys., Vol. 6, 1-50.

[301] S. M. Lozinskii, Error estimates for the numerical integration of ordinary differential equations, Izv. Vyss. Ucebn. Zaved. Mathematika 5 (1958), 52-90.

[302] S. M. Lozinskii, Matrix norms and their applications to the estimation of eigenvalues of matrices and to differential equations, Vestnik Leningrad Univ. 23 (1968), 51-61.

[303] G. Lumer, Semi-inner-product spaces, Trans. Amer. Math. Sco. 100 (1961), 29-43.

[304] G. Lumer and R. S. Phillips, Dissipative operators in a Banach space, Pacific J. Math. 11 (1961), 679-698.

[305] R. H. Martin, Jr., Existence and bounds of solutions to ordinary differential equations, Doctoral Dissertation, Georgia Institute of Technology, Atlanta, 1970.

[306] R. H. Martin, Jr., The logarithmic derivative and equations of evolution in a Banach space, J. Math. Soc. Japan 22 (1970), 411-429.

[307] V. B. Melamed and A. I. Perov, A generalization of a theorem of M. A. Krasnoselskii on the complete continuity of the Fréchet derivative of a completely continuous operator, Sibirsk. Mat. Z. 4 (1963), 702-704.

[308] S. G. Mikhlin, The Problem of the Minimum of a Quadratic Functional, Holden-Day, San Francisco, 1965.

[309] S. G. Mikhlin, Variational Methods in Mathematical Physics, Pergamon Press, London, 1964.

[310] S. G. Mikhlin and K. L. Smolitskiy, Approximate Methods for Solution of Differential and Integral Equations, American Elsevier, New York, 1967.

[311] G. J. Minty, Monotone (nonlinear) operators in Hilbert space, Duke Math. J. 29 (1962), 341-346.

[312] G. J. Minty, On the monotonicity of the gradient of a convex function, Pacific J. Math. 14 (1964), 243-247.

[313] G. J. Minty, Integrability conditions for vector fields in Banach spaces, Doctoral Dissertation, University of Michigan, Ann Arbor, 1958.

[314] R. H. Moore, Differentiability and convergence for compact nonlinear operators, J. Math. Anal. Appl. 16 (1966), 65-72.

[315] R. H. Moore, Approximations to nonlinear equations and Newton's method, Numer. Math. 12 (1968), 23-34.

[316] J.J. Moreau, Fonctionelles sous-differentiables, C. R. Acad. Sci. 257 (1963), 4117-4119.

[317] J. J. Moreau, Fonctionelles convexes, multilith notes, Collège de France, Montpellier, 1966-1967.

[318] J. J. Moreau, Sous-differentiabilité, Proc. Colloquium on convexity (Copenhagen, 1965), Københavns Univ. Mat. Inst., Copenhagen, 1967, pp. 185-201.

[319] M. Z. Nashed, Supportably and weakly convex functionals with applications to approximation theory and nonlinear programming, J. Math. Anal. Appl. 18 (1967), 504-521.

[320] M. Z. Nashed and J. S. W. Wong, Some variants of a fixed point theorem of Krasnoselskii and applications to nonlinear integral equations, J. Math. Mech. 18 (1969), 767-778.

[321] M. Z. Nashed, The convergence of the method of steepest descents for nonlinear equations with variational or quasi-variational operators, J. Math. Mech. 13 (1964), 765-794.

[322] J. Nečas, Les équations elliptiques non lineaires, Czechoslovak Math. J. 19 (94) (1969), 252-274.

[323] J. Nečas and Z Poracká, On extrema of functionals, Comment. Math. Univ. Carolinae 7 (1966), 509-520.

[324] L. W. Neustadt, An abstract variational theory with applications to a broad class of optimization problems. I and II. SIAM J. Control 4 (1966), 505-527 and 5 (1967), 90-137.

[325] L. W. Neustadt, A general theory of extremals, J. Computer and System Sciences 3 (1969), 57-92.

[326] L. J. Nicolescu, On the integrability of Fréchet differentials and applications, Rev. Math. Pures Appl. 5 (1960), 625-653.

[327] L. J. Nicolescu, Montel's equation in Banach spaces, Mathematica 3 (1961), 159-169.

[328] J. M. Ortega and W. C. Rheinboldt, Iterative Solution of Nonlinear Equations in Several Variables, Academic Press, New York, 1970.

[329] A. M. Ostrowski, Solution of Equations and Systems of Equations, 2nd ed., Academic Press, New York, 1966.

[330] K. J. Palmer, On the complete continuity of differentiable mappings. J. Austral. Math. Soc. 9 (1969), 441-444.

[331] K. J. Palmer and S. Yamamuro, A note on finite-dimensional differentiable mappings, J. Austral. Math. Soc. 9 (1969), 405-408.

[332] A. I. Perov, On certain aspects of the general theory of convex functions, Ukrain. Mat. Ž. 18 (1966), 129-132.

[333] W. V. Petryshyn, Further remarks on nonlinear P-compact operators in Banach spaces, J. Math Anal. Appl. 16 (1966), 243-253.

[334] W. V. Petryshyn, Nonlinear equations involving noncompact operators, Proc. Amer. Math. Soc. Sympos. Pure Math., Vol. 18, Part 1, Providence, R. I., 1970, pp. 206-233.

[335] B. T. Poljak, Existence theorems and convergence of minimizing sequences in extremum problems with restrictions, Soviet Math. Dokl., 7 (1966), 72-75.

[336] B. T. Poljak, Semicontinuity of integral functionals and existence theorems on extremal problems, Math. USSR Sbornik 7 (1969), 59-77.

[337] L. B. Rall, Variational methods for nonlinear integral equations, pp. 155-189 in [232].

[338] R. T. Rockafellar, Characterization of the subdifferential of convex functions, Pacific J. Math. 17 (1966), 497-510.

[339] R. T. Rockafellar, Convex Analysis, Princeton Univ. Press, Princeton, N.J., 1970.

[340] R. T. Rockafellar, On the maximal monotonicity of subdifferential mappings, Pacific J. Math. 33 (1970), 209-216.

[341] E. L. Roetman, Tangent planes and differentiation, Math. Mag. 43 (1970), 1-7.

[342] E. H. Rothe, Completely continuous scalars and variational methods, Annals of Math. 49 (1948), 265-278.

[343] E. H. Rothe, Gradient mappings and extrema in Banach spaces, Duke Math. J. 15 (1948), 421-431.

[344] E. H. Rothe, A note on the Banach spaces of Calkin and Morrey, Pacific J. Math. 3 (1953), 433-439.

[345] E. H. Rothe, Some applications of functional analysis to the calculus of variations, Amer. Math. Soc. Proceedings of the Symposia in Appl. Math., VIII (1958), 143-151.

[346] E. H. Rothe, A note on gradient mappings, Proc. Amer. Math. Soc., 10 (1959), 931-935.

[347] E. H. Rothe, Remarks on the application of gradient mappings to the calculus of variations and connected boundary-value problems, Comm. Pure Appl. Math. 9 (1956), 551-568.

[348] E. H. Rothe, An existence theorem in the calculus of variations based on Sobolev's imbedding theorems, Arch. Rational Mech. Anal. 21 (1965), 151-162.

[349] E. H. Rothe, Weak topology and calculus of variations, pp. 207-237 in "Calculus of Variations, Classical and Modern, I Ciclo, Centro Internazionale Matematico Estivo (C.I.M.E.) Edizioni Cremonese, 1967.

[350] T. L. Saaty, Modern Nonlinear Equations, McGraw-Hill, New York, 1967.

[351] J. Schwartz, Nonlinear Functional Analysis, Gordon and Breach, New York, 1969.

[352] K. Srinivasacharayulu, On some non-linear problems, Canad. J. Math. 20 (1968), 394-397.

[353] F. H. Szafarniec, (a) A method of localization of implicit functions, Bull. Acad. Polon. Sci. Sér. Sci. Math. Astronom. Phys., 16 (1968), 937-942. (b) A method of estimation for the solution of integral and differential equations depending on a parameter, ibid., 16 (1968), 947-950. (c) A numerical approach to a localization problem of implicit functions, ibid., 16 (1968), 943-946.

[354] M. M. Vaĭnberg and R. I. Kačurovskiĭ, On the variational theory of nonlinear operators and equations, Dokl. Akad. Nauk 129 (1959), 1199-1202. An English translation appeared as an MRC Report No. 666, Mathematics Research Center, University of Wisconsin, Madison, August, 1966.

[355] P. P. Varaiya, Nonlinear programming in Banach space, SIAM J. Appl. Math. 15 (1967), 284-293.

[356] S. Yamamuro, The adjoints of differentiable mappings, J. Austr. Math. Soc. 8 (1968), 397-409.

[357] E. H. Zarantonello, Solving functional equations by contractive averaging, MRC Report No. 160, Mathematics Research Center, University of Wisconsin, Madison, 1960.

[358] E. H. Zarantonello, The closure of the numerical range contains the spectrum, Bull. Amer. Math. Soc. 70 (1964), 781-787. (The original expanded version appeared as a technical report, Univ. of Kansas, 1964).

[359] E. H. Zarantonello, The closure of the numerical range contains the spectrum, Pacific J. Math. 22 (1967), 575-595.

[360] E. H. Zarantonello, editor, Advances in Nonlinear Functional Analysis, Proceedings of a Symposium held at the Mathematics Research Center, April 12-14, 1971, Academic Press, New York, 1971.

[361] V. Zizler, Banach spaces with differentiable norms, Comment. Math. Univ. Carolinae 8 (1968), 415-440.

[362] V. Zizler, Some notes on various rotundity and smoothness properties of separable Banach spaces, Comment. Math. Univ. Carolinae 10 (1969), 195-206.

[363] R. A. Bonic, Four brief examples concerning polynomials on certain Banach spaces, J. Differential Geometry 2 (1968), 391-392.

Generalized Inverses, Normal Solvability, and Iteration for Singular Operator Equations

M. Z. NASHED

Introduction

The operator equation $Tx = y$ where T is a mapping on some space into another has a solution if and only if y is in the range of T. This embodies the notion of a solution in the traditional sense; it is an ideal situation.

On the other hand, one may look at the problem from a different angle. May be the reason that a certain equation does not have a solution is that our adopted <u>notion of solution</u> is not broad enough. In other words, we may not necessarily be looking at the problem in the "right" setting. The search for the "natural setting" is a recurring phenomenon in the development of mathematics. This has led, for instance, to the notions of weak and strong solutions of boundary value problems of differential equations, weak and strong extrema in the calculus of variations, etc.

The necessity to consider <u>least squares solutions</u> of linear operator equations arises in many contexts. In this paper we consider some aspects of least squares solutions of linear operator equations, both for the case when the range of the operator is closed and for the case when the range is arbitrary. Related questions for nonlinear operators are also briefly treated.

In Part I we introduce the generalized inverse for a bounded linear operator, whose range is not necessarily closed, from a point of view which depends heavily on projection operators and which preserves (and lends itself readily to) the "best approximate solution" property in the case of Hilbert spaces. Bounded generalized inverses of closed linear operators with closed range in a Banach space and, more generally, of topological homomorphisms are also defined under certain conditions.

In Part II we state several results on the convergence of the method of successive approximations (with a fixed averaging parameter), and the steepest descent and conjugate gradient methods to least squares solutions of the linear operator equation $Tx = y$.

In Part III we consider some recent results on normal solvability of nonlinear operator equations, which were obtained by Pohožaev [68-70], and indicate some directions for extending generalized inverses to nonlinear operators under certain conditions.

The results stated in Sections 5 and 6 of Part II are based on a joint work with W. J. Kammerer, whom I would like to thank for many interesting discussions.

I. Generalized Inverses of Linear Operators

1. Preliminary Remarks and a Fairy Tale

Let us consider the system of linear equations

(1.1) $\quad Ax = b, \quad \text{where } A = \begin{pmatrix} 1 & 1 \\ 1 & 1 \end{pmatrix}, \quad x = \begin{pmatrix} x_1 \\ x_2 \end{pmatrix}, \quad b = \begin{pmatrix} 1 \\ 2 \end{pmatrix}.$

This system obviously has no solution in the traditional sense; the vector b is not in the range of the matrix A. One may broaden the notion of a solution so that (1.1), and indeed any system of linear algebraic equations, has a solution in a meaningful sense. One way to do this is to replace the vector b in (1.1) by another vector $\tilde{b}$ which satisfies two requirements: $\tilde{b} \in \mathcal{R}(A)$, the range of A, and $\tilde{b}$ approximates b in some sense; for instance we may take $\tilde{b}$ to be the point in $\Re(A)$ "nearest" to b. If the usual Euclidean distance is used, then $\tilde{b}$ is obviously the orthogonal projection of b on $\Re(A)$. Then the equation

(1.2) $$Ax = P_\Re b = \tilde{b},$$

where $P_\Re$ denotes the orthogonal projection on $\Re(A)$, is solvable but the solution is not unique. Note that $\Re(A) = \mathrm{span}\{(1,1)\}$, $P_\Re b = (\frac{3}{2}, \frac{3}{2})$, and thus the set of all solutions of (1.2) is given by $S = \{(x_1, x_2): x_1 + x_2 = \frac{3}{2}\}$.

Another way of attaching a notion of solution to (1.1) is to seek a solution in the least squares sense. A vector $w \in \mathbb{R}_2$ is called a <u>least squares solution</u> of (1.1) if

(1.3) $$\inf\{\|Ax - b\|: x \in \mathbb{R}_2\} = \|Aw - b\|,$$

where $\|\cdot\|$ is some norm on $\mathbb{R}_2$. It is easy to show that corresponding to the choice of the <u>Euclidean norm</u> $\|x\| = (x_1^2 + x_2^2)^{\frac{1}{2}}$, a vector $w \in \mathbb{R}_2$ is a solution of (1.2) if and only if w is a least squares solution of (1.1). The problem of minimizing $\|Ax - b\|^2$ in the Euclidean norm is also equivalent to solving the equation

(1.4) $$A^*Ax = A^*b,$$

where A^* is the adjoint of A. ((1.4) is often called the "normal" equation by analogy with the normal equation arising in least squares problems in statistics).

If we seek a vector of minimal Euclidean norm which minimizes $\|Ax - b\|$, we get the unique solution $\hat{x} = (\frac{3}{4}, \frac{3}{4})$. This analysis leads to an analytic definition of the Moore-Penrose generalized inverse $A^\dagger$. More explicitly, $A^\dagger$ is the map on $\mathbb{R}_2$ into $\mathbb{R}_2$ which assigns to each $b \in \mathbb{R}_2$, the vector $\hat{x}$ which minimizes $\|Ax-b\|$ and has the property that $\|\hat{x}\| < \|w\|$ for all w with $\|A\hat{x} - b\| = \|Aw-b\|$. For the example considered, we thus have

$$A^\dagger = \begin{pmatrix} \frac{1}{4} & \frac{1}{4} \\ \frac{1}{4} & \frac{1}{4} \end{pmatrix}.$$

The Moore-Penrose inverse is characterized also as the unique solution X of the following equations: $XAX = X$, $(AX)^* = AX$, $AXA = A$, $(XA)^* = XA$.

The preceding simple ideas, together with geometric and algebraic manifestations thereof, constitute the pivotal point of the generalizations to linear operators on infinite dimensional spaces. According to a fairy tale, a group of matrices demanded in the year 1920 that the mathematical community put an end to the discrimination between "singular" and "nonsingular" matrices, whether they are square or rectangular. A committee was appointed and its dictum was that all matrices shall have "inverses" (an abstract of the dictum, but not the tale was given in [1]). Moore's various results on the general reciprocal were incorporated later in [2]. However, these results were not too well known; generalized inverses were defined later, independently, by Penrose [3], [4], Tseng [5], [6], [7], [8] and others. Greville [9], [10] gave an impetus to the study of pseudoinverses following the original approach of Moore. In the last few years there has been considerable interest not only in new definitions of generalized inverses of rectangular matrices and their applications to various fields, but also in the extension of the concept of a generalized inverse to various algebraic and topological structures. The literature on generalized inverses has proliferated rapidly; as a result we shall limit our bibliographical indications in the following sections to papers dealing with generalized inverses of linear operators in the setting of functional analysis, with the exception of a few references on generalized inverses of matrices which are included for historical considerations or particular applications. A comprehensive bibliography on generalized inverses of matrices and related applications was compiled recently by Lewis, Boullion, and Odell [11].

2. Generalized Inverses of Bounded Linear Operators on Hilbert Spaces

Let X and Y be two Hilbert spaces over the same field, either real or complex, and let T be a bounded linear operator mapping X into Y, whose range is not necessarily closed. We denote the range of T by $\Re(T)$, the null space of T by $n(T)$ and the adjoint of T by T^*, i.e. $<Tx,y> = <x,T^*y>$ for all $x \in X$, $y \in Y$. For any subspace S of a Hilbert space H we denote by $S^\perp$ the orthogonal complement of S and by $\bar{S}$ the closure of S. Since T is continuous, $n(T)$ is closed and thus we have the following well known and easily deduced (see, e.g., [12], [13]) orthogonal decomposition

$$(2.1) \qquad X = n(T) \oplus n(T)^\perp .$$

The adjoint T^* is also bounded and we have similarly

$$(2.2) \qquad Y = n(T^*) \oplus n(T^*)^\perp .$$

The following relations are also valid

$$(2.3) \qquad \overline{\Re(T)} = n(T^*)^\perp, \quad \Re(T)^\perp = n(T^*) ,$$

$$(2.4) \qquad \overline{\Re(T^*)} = n(T)^\perp ,$$

$$(2.5) \qquad \overline{\Re(T)} = \overline{\Re(TT^*)} ,$$

and the Closed Range Theorem holds (see Yosida [12]):

$(2.6) \qquad \Re(T)$ is closed in $Y \iff \Re(T^*)$ is closed in X.

For a closed subspace S of a Hilbert space H, we let P_S denote the <u>orthogonal projection</u> of H onto S. For simplicity, we use P for $P_{N(T)^\perp}$, the orthogonal projection of X onto $N(T)^\perp = \overline{\mathcal{R}(T^*)}$, and Q for the orthogonal projection of Y onto $N(T^*)^\perp = \overline{\mathcal{R}(T)}$. Then $Tx = TPx$ for $x \in X$ and $T^*y = T^*Qy$ for $y \in Y$. We denote the restriction of an operator L to a subset $\mathcal{K}$ of its domain by $L|\mathcal{K}$. It follows immediately that the restriction of T to $n(T)^\perp$ has an inverse, which is not necessarily continuous. We are now in a position to define the generalized inverse of a bounded linear operator whose range is not necessarily closed.

<u>Definition 1.1</u>. The generalized inverse $T^\dagger$ of the bounded linear operator $T: X \to Y$ is the linear extension of $\{T|n(T)^\perp\}^{-1}$ so that its domain of definition $\mathfrak{D}(T^\dagger)$ is $\mathcal{R}(T) + \mathcal{R}(T)^\perp$ and its null space is $n(T^*) = \mathcal{R}(T)^\perp$.

Clearly this extension is unique.
If $\mathcal{R}(T)$ is not closed, then $\mathcal{R}(T^\dagger)$ is $\overline{\mathcal{R}(T^*)}$, with $T^\dagger y \in \mathcal{R}(T^*)$ if and only if $Qy \in \mathcal{R}(TT^*)$.
On the other hand if we assume that the range of T is closed, then by the closed range theorem (2.6), $\mathcal{R}(T^*)$ is also closed and

(2.7) $\qquad \mathcal{R}(T) = n(T^*)^\perp = \mathcal{R}(TT^*)$

and

(2.8) $\qquad \mathcal{R}(T^*) = n(T)^\perp = \mathcal{R}(T^*T)$.

In this case, by the Banach open mapping theorem (see e.g., [12] or [13]), the operator $\{T|n(T)^\perp\}^{-1}$ is bounded. Also from (2.2)-(2.3), $\mathcal{R}(T) \oplus \mathcal{R}(T)^\perp = Y$ and thus the domain of $T^\dagger$ in the case of a bounded linear operator T with a closed range is all of Y, $n(T^\dagger) = \mathcal{R}(T)^\perp = n(T^*)$, $\mathcal{R}(T^\dagger) = \mathcal{R}(T^*)$, $T^\dagger$ is bounded, and the following relations are valid:

(2.9) $\qquad P = T^\dagger T$,

SINGULAR OPERATOR EQUATIONS

(2.10) $$Q = TT^\dagger ,$$

and

(2.11) $$T^\dagger \{\mathfrak{R}(T)^\perp\} = \{0\} .$$

Definition 1.1 is the same as Hestenes' definition [26] of a reciprocal. In the case of a bounded linear operator with a closed range, Definition 1.1 is the same as used by Petryshyn [14] and is obviously equivalent to the definition given by Desoer and Whalen [15], i.e. $T^\dagger$ is the operator which is uniquely characterized by the following three conditions:

(i) $T^\dagger T x = x$ for all $x \in \mathfrak{R}(T^*)$,

(ii) $T^\dagger y = 0$ for all $y \in \mathfrak{n}(T^*)$,

(iii) if $y_1 \in \mathfrak{R}(T)$ and $y_2 \in \mathfrak{n}(T^*)$, then

$$T^\dagger(y_1 + y_2) = T^\dagger y_1 + T^\dagger y_2 .$$

Petryshyn [14; Theorem 1] summarized and succinently proved the equivalence of several definitions of the generalized inverse of a bounded linear operator with closed range. In particular, the preceding definition is equivalent to the Moore-Penrose generalized inverse [3], [14]. The generalized inverse $T^\dagger$ of a bounded linear operator T with closed range is characterized as the unique solution V of the following equations:

(2.12) $$VTV = V$$

(2.13) $$(TV)^* = TV$$

(2.14) $$TVT = T$$

(2.15) $$(VT)^* = VT .$$

We now return to the case when the range of T is not closed. Then (2.7) and (2.8) are no longer valid. However, $\mathfrak{R}(TT^*) \subset \mathfrak{R}(T) \subset \overline{\mathfrak{R}(T)} = \mathfrak{n}(T^*)^\perp$ and

$$\{y \in Y: Qy \in \mathfrak{R}(T)\} = \mathfrak{R}(T) + \mathfrak{R}(T)^\perp$$

is a proper dense linear manifold in Y on which the generalized inverse is defined. The relations (2.9) and (2.11) are still valid, but the projection Q is now the <u>unique continuous extension</u> of $TT^\dagger$ to all of Y. A natural question to ask at this point is the following: What is the significance of the generalized inverse as far as the "solvability" of the operator equation

(2.16) $\qquad Tx = y, \quad y \in Y$

is concerned?

As in the case of the Moore-Penrose inverse for a bounded linear operator with closed range (see, e.g., [15], [16], [17]) a very important property of $T^\dagger$ is its relationship to the least squares solution of minimal norm of the operator equation (2.16).

<u>Definition 2.2</u>. A vector $u \in X$ is a <u>least squares solution</u> of the linear operator equation (2.16) if and only if

$$\|Tu - y\| = \inf\{\|Tx - y\|: x \in X\} \ .$$

The vector $\hat{u}$ is a <u>least squares solution of minimal norm</u> of (2.16) if and only if $\hat{u}$ is a least squares solution of (2.16) and $\|\hat{u}\| \leq \|u\|$ holds for all least squares solutions u of (2.16). A least squares solution of minimal norm is also called a <u>best approximate solution</u> (of minimal norm).

When the range of T is closed, the set of all least squares solutions of the operator equation $Tx = y$ is nonempty for any $y \in Y$, and is closed and convex, and hence has a unique element $\hat{u}$ of minimal norm; $\hat{u}$ is related to $T^\dagger$ in the following important result (which was established by Desoer and Whalen [15], Beutler [16] and others).

<u>Theorem 2.1</u>. If T is a bounded linear operator with closed range mapping X into Y, then the least squares solution $\hat{u}$ of minimal norm of the equation (2.16) is given by $\hat{u} = T^{\dagger}y$.

We now show that Theorem 2.1 generalizes to the case when the range of T is not necessarily closed and establish other characterizations of $T^{\dagger}$.

When the range of T is not closed the linear operator equation $Tx = y$ need not have a least squares solution. This will be the case when Qy is in $\overline{\mathfrak{R}(T)}$ but not in $\mathfrak{R}(T)$, because in this case $\inf\{\|Tx - y\| : x \in X\}$ is not attained for any $x \in X$. However, we have

<u>Theorem 2.2</u>. Let T be a bounded linear operator on X into Y, whose range is not necessarily closed. For each $y \in \mathfrak{D}(T^{\dagger}) = \mathfrak{R}(T) + \mathfrak{R}(T)^{\perp}$, the equation $Tx = y$ has a unique least squares solution $\hat{u}$ of minimal norm and is given by $\hat{u} = T^{\dagger}y$. The set of all least squares solutions for each $y \in \mathfrak{D}(T^{\dagger})$ is given by $T^{\dagger}y \oplus \mathfrak{n}(T)$.

<u>Proof</u>: For $y \in \mathfrak{D}(T^{\dagger})$, i.e. $Qy \in \mathfrak{R}(T)$, the equation (2.16) has a least squares solution, namely any element of the set $S = \{z \in X : Tz = Qy\}$, since for any $z \in S$,

$$\|Tz - y\| = \|Qy - y\| \le \|Tx - y\|$$

holds for all $x \in X$. The set S is also convex and closed and hence contains a unique element $\hat{u}$ of minimal norm. Noting that $S = \hat{u} \oplus \mathfrak{n}(T)$ and $\|\hat{u}\| \le \|\hat{u} + x\|$ for all $x \in \mathfrak{n}(T)$ shows that $\hat{u} \in \mathfrak{n}(T)^{\perp}$ and $Pz = \hat{u}$ for all $z \in S$. Thus

$$T^{\dagger}y = T^{\dagger}Qy = T^{\dagger}T\hat{u} = P\hat{u} = \hat{u}.$$

This proves the theorem.

Thus if u is a least squares solution of (2.16), then

(2.17) $$u = T^{\dagger}y + (I - P)x_0$$

for some $x_0 \in X$ and, conversely, every vector of the form (2.17) is a least squares solution of (2.16)

We now state a well known characterization of the set of all least squares solutions of (2.16).

Proposition 2.1. A vector u is a least squares solution of (2.16) if and only if u is a solution of the "normal" equation

(2.18) $$T^*Tx = T^*y .$$

Proof: The usual proof for both the necessity and sufficiency is based on simple algebraic manipulations (see, e.g., [18]). However, it is perhaps more instructive to appeal to the geometry of the situation. Clearly the problem of finding the least squares solutions of (2.16) is equivalent to minimizing $\|w - y\|$ over $w \in \Re(T)$. A minimizing element $\hat{w} \in \Re(T)$ is obviously characterized by the condition $y - \hat{w} \in \Re(T)^\perp$, or equivalently $y - \hat{w} \in \mathfrak{n}(T^*)$, i.e. $T^*y = T^*\hat{w}$. But $\hat{w} = Tu$ for some $u \in X$ and u is a least squares solution of (2.16). Thus $T^*y = T^*Tu$.

The set of all least squares solution can be characterized also in a different way. We associate with (2.16) the projectional equation

(2.19) $$Tx = Qy \text{ for } y \in Y .$$

Obviously, (2.19) is not equivalent to (2.16) only if $\Re(T)$ is not dense in Y, in which case the solvability of (2.19) does not necessarily imply the solvability of (2.16).

Proposition 2.2. For a fixed $y \in Y$, let $S = \{x \in X: Tx = Qy\}$ and $N = \{x \in X: T^*Tx = T^*y\}$. Then $S = N$.

Proof: Let $u \in S$, then $T^*Tu = T^*Qy = T^*y$ by definition of Q. Conversely, if $v \in N$, then $T^*Tv = T^*y = T^*Qy$, which means that $Tv - Qy \in \mathfrak{n}(T^*)$. But $Tv - Qy$ is also in $\overline{\Re(T)}$. Therefore, $Tv - Qy \in \mathfrak{n}(T^*) \cap \overline{\Re(T)} = \{0\}$.

Theorem 2.2 and Propositions 2.1 and 2.2 provide three equivalent definitions of $T^\dagger$ which may be more appropriate for certain applications than Definition 2.1.

Remark 2.1. Let X and Y be two vector spaces over the same scalars. Let T be a linear transformation on X into Y. A <u>linear</u> transformation M is called a <u>partial inverse</u> of T if TMT = T. The notion of a partial inverse appears in the work of Friedrichs [19], Hamburger [20], Sheffield [21] and others. It is an algebraic notion and, in contrast with the notion of the generalized inverse, is not connected with any topological structure of the vector spaces X and Y, or continuity properties of the operator. Moreover, every linear transformation has a partial inverse. If the equation $Tx = y$ is consistent (i.e. $Tx_0 = y$ for some x_0), then My is a solution of the given equation, since $TMy = TM(Tx_0) = (TMT)x_0 = Tx_0 = y$. Also the vector $(I - MT)z$ is a solution of the homogeneous equation $Tx = 0$ for any $z \in X$; thus $I - MT$ projects (not necessarily orthogonally) the space X onto the null space of T. Obviously every generalized inverse is a partial inverse. Properties of partial inverses and connections with generalized inverses are reviewed in [23]; we mention here the following theorem which sheds more light on the generalized inverse of a bounded linear operator with closed range.

Theorem 2.3. Let X and Y be two Hilbert spaces and T be a bounded linear operator on X into Y. Then the following statements are equivalent:

(a) T has a bounded generalized inverse;

(b) T has a bounded right inverse;

(c) all partial inverses of T are bounded;

(d) the restriction of T to $n(T)^\perp$ has a bounded inverse;

(e) the quotient space $X/n(T)$ is isomorphic with $\Re(T)$;

321

(f) the number $\gamma(T)$ defined by (*)

$$\gamma(T) = \sup\{\inf\{\|x\|: Tx = y\}: y \in \mathfrak{R}(T), \|y\| = 1\}$$

is finite;

(g) $\mathfrak{R}(T)$ is closed;

(h) $Tx = y$ is <u>normally solvable</u> in the sense of Hausdorff, i.e. for a given $y \in Y$, the necessary and sufficient condition for the equation $Tx = y$ to be solvable is that $y \in \mathfrak{n}(T^*)^\perp$;

(i) $\inf\{\|Tx - y\|: x \in X\}$ is attained for each $y \in Y$;

(j) there exists a unique operator $T^\dagger: Y \to X$ such that $T^\dagger T T^\dagger = T^\dagger$, $T^\dagger T = P$, and $TT^\dagger = Q$.

Remark 2.2. The set of all bounded linear operators with closed range includes the following classes of operators: (a) all operators which are bounded below, i.e., $\|Tx\| \geq m\|x\|$, $m > 0$; (b) all operators of the form $T = T_1 + T_2$, where $\mathfrak{R}(T_1)$ is closed and $\mathfrak{R}(T_2)$ is finite dimensional; (c) all operators of the form $T = A - \lambda L$, $\lambda \neq 0$, where A is completely continuous and L has a bounded inverse; (d) all continuous normal operators of finite descent. (Recall that an operator is said to have a finite descent if for some nonnegative integer r for which $\mathfrak{R}(T^r) = \mathfrak{R}(T^{r+k})$ for $k = 1, 2, \ldots$). On the other hand, many operators that arise in applications have nonclosed range; their prototype is a completely continuous linear operator with an infinite dimensional range. The integral operator $Tx = \int_0^1 K(\cdot, t)x(t)dt$, where $K(s, t)$ is a square integrable function on $[0,1] \times [0,1]$ does not have a closed range in $L_2[0,1]$ unless $K(s, t)$ is a

---

(*) Note that if T has a bounded generalized inverse, then
$\gamma(T) = \|T^\dagger\|$.

separable kernel.

Minamide and Nakamura [22] introduced recently the concept of the restricted generalized inverse which possesses a "constrained best approximation property" and which has applications to certain constrained minimization problems.

Let X, Y and Z be Hilbert spaces over the same scalars and let $T: X \to Y$, $S: X \to Z$ be bounded linear operators with closed ranges. Consider the product transformation (T,S) on X into $Y \times Z$ defined by $(T,S)u = (Tu, Su)$, where $Y \times Z$ is the product Hilbert space equipped with the usual inner product. Assume that the transformation (T,S) has a closed range (*) in $Y \times Z$. Since S is continuous, the null space $n(S)$ is a closed linear subspace of X and hence a Hilbert space. Denote the restriction of T to $n(S)$ by T_S. Then T_S has a closed range in Y (see footnote), and the generalized inverse $T_S^\dagger: Y \to n(S)$ is well defined. $T_S^\dagger$ is called the <u>restricted generalized inverse</u> of T with respect to $n(S)$.

The restricted generalized inverse $T_S^\dagger$ may be characterized algebraically as the unique solution C of the following equations:

$$SC = 0$$

$$CTC = C$$

$$(TC)^* = TC$$

$$TCT = T \text{ on } n(S)$$

$$P_{n(S)}(CT)^* = CT \text{ on } n(S)$$

(Compare with (2.12) - (2.15)).

(*) This is equivalent to assuming that the image of the null space $n(S)$ (respectively $n(T)$) under T (respectively S) is closed in Y (respectively Z). This assumption is redundant if, for example, the inclusion $n(S) \subset n(T)$ or $n(T) \subset n(S)$ holds.

The restricted generalized inverse possesses the following constrained best approximate solution property. Let $y \in Y$ and $\hat{u} = T_S^\dagger y$. Then

(2.20) $\quad S\hat{u} = 0$

(2.21) $\quad \|T\hat{u} - y\| \leq \|Tu - y\|$ for all $y \in n(S)$

(2.22) $\quad \|\hat{u}\| \leq \|u\|$ for all $u \in n(S)$ such that

$$\|T\hat{u} - y\| = \|Tu - y\|.$$

(2.20)-(2.22) show that $\hat{u}$ possesses the best approximate solution property <u>over $n(S)$</u>. As an application, consider the following constrained minimization problem: For $z \in \mathcal{R}(S)$ and $y \in Y$, find an element $\hat{u} \in X$ satisfying $S\hat{u} = z$ such that (i) $\|T\hat{u} - y\| \leq \|Tu - y\|$ for all $u \in \{u: Su = y, u \in X\}$ and (ii) $\|\hat{u}\| \leq \|u\|$ for all such u with $\|T\hat{u} - y\| = \|Tu - y\|$. It is easy to show that this problem has a unique solution which is given by

$$\hat{u} = T_S^\dagger(y - TS^\dagger z) + S^\dagger z.$$

Some results dealing with explicit computation of the restricted generalized inverse, as well as an application to control problems are given in [22].

Restricted generalized inverses for bounded linear operators with a nonclosed range can also be defined. The development of their properties, and in particular the "constrained best approximate property", require only minor technical modifications in view of the framework in which generalized inverses were developed in this section. (See also [23]).

3. Generalized Inverses of Linear Topological Homomorphisms and Closed Linear Operators with Closed Range on Banach Spaces

The theory of differential operators gives rise usually to closed unbounded operators on a Banach space. Thus the preceding development of the generalized inverse does not apply to such operators. In this section, the notion of a generalized inverse is extended so that under certain conditions a closed transformation with a closed range on a Banach space has a <u>continuous</u> generalized inverse defined on the whole space. We also consider generalized inverses of linear topological homomorphisms. The generalized inverse for a continuous linear operator from one topological space to another has been first studied by Votruba [17]. Some aspects of generalized inverses in Banach spaces have been considered by Sheffield [21], Beutler [16; I], Reid [24; pp. 19-21], and implicitly by Wyler [25] in the setting of so-called "Green's operators". Various routes to generalized inverses of unbounded linear operators in a Hilbert space setting were taken by Tseng [5], [6], [7], [8], Hestenes [26], Beutler [16; II] and others. The material of this and the following section is not needed for the rest of this paper; consequently we shall only indicate the main results referring the interested reader to [23] and the cited literature.

Let X and Y be <u>linear topological spaces</u> over the real or complex numbers. If X is a <u>direct sum</u> of $\mathfrak{m}$ and $\mathfrak{n}$, the function $(x,y) \to x+y$ mapping $\mathfrak{m} \times \mathfrak{n}$ into X is continuous but is not necessarily a homeomorphism. X is said to be a <u>topological direct sum</u> of $\mathfrak{m}$ and $\mathfrak{n}$ if this function is a homeomorphism. In this case, we write $X = \mathfrak{m} \oplus \mathfrak{n}$ and call $\mathfrak{n}$ a <u>topological supplement</u> of $\mathfrak{m}$ in X. It is easy to show that if X is Hausdorff, then $\mathfrak{m}$ and $\mathfrak{n}$ are closed.

A <u>projector</u> on a linear topological space is a continuous linear idempotent (i.e., $P^2 = P$) operator. Every projector P on X induces a decomposition of X into two topological supplements $P(X)$ and $(I-P)(X)$ and

$$X = P(X) \oplus (I-P)(X) ,$$

where I is the identity operator. On the other hand, a subspace $\mathfrak{M}$ of X has a (not necessarily unique) topological supplement if and only if there exists a projector P: $X \to X$ such that $P(X) = \mathfrak{M}$. (See [12, p. 242], [13]).

A linear mapping T: $X \to Y$ is called a <u>topological homomorphism</u> if the image of every open set in X is an open set in T(X) in the induced topology.

Let T: $X \to Y$ be a topological homomorphism with the property that $\mathfrak{n}(T)$ has a topological supplement in X and $\mathfrak{R}(T)$ has a topological supplement in Y. Let P and Q be given projectors onto $\mathfrak{n}(T)$ and $\mathfrak{R}(T)$ respectively. Then it can be shown that there exists a unique topological homomorphism $T^\dagger$ on Y into X satisfying the following conditions:

(3.1) $\qquad TT^\dagger = Q \quad$ on Y,

(3.2) $\qquad T^\dagger T = I - P \quad$ on $\mathfrak{D}(T)$,

and

(3.3) $\qquad T^\dagger TT^\dagger = T^\dagger \quad$ on Y.

Furthermore, $T^\dagger$ is continuous. Conditions (3.1)-(3.3) imply that

(3.4) $\qquad T^\dagger TT^\dagger = T \quad$ on $\mathfrak{D}(T)$.

It should emphasized that the uniqueness of $T^\dagger$ is <u>relative to given</u> P and Q.

<u>Definition 3.1</u>. The unique continuous linear mapping $T^\dagger$ determined by (3.1)-(3.3) is called <u>the generalized inverse of</u> T <u>relative to the projectors</u> P <u>and</u> Q <u>and is denoted by</u> $T^\dagger_{PQ}$.

It follows easily that T is onto if and only if $T^\dagger_{PQ}$ is a right inverse for T and that $T^\dagger_{PQ}$ is a left inverse for T if and only if $\mathfrak{n}(T) = \{0\}$. In particular if T is a topological isomorphism, then $T^\dagger = T^{-1}$. If $T^\dagger_{PQ}$ exists, then $(T^\dagger_{PQ})^\dagger_{I-Q,I-P}$ also exists and is equal to T. In particular, in the Hilbert space setting if we take P and Q to be the <u>orthogonal</u> projectors onto $\mathfrak{n}(T)$ and $\mathfrak{R}(T)$ respectively, then $(T^\dagger)^\dagger = T$.

As an example, suppose T is a projection mapping, then since $T^2 = T$, we may take $P = I - T^2$ as a projector on $\mathfrak{n}(T)$ and $Q = T^2$ as a projector on $\mathfrak{R}(T)$. Relations (3.1)-(3.3) are then satisfied by $T^\dagger = T$. As another example, suppose f is a continuous linear functional on a Banach space X. Then $\mathfrak{n}(f)$ is a closed subspace and $X = \mathfrak{n}(f) \oplus \text{span}\{x_0\}$, where $f(x_0) = 1$. The choice of x_0 determines a projection onto $\mathfrak{n}(f)$ and relative to this projection $f^\dagger$ is the map from $\mathbb{R}$ into X taking λ to λx_0.

We note that in Banach spaces, closed linear operators with closed ranges are homomorphism ([27], Theorem 11), but homomorphism with closed ranges may be nonclosed operators. Now let X and Y be Banach spaces. A linear operator with domain $\mathfrak{D}(T)$ in X and range in Y is said to be closed if its graph $\{(x, Tx): x \in \mathfrak{D}(T)\}$ is closed in $X \times Y$. It follows immediately that the null space of a closed linear operator is a closed subspace in X. We then have the following theorem (see also [24, p. 19]).

<u>Theorem 3.1.</u> Let X and Y be Banach space, and T be a closed linear operator with domain $\mathfrak{D}(T) \subset X$ and range $\mathfrak{R}(T) \subset Y$ such that

(a) $\mathfrak{R}(T)$ is a closed subspace in Y;

(b) there exists a closed subspace $\mathfrak{S}$ in Y such that $Y = \mathfrak{R}(T) + \mathfrak{S}$;

(c) there exists a closed subspace $\mathfrak{M}$ in X such that $X = \mathfrak{n}(T) \oplus \mathfrak{M}$.

Then T has a continuous generalized inverse $T^\dagger$ on Y into X.

Remarks. 1. The setting of this section applies in particular to the so-called <u>Fredholm operators</u>, i.e. closed linear operators $T: X \to Y$ with closed range for which both $n(T)$ and the quotient space $Y/\Re(T)$ are finite dimensional. In particular, the preceding theory applies to operators of the form $T = \lambda I - \mathcal{K}$, where $\mathcal{K}$ is a compact linear operator on a Banach space X into itself. On the other hand, Theorem 3.1 does not apply for instance to integral operators of the first kind since, if the kernel is nondegenerate, the range of such operators is nonclosed. Applications of generalized inverses to integral equations of the first and second kinds are given in Kammerer and Nashed [28].

2. The generalized inverse in Banach spaces does not possess the "best approximate solution" property which gives the generalized inverse in Hilbert spaces its unique role in problems of prediction theory, optimal control, etc. The "best approximate solution" property is intimately connected with the inner product structure <u>and the choice of orthogonal</u> (equivalently self-adjoint) <u>projections</u> for P and Q; it does not even hold for matrices if one uses a norm which is not induced by the inner product. On the other hand, for a given linear operator $T: X \to Y$, where X and Y are Banach spaces, if we can embed X and Y in two Hilbert spaces H_1 and H_2 respectively, such that T considered as a mapping on the domain $X \subset H_1$ into H_2 has a generalized inverse $T^\dagger$ with $\Re(T^\dagger) \subset X$, then we can still consider least squares solutions of the equation $Tx = y$, $y \in Y$ with the least squares property taken in the norm of H_2.

3. The definition of a generalized inverse relative to projections P and Q, even in the case of operators on Hilbert spaces, seems to be a natural general definition leading to a <u>class</u> of generalized inverses. Particular generalized inverses, corresponding to specific choices of P and Q, may then be chosen which have the desired properties for concrete applications. For an interesting and well motivated application to dynamics which calls for a specific choice of nonorthogonal P and Q, see Milne [29]. It may be noted

in this connection that the setting of this section includes what is called in [29] an oblique generalized inverse, and, in view of Section 5.2 in [29], the generalized inverse defined by Scroggs and Odell [30].

4. The existence of complementary subspaces plays a decisive role in the definitions of generalized inverses. Fundamental questions in the theory of complementary manifolds and projections were first considered in a classic paper of Murray [31].

4. Other Definitions of Generalized Inverses and Miscellaneous Remarks

The remarks made on page 140 of the preceding paper in this volume [36] seem to apply also to the area of generalized inverses. In most of the papers on generalized inverses of linear operators that have appeared in the last two decades, the topic has been presented de novo, as observed also in [24], leading to different definitions of generalized inverses. Most of these definitions turned out to be equivalent in the case of bounded linear operators with closed range. In this section, we comment briefly on some of the approaches to generalized inverses which have not been considered in the preceding two sections, with particular reference to the case of unbounded operators.

4.1. The earliest work explicitly devoted to the study of generalized inverses of linear operators is that of Tseng [5], [6], [7], [8] (see also Ben-Israel and Charnes [32] for a summary of some of Tseng's results and a comprehensive treatment of several aspects of generalized inverses, together with an extensive bibliography up to 1962). Let X and Y be two Hilbert spaces. Tseng defined a generalized inverse of a linear operator $T: \mathcal{D}(T) \to Y$, where $\overline{\mathcal{D}(T)} = X$, as an operator T^g with $\overline{\mathcal{D}(T^g)} = Y$, satisfying the following relations

(4.1) $\qquad \mathcal{R}(T) \subseteq \mathcal{D}(T^g), \quad \mathcal{R}(T^g) \subseteq \mathcal{D}(T)$,

(4.2) $$TT^g = P_{\overline{R(T)}}, \quad T^gT = P_{\overline{R(T^g)}}.$$

A necessary and sufficient condition for a linear operator T with $\mathcal{D}(T) = X$ to have a generalized inverse is that

(4.3) $$\mathcal{D}(T) = \mathfrak{n}(T) \oplus \{\mathfrak{n}(T)^\perp \cap \mathcal{D}(T)\}.$$

If T has a generalized inverse, then it has a unique <u>maximal generalized inverse</u>, denoted by $T^\dagger$ (i.e. every other generalized inverse T^g is a restriction of $T^\dagger$) with

(4.4) $$\mathcal{D}(T^\dagger) = \mathfrak{R}(T) \oplus \mathfrak{R}(T)^\perp,$$

(4.5) $$\mathfrak{n}(T^\dagger) = \mathfrak{R}(T)^\perp.$$

Furthermore, $T^\dagger$ is the only generalized inverse of T with a closed null space.

Tseng [7,8] also considered the relationship between the generalized inverse of T and least squares solutions (which he called virtual solutions) of the linear equation

(4.6) $$Tx = y, \quad \overline{\mathcal{D}(T)} = X, \quad y \in Y,$$

and gave the following criterion: A necessary and sufficient condition for (4.6) to have a least squares solution (Definition 2.2) is that there exists a constant γ such that

(4.7) $$|<y, u>|^2 \leq \gamma <u, TT^*u>$$

for every $u \in \mathcal{D}(TT^*) \ominus \mathfrak{n}(TT^*)$. If (4.7) is satisfied, then $y \in \mathcal{D}(T^\dagger)$ and the vector $\hat{x} = T^\dagger y$ is a least squares solution of minimal norm.

It may be noted in passing that in the case of a bounded linear operator, condition (4.3) holds, and (4.7) is satisfied for any $y \in \mathcal{R}(T) + \mathcal{R}(T)^\perp$. Tseng's maximal generalized inverse in this case is equivalent to our Definition 1.1. The author feels, however, that the approach taken in Section 1 is more beneficial, and has the same advantages as the approach taken in [15], for bounded linear operators with closed range, in that the definition has a strong motivation, the concepts are illuminated geometrically and the proofs are quite simple.

4.2. An interesting function-analytic approach to generalized inverses of bounded linear operators on a Banach space as well as unbounded closed linear operators, whose range is not necessarily closed, is given in the comprehensive papers of Beutler [16]. The basic tool in Beutler's approach is a representation theorem giving a necessary and sufficient condition for a bounded linear operator $T: B_1 \to B_2$, where B_1 and B_2 are Banach spaces, to admit a decomposition in the form

$$(4.8) \qquad T = P_{\mathcal{R}(T)} \widetilde{T}$$

where $\widetilde{T}$ is a bounded linear operator from B_1 onto B_2 with a bounded inverse. The existence of such a representation of T implies that $\mathcal{R}(T)$ is closed and has a closed complement $\mathcal{S}$, and $\mathfrak{n}(T)$ has a closed complement $\mathfrak{M}$, such that $\widetilde{T}$ maps $\mathfrak{n}$ onto $\mathcal{S}$ and is one-to-one. If T has a representation (4.8), Beutler calls $\widetilde{T}^{-1}$ a <u>semi-inverse</u> of T, which we denote by T^∂; for a given operator T, there is a (nondenumerable) infinity of semi-inverses, unless T is itself invertible. It is easy to verify that a semi-inverse is a partial inverse, i.e. it satisfies the identity $TT^\partial T = T$. The operator $T^\dagger = P_\mathfrak{M} T^\partial$ is called in [16] a <u>pseudo-inverse</u> of T. In the case of Hilbert spaces, T^∂ and $T^\dagger$ are related to least squares solutions of the equation $Tx = y$. Let H_1 and H_2 be Hilbert spaces and T a bounded linear operator from

H_1 into H_2. If T has a representation as in (4.8), then for any $y \in H_2$, $T^\partial y$ is a least squares solution. The set of all least squares solutions is given by $S = \{x: Tx = TT^\partial y\}$, and $\inf\{\|x\|: x \in S\} = \|T^\dagger y\| = \|P_\mathfrak{m} T_y^\partial\|$.

Turning to the general case of a linear operator $T: H \to H$, on the Hilbert space H, $T^\dagger$ is said to be a generalized inverse if (a) $\mathfrak{D}(T^\dagger)$ is dense in H; (b) for every $y \in \mathfrak{D}(T^\dagger)$, $\inf\{\|Tx-y\|: x \in \mathfrak{D}(T)\}$ is attained by $\hat{x} = T^\dagger y$ and $\|\hat{x}\| < \|u\|$ for all $u \in \mathfrak{D}(T)$ which attain the preceding infimum (for given $y \in \mathfrak{D}(T^\dagger)$), $u \neq \hat{x}$. Beutler [16; II, Theorems 8 and 9] establishes the following two theorems:

<u>Theorem 4.1.</u> Let T be a closed linear operator with domain $\mathfrak{D}(T) \subset H$ and whose range $\mathfrak{R}(T)$ is closed. Let $H = \mathfrak{R}(T) \oplus S$, and let $\mathfrak{m}$ denote the orthogonal complement of $\mathfrak{n}(T)$. Let $\mathfrak{n}$ and S be of the same dimension. Then T has a generalized inverse $T^\dagger = P_\mathfrak{m} \tilde{T}^{-1}$, where $\tilde{T}$ may be taken as any operator having a bounded everywhere defined inverse and satisfying $T = P_{\mathfrak{R}(T)} \tilde{T}$ (the existence of such representations of T are guaranteed by the hypotheses of the theorem). Furthermore, $T^\dagger$ is bounded, everywhere defined, and satisfies the equations

$$TT^\dagger = P_{\mathfrak{R}(T)} \text{ for all } y \in H,$$

and

$$T^\dagger T = P_\mathfrak{m} \text{ for all } x \in \mathfrak{D}(T).$$

<u>Theorem 4.2.</u> Let T be a closed linear operator with domain $\mathfrak{D}(T) \subset H$ and whose range $\mathfrak{R}(T)$ is not necessarily closed. Let $\mathfrak{n}(T)$ and $\mathfrak{R}(T)^\perp$ be of the same dimension. Then there exists a generalized inverse of T, and an operator $T^\dagger = P_\mathfrak{m} \tilde{T}^-$ is a generalized inverse whenever $\tilde{T}$ is an invertible closed operator with dense range satisfying $T = P_{\mathfrak{R}(T)} \tilde{T}$. Furthermore, $T^\dagger$ is unbounded and satisfies

$$TT^\dagger = P_{\mathfrak{R}(T)} \text{ for all } y \in \mathfrak{D}(T^\dagger).$$

As remarked in Section 2, Beutler's definition is equivalent to Definition 1.1 in the case of a bounded linear operator with closed range. Beutler's definition of a generalized inverse for an operator with nonclosed range does not lead to a unique generalized inverse since it does not preclude various generalized inverses defined on different dense sets. However, if T' and T'' are two generalized inverses of T, then

$$T'y = T''y \text{ for } y \in \mathcal{D}(T') \cap \mathcal{D}(T'').$$

In view of Theorem 2.2, Definition 2.1 is equivalent to Beutler's definition if one takes $\mathcal{D}(T^\dagger) = \mathcal{R}(T) + \mathcal{R}(T)^\perp$ in the latter definition. Connections with other definitions are given in [23], where $\mathcal{D}(T)$ is not required to be dense.

4.3. It is perhaps appropriate to mention in this Advanced Seminar on Nonlinear Functional Analysis that historically notions of generalized or pseudo inverses appeared first in the context of analysis (differential and integral equations) long before their widespread use in matrix theory. More specifically, the germ of these notions may be found in the celebrated paper of Fredholm [33] "pseudo inverse"[*] of a transformation ..., the work of Hurwitz [34] on "pseudo-resolvent to the kernel of an integral equation", the work of Hilbert [35], Elliott [37], Reid [38] and others on generalized Green's functions and generalized Green's matrices. See Reid [24] for a vivid historical account of these aspects of generalized inverses. We refer also to [24] for a motivated and informal discussion of the interrelations between the concepts of a generalized Green's matrix for a differential system and the generalized inverse of a finite matrix. Generalized Green's functions and Green's matrices for differential systems were also studied by Greub and Rheinboldt [39], Wyler [40], Bradley [41], [42]. Generalized inverses of linear operators have been used recently in connection with generalized Green's functions and other aspects of differential

---

[*] The terms in quotations following the references are terms used by these authors.

equations by Loud [43], [44], Reid [45], [46], Wyler [25], Landesman [47], Kallina [48], Halany and Moro [49] and others.

II. Some Iterative Methods for Least Squares Solutions of Singular Linear Operator Equations

In the following three sections we shall use the terminology and notation of Section 2. Therefore it is convenient to distinguish the two cases (i) $\Re(T)$ is closed and (ii) $\Re(T)$ is not closed; in the former, stronger results and sharper error bounds are obtained.

5. The Method of Steepest Descent

Let X and Y be Hilbert spaces and T a bounded linear operator on X into Y. We associate with the linear operator equation

(5.1) $$Tx = y$$

the quadratic functional

(5.2) $$J(x) = \frac{1}{2} \|Tx - y\|^2 .$$

The method of steepest descent for minimizing $J(x)$ (see, for instance, Kantorovich and Akilov [51]) is defined by the sequence

(5.3) $$x_{n+1} = x_n - \alpha_n \text{ grad } J(x_n) ,$$

where α_n is chosen to minimize $J(x_{n+1})$. Since grad $J(x) = T^*Tx - T^*b$, the sequence (5.3) may be written in the form

(5.4) $$x_{n+1} = x_n - \alpha_n r_n \quad \text{for} \quad n = 1, 2, \ldots$$

where x_0 is any initial approximation,

(5.5) $$r_n = T^*(Tx_n - y)$$

and

(5.6) $$\alpha_n = \frac{\|r_n\|^2}{\|Tr_n\|^2}.$$

The first proof of the convergence of the method of steepest descent in the setting of functional analysis was given by Kantorovich [50] for positive definite bounded linear operators in Hilbert space. Kantorovich also generalized the method to unbounded self-adjoint and positive definite operators. In [52] the author extended Kantorovich's result to <u>least squares</u> solutions of linear operator equations (which need not necessarily have solutions) and proved the following

Theorem 5.1. Let $T: X \to Y$ be a bounded linear operator which has a closed range. Then the sequence $\{x_n\}$ generated by the method of steepest descent (5.4) - (4.6) converges for $y \in Y$ and any initial approximation $x_0 \in X$ to the vector $T^\dagger y + (I-P)x_0$, a least squares solution of (5.1) and

(5.7) $$\|T^\dagger y + (I-P)x_0 - x_n\| \leq C\left[\frac{M-m}{M+m}\right]^n,$$

where C is a constant and

(5.8) $$m\|x\|^2 \leq \langle T^*Tx, x\rangle \leq M\|x\|^2, \quad x \in \mathcal{R}(T^*)$$

with $0 < m \leq M < \infty$ [#]

[#] If $\mathcal{R}(T)$ is closed, then it follows easily from the open mapping theorem that the restriction of T to $\mathcal{R}(T^*)$ has a bounded inverse; equivalently (5.8) holds. {In Remark 1, p. 362 in [52], $\langle Tx, x\rangle$ should be $\langle T^*Tx, x\rangle$}.

Crucial to the proof of this theorem is the fact that the restriction of T^*T to the closed subspace $\Re(T^*)$ is positive definite. However $T^*T|\Re(T^*)$ need not be positive definite when $\Re(T)$ is not closed. The next theorem (see Kammerer and Nashed [59] for a proof) states that with the restriction $Qy \in \Re(TT^*)$, the method of steepest descent will still converge, starting with $x_0 = 0$, to the least squares solution of minimal norm even when $\Re(T)$ is not closed.

Theorem 5.2. Let X and Y be Hilbert spaces over the real or complex numbers and let T be a bounded linear operator on X into Y. Let Q denote the orthogonal projection of Y onto $n(T^*)^\perp = \overline{\Re(T)}$. If $Qy \in \Re(TT^*)$, then the sequence $\{x_n\}$ generated by the method of steepest descent (5.4)–(5.6) with initial vector $x_0 = 0$ converges monotonically to the least squares solution of minimal norm $\hat{x} = T^\dagger y$. In fact

$$\|x_n - T^\dagger y\|^2 \leq \frac{\|T\|^2 \|T^\dagger y\|^2 \|(TT^*)^\dagger y\|^2}{\|T\|^2 \|(TT^*)^\dagger y\|^2 + n \|T^\dagger y\|^2}, \quad n = 1, 2, \ldots.$$

Note that $\Re(TT^*)$ is a dense linear manifold in the domain of $T^\dagger$ since $\mathcal{D}(T^\dagger) = \Re(T) + \Re(T)^\perp$. We conjecture that the method of steepest descent converges for any y such that $Qy \in \mathcal{D}(T^\dagger)$, to a least squares solution of (5.1).

6. The Conjugate Gradient Method

We now consider the conjugate gradient method of Hestenes and Stiefel [53], [54] for minimizing the functional $J(x)$ defined by (5.2). That is, we choose an initial approximation $x_0 \in X$ and compute $r_0 = p_0 = T^*(Tx_0 - y)$. If $p_0 \neq 0$, we compute $x_1 = x_0 - \alpha_0 p_0$, where $\alpha_0 = \dfrac{\|r_0\|^2}{\|Tp_0\|^2}$. For $i = 1, 2, \ldots$ we compute

(6.1) $$r_i = T^*(Tx_i - y) = r_{i-1} - \alpha_{i-1} T^* T p_{i-1},$$

where

(6.2) $$\alpha_{i-1} = \frac{\langle r_{i-1}, p_{i-1} \rangle}{\|Tp_{i-1}\|^2},$$

and if $r_i \neq 0$, then we compute

(6.3) $$p_i = r_i + \beta_{i-1} p_{i-1}, \quad \text{where} \quad \beta_{i-1} = \frac{\langle r_i, T^* T p_{i-1} \rangle}{\|Tp_{i-1}\|^2}$$

and set

(6.4) $$x_{i+1} = x_i - \alpha_i p_i.$$

The convergence of this method has been established by Hestenes and Stiefel for matrix equations and by Hayes [55], Antosiewicz and Rheinboldt [56] and Daniel [57] when T is a bounded linear operator on X into Y, with a bounded inverse. Kammerer and the author have investigated the convergence of the conjugate gradient method when the assumptions that T be onto and that T^{-1} exist have been removed. In particular the following two theorems are established in [58].

Theorem 6.1. Let X and Y be Hilbert spaces over the real field and let T be a bounded linear operator on X into Y. If the range of T is closed, then the conjugate gradient method (6.1)-(6.4) converges monotonically to $u = T^\dagger y + (I - P)x_0$, a least squares solution of (5.1), for each initial vector $x_0 \in X$. Moreover, if m and M are as in (5.8), then

$$\|x_i - u\|^2 \le \frac{g(x_0)}{m}\left(\frac{M-m}{m+m}\right)^{2i}, \quad i = 1, 2, \ldots$$

where $g(x) = \|Tx - Qy\|^2$.

Theorem 6.2. Let $T: X \to Y$ be a bounded linear operator whose range is not necessarily closed. Let Q denote the orthogonal projection of Y onto $\mathfrak{n}(T^*)^\perp$. If $Qy \in \mathfrak{R}(TT^*T)$, then the conjugate gradient method with initial approximation $x_0 \in \mathfrak{R}(T^*T)$ converges monotonically to the least squares solution of minimal norm $u = T^\dagger y$. In fact

$$\|x_n - T^\dagger y\|^2 \le \frac{\|T\|^2 \|x_0 - T^\dagger y\|^2 \|(T^*)^\dagger x_0 - (TT^*)^\dagger y\|^2}{\|T\|^2 \|(T^*)^\dagger x_0 - (TT^*)^\dagger y\|^2 + n\|x_0 - T^\dagger y\|^2}.$$

As a corollary to Theorem 6.1 we may easily prove the following

Theorem 6.3. Let X and Y be Hilbert spaces over the same field. Let T be a linear transformation on X into Y with rank $T = \dim \mathfrak{R}(T) = r$. Then the conjugate gradient method associated with the equation $Tx = y$ converges in at most r steps to the least squares solution $u = T^\dagger y + (I - P)x_0$ for any $x_0 \in X$.

7. Series and Integral Representations of Generalized Inverses, and the Successive Approximations Method for Singular Linear Operator Equations

The method of successive approximations is a "medieval iterative algorism" (see Kennedy and Transue [60] for some interesting historical aspects) which was a standard expedient of ancient astronomy. It was used for extraction of roots of simple polynomial equations about 200 B.C. In a more complicated setting, the method had cropped up in the construcstruction of astronomical tables which are based on the computation of $\Phi(t) = k \sin \theta(t)$, where θ is a solution of the equation $\theta - m \sin \theta = t$. It is stated in [60] that a recursion relation for computing Φ was given in the ninth century by Habash al-Hāsib al-Marwazī (i.e., The Computer from Merv, in Turkestan) in the form $\theta_0(t) = t + m \sin t$, ..., $\theta_n(t) = t + m \sin \theta_{n-1}(t)$, $n = 1, 2, ...$ (of course, the processes were written out as verbal statements). Habash gave no proof of convergence, contenting himself with the statement that the solution is $\Phi = k \sin \theta_3(t)$.

We shall indicate in this section some applications of the method of successive approximations (with a fixed averaging parameter) to the computations of generalized inverses and least squares solutions of linear operator equations.

Let X and Y be Hilbert spaces over the same scalars and let T be a bounded linear operator on X into Y. Assume that $n(T) \neq \{0\}$ and let α be any positive real number. Then $(I - \alpha T^*T)^n$ converges to $I - P$ in the operator norm if and only if $\mathcal{R}(T)$ is closed and $0 < \alpha < \dfrac{2}{\|T\|^2}$ (see [14]).

The optimal value of α is $\alpha_b = \dfrac{2}{\gamma^2 + \|T\|^2}$, where

$$\gamma = \text{g.l.b}\left\{\dfrac{\|Tx\|}{\|x\|} : x \in n(T)^\perp, x \neq 0\right\} = \dfrac{1}{\|T^\dagger\|},$$

with the error estimate

(7.1) $\quad \|(I - \alpha_b T^*T)^n - (I - P)\| \leq \left\{\dfrac{\gamma^2 - \|T\|^2}{\gamma^2 + \|T\|^2}\right\}^n.$

Furthermore, if T has a <u>closed range</u>, one easily obtains a Neumann-type series expansion for $T^\dagger$, namely,

(7.2) $\quad T^\dagger = \sum_{n=0}^{\infty} \alpha(I - \alpha T^*T)^n T^*,$ for $0 < \alpha < \dfrac{2}{\|T\|^2}.$

For details see [14], [17], [61], [32]. Note that T^* may not be factored out from the series in (7.2).

Remarks. 1. The series representation (7.2) is analogous to the Neumann expansion for the inverse of a nonsingular bounded linear operator T on a Hilbert space:

$$T^{-1} = \sum_{n=0}^{\infty} \alpha(I - \alpha T^*T)^n T^*, \quad \text{for} \quad 0 < \alpha < \frac{2}{\|T\|^2}.$$

2. The Neumann series expansion (7.2) for $T^\dagger$ can be replaced by a more rapidly convergent series, making use of Euler's identity

$$\frac{1}{1-x} = (1+x)(1+x^2)(1+x^4)\ldots(1+x^{2^k})\ldots, \quad \text{for} \quad |x| < 1.$$

In particular the nth partial product

$$\prod_{k=0}^{n-1}(1+x^{2^k}) = \sum_{k=0}^{2^n-1} x^k,$$

so that if $0 < \alpha < 2\|T\|^{-2}$ and

$$T_n^\dagger = \alpha\{I + (I - \alpha T^*T)\} \prod_{k=1}^{n-1}\{I + (I - \alpha T^*T)^{2^n}\} T^*,$$

then it is not hard to show using (7.2) that

$$\|T^\dagger - T_n^\dagger\| \leq \frac{(1 - \alpha\|T\|^2)^{2^n}}{\|T\|}.$$

A hierachy of other rapidly convergent series representations for $T^\dagger$ may be constructed by using generalizations of Euler's identity as suggested by Lonseth [62] for representations of T^{-1}, when T is invertible.

3. A rapidly convergent series representation for $T^\dagger$ can also be constructed as follows. Pick $0 < \alpha < \dfrac{2}{\|T\|^2}$, and define $A_0 = \alpha T^*$, $A_{n+1} = 2A_n - A_n T A_n$. Then

(7.3) $$\|T^\dagger - A_n\| \leq \|T^\dagger\| \beta^{2^n},$$

where

$$\beta = \max\{|1 - \alpha\|T\|^2|,\ |1 - \alpha\|T^\dagger\|^{-2}|\}$$

(see also [61]).

4. An asymptotic integral representation for $T^\dagger$ was given by Showalter [61]. Define for $t \geq 0$,

(7.4) $$T^\dagger(t) = \int_0^1 \exp[-T^*T(t-s)] T^* ds,$$

then

(7.5) $$\|T^\dagger - T^\dagger(t)\| \leq \|T^\dagger\| \exp(-t\|T^\dagger\|^{-2}).$$

This result remains valid if T is an unbounded linear operator such that T^*T is a closed operator with dense domain and $\|T^\dagger\| < \infty$.

5. The generalized inverse of a bounded linear operator $T: X \to Y$ with closed range can also be represented as a limit:

(7.6) $$T^\dagger = \lim_{\lambda \to 0^+} T^*(TT^* + \lambda I)^{-1} = \lim_{\lambda \to 0^+} (T^*T + \lambda I)^{-1} T^*$$

where convergence is in the operator norm. This form is quite suitable for regularization of ill-conditioned problems and other applications.

We consider the sequence

$$(7.7) \qquad x_{n+1} = (I - \alpha T^*T)x_n + \alpha T^*y$$

where y is a given element in Y and α is chosen as before. Using the relation $T^*TT^\dagger y = T^*Qy = T^*y$, it follows by recursion that

$$x_n - T^\dagger y = (I - \alpha T^*T)x_{n-1} + \alpha T^*TT^\dagger y - T^\dagger y$$

$$= (I - \alpha T^*T)(x_{n-1} - T^\dagger y) = (I - \alpha T^*T)^n (x_0 - T^\dagger y).$$

Hence,

$$\lim_{n \to \infty} (x_n - T^\dagger y) = \lim_{n \to \infty} (I - \alpha T^*T)^n (x_0 - T^\dagger y)$$

$$= (I - P)(x_0 - T^\dagger y) = (I - P)x_0.$$

Thus we have

Theorem 7.1. Let T be a bounded linear operator with closed range. The sequence $\{x_n\}$ defined by (7.7), for any initial vector x_0, converges to $T^\dagger y + (I - P)x_0$, which is a least squares solution of $Tx = y$. In particular, the optimal choice $\alpha = \alpha_b$ yields

$$\|T^\dagger y + (I - P)x_0 - x_n\| \le \left(\frac{1-c^2}{1+c^2}\right)^n \|x_0 - T^\dagger y\|,$$

where $c = \|T\| \|T^\dagger\|$ is the <u>pseudocondition number</u> of T.

Now we turn to the case when the range of T is not closed. Since convergence of $(I - \alpha T^* T)^n$ in the <u>operator norm</u> to $I - P$ implies that $\Re(T)$ is closed, we consider <u>pointwise</u> convergence of $(I - \alpha T^* T)^n$. For each $x \in X$, we have (see [63])

$$\lim_{n \to \infty} (I - \alpha T^* T)^n x = (I - P)x \quad \text{for} \quad 0 < \alpha < \frac{2}{\|T\|^2}.$$

The series $\sum_{k=0}^{n} \alpha (I - \alpha T^* T)^k T^* y$ for $0 < \alpha < \frac{2}{\|T\|^2}$ converges in norm monotonically to $T^\dagger y$ for each $y \in \mathcal{D}(T^\dagger) = \Re(T) + \Re(T)^\perp$. Moreover if $Qy \in \Re(TT^*)$, then

$$\|T^\dagger y - \sum_{k=0}^{n} \alpha (I - \alpha T^* T)^k T^* y\|^2 \leq \frac{\|T^\dagger y\|^2 \|(TT^*)^\dagger y\|^2}{\|(TT^*)^\dagger y\|^2 + n \alpha (2 - \alpha \|T\|^2) \|T^\dagger y\|^2}.$$

Rephrased in the setting of the iterative process (7.7) the preceding estimate yields

Theorem 7.2. Let X and Y be Hilbert spaces over the same field and T be a bounded linear operator on X into Y, with the range of T not necessarily closed. The sequence (7.6), starting with $x_0 = 0$, converges in norm monotonically to $T^\dagger y$ for each $y \in \mathcal{D}(T^\dagger)$ and each fixed number α in the range $0 < \alpha < \frac{2}{\|T\|^2}$. Moreover if $Qy \in \Re(TT^*)$, then

$$\|x_n - T^\dagger y\|^2 \leq \frac{\|T^\dagger y\|^2 \|(TT^*)^\dagger y\|^2}{\|(TT^*)^\dagger y\|^2 + n \alpha (2 - \alpha \|T\|^2) \|T^\dagger y\|^2}.$$

Hyperpower iterative methods for computing the generalized inverse of a bounded linear operator T on a Hilbert space have been considered by Petryshyn [14] when $\Re(T)$ is closed and by Showalter and Ben-Israel [63] for the case when $\Re(T)$ is arbitrary.

The uniform convergence of $(I-\beta K)^n$ to a projection can also be asserted in Banach spaces under certain conditions, leading to a series representation for $K^\dagger y$. For example we have the following theorem ([17], [23]).

Theorem 7.3. Let L be a bounded linear operator on a Banach space X into X. Assume the spectrum of L is contained in some closed disk of radius r whose center lies on the interval $(-\infty, 1)$ and whose boundary passes through the point 1. Assume further that 1 is not a limit point of the spectrum of L, that $K = I - L$ has a generalized inverse, and that $X = \mathfrak{n}(K) \oplus \mathfrak{R}(K)$. Let $P: X \to \mathfrak{n}(K)$, $Q: X \to \mathfrak{R}(K)$ be the projectors determined by this direct sum decomposition, and let β be a real number such that $0 < \beta < \frac{1}{r}$. Then

(a) $(I - \beta K)^n$ converges in the operator norm to P;

(b) for all $y \in \mathfrak{R}(K)$, the series $\sum_{n=0}^{\infty} \beta(I - \beta K)^n y$ converges strongly to $K^\dagger y$, where $K^\dagger$ is the generalized inverse of K with respect to P and Q, and the convergence is uniform on every bounded subset of $\mathfrak{R}(K)$;

(c) the sequence $x_{n+1} = \beta L x_n + \beta y + (1-\beta) x_n$ converges for any initial approximation x_0 and for any $y \in \mathfrak{R}(K)$ to a solution of the equation $x - Lx = Kx = y$, and every such solution is attainable by a suitable choice of x_0.

Finally let us mention that several authors have studied, without using generalized inverses, iterative methods which converge to some solution of bounded linear operator equations which may have infinitely many solutions. Let T be a nonnegative bounded linear operator on a Hilbert space H into H, i.e. $\langle Tx, x \rangle \geq 0$ for all $x \in H$. For $y \in H$, consider the iteration procedure

$$x_{n+1} = x_n + \alpha(y - Tx_n), \qquad 0 < \alpha < 2\|T\|^{-1}.$$

Bialy [64] has shown that $Tx_n \to Qy$, where Q is the orthogonal projection on $\overline{\mathcal{R}(T)}$. The sequence $\{x_n\}$ converges if and only if the equation $Tx = y$ has a solution, in which case $x_n \to (I-P)x_0 + \hat{x}$, where $\hat{x}$ is the solution of minimal norm, i.e. $x_n \to (I-P)x_0 + T^{\dagger}y$. Keller [65] studied block successive and simultaneous relaxations with a matrix overrelaxation parameter for the solutions of singular and semidefinite linear systems. Most of the results in [65] can be recast in the terminology of generalized inverses for bounded linear operators with closed range. The alternating direction method for singular matrices was also considered by Douglas and Pearcy [66] and Kellogg and Spanier [67]. Iterative methods for least squares solutions of singular linear operator equations of the first and second kinds with particular applications to integral equations are considered in [28].

III. Normal Solvability of Nonlinear Operator Equations and Remarks on Generalized Inverses

8. Normal Solvability of Nonlinear Operator Equations

In a series of recent papers ([68],[69],[70]), Pohožaev has studied the concept of <u>normal solvability</u> of nonlinear operator equations for mappings which have continuous Fréchet derivative. We shall use this concept to provide in Section 9 a motivation for a meaningful definition of a generalized inverse of a nonlinear operator which reduces to Definition 1.1 in the case of a bounded linear operator with closed range. We first recall the concept of normal solvability of linear operator equations.

Let X and Y be Banach spaces over the real (or complex) numbers, and let L be a bounded linear operator from X to Y. Let X^* and Y^* denote the dual spaces of X and Y respectively and let $L^*: Y^* \to X^*$ be the adjoint of L. Consider the equation

(8.1) $$Lx = y .$$

If for some $y \in Y$, equation (8.1) is solvable, then $y \in \mathfrak{n}(L^*)^{\perp} = \{y \in Y: <y^*,y> = 0 \text{ for any } y^* \in \mathfrak{n}(L^*)\}$. Here $<y^*,y>$

is the value of the functional $y^* \in Y^*$ at the element $y \in Y$. The linear equation (8.1) is called <u>normally (Hausdorff) solvable</u> (see [71]) if this necessary condition for solvability is also sufficient. It follows immediately from the relation $\mathfrak{n}(L^*)^\perp = \overline{\mathfrak{R}(L)}$ that a necessary and sufficient condition for (8.1) to be normally (Hausdorff) solvable is that $\mathfrak{R}(L)$ be closed in Y. As noted earlier, normal solvability is equivalent to the "best approximate solution property" (see Theorem 2.3).

Now let the space Y be uniformly convex. This means that for each $\varepsilon > 0$ there is a $\delta > 0$ such that $\|u-v\| < \varepsilon$ whenever $\|u\| = \|v\| = 1$ and $\|\frac{1}{2}(u+v)\| > 1 - \delta$. Let $F: X \to Y$ be a (nonlinear) operator which is continuously Fréchet differentiable at each $x \in X$. Consider the equation

(8.2) $$F(x) = y.$$

<u>Definition 8.1.</u> Equation (8.2) is called normally solvable if

(i) for any $y \in Y$ there exists a sequence $\{y_n\}$ (which may be the constant sequence) such that $y_n \to y$ strongly as $n \to \infty$ and for any y_n $(n = 1, 2, \ldots)$ there exists an $x_n \in X$ minimizing the functional $\|F(x) - y_n\|$;

(ii) for any such sequence $\{y_n\}$ if $F(x_n) - y_n \in \{\mathfrak{n}\,[F'(x_n)^*]\}^\perp$ for $n = 1, 2, \ldots$, then $y \in \mathfrak{R}(F)$.

The main theorem in [70] is the following characterization of normal solvability.

<u>Theorem 8.1.</u> Let X be a Banach space, Y a uniformly convex Banach space and let $F: X \to Y$ be continuously Fréchet differentiable on X. A necessary and sufficient condition for the operator equation (8.2) to be normally solvable is that the range of F be closed in Y.

In particular if F is a bounded linear operator, then normal solvability in the sense of Definition 8.1 is equivalent to normal Hausdorff solvability when the space Y is uniformly convex.

A uniformly convex Banach space is reflexive but the converse is not necessarily true. Some results on normal solvability also hold when the space Y is reflexive but not necessarily uniformly convex. Note that if some $x_0 \in X$, $F(x_0) = y$, then obviously $y - F(x_0) \in \{n[F'(x_0)^*]\}^\perp$. Let us call the equation (8.2) <u>strictly normally solvable</u> if the constant sequence $y_n = y$, $n = 1, 2, \ldots$ satisfies Definition 8.1.

Definition 8.2. For any continuously Fréchet differentiable operator F, the equation (8.2) is <u>strictly normally solvable</u> if for any $y \in Y$ there exists an x_0 minimizing $\|F(x) - y\|$, and if for any such x_0, the condition $y - F(x_0) \in \{n[F'(x_0)^*]\}^\perp$ holds, then $y = F(x_0)$.

Pohožaev [69] gave a sufficient condition for strict normal solvability.

Theorem 8.2. Let X be a Banach space, Y a reflexive Banach space and let $F: X \to Y$ be continuously Fréchet differentiable on X. If the range of F is <u>weakly closed</u> in Y then (8.2) is strictly normally solvable.

In particular if F is a bounded linear operator and Y is a reflexive Banach space, then the notion of strict normal solvability is equivalent to normal Hausdorff solvability.

We remark in passing that as corollaries to Theorem 8.1 and 8.2 Pohožaev obtains the following theorem on <u>solvability</u> of the nonlinear equation (8.2)

Theorem 8.3. Let X and Y be Banach spaces and let $F: X \to Y$ be continuously differentiable on all of X. If $n([F'(x)]^*) = \{0\}$ for every x in X, and if one of the two following hypotheses hold:

(i) Y is reflexive and $\Re(F)$ is weakly closed in Y;

(ii) Y is uniformly convex and $\Re(F)$ is closed in Y;

then (8.2) is solvable for any $y \in Y$.

The results of Theorem 8.3 are considerably sharpened and generalized in a recent paper of Browder [72]. Applications of Theorem 8.3 to Fredholm alternative theory of nonlinear operator equations as well as to boundary value problems for quasilinear elliptic differential equations are given in [69], [70]. Several interesting examples of normally solvable (but not necessarily solvable) nonlinear boundary value problems are given in [69].

9. Remarks on Generalized Inverses of Arbitrary Linear Transformations Between Hilbert Spaces and Generalized Inverses of Nonlinear Operators

There are several (nonequivalent) approaches to generalized inverses of an arbitrary linear transformation between Hilbert spaces, or between inner product spaces which are not necessarily complete. Two unifying approaches using projectional and domain-range properties and the graphs of the operators are studied in [23]; various other definitions of generalized inverses of unbounded operators are related to this setting. Here we only make a few remarks on some approaches to generalized inverse under the assumption that the domain $\mathcal{D}(T) \subset X$ of T is such that

(9.1) $$\mathcal{D}(T) = n(T) \oplus (\mathcal{D}(T) \cap n(T)^{\perp}) .$$

In this connection and for other works related to these remarks see Hestenes [26], [78], Arghiriade [74], Arghiriade and Dragomir [75], Arghiriade and Boros [76]. If (9.1) holds, we say that T has a <u>decomposable domain</u> and following Hestenes [26] we call $\mathcal{D}(T) \cap n(T)^{\perp}$ the <u>carrier</u> of T and denote it by $C(T)$. Note that we do not require T to be closed or densely defined. The operator $T: \mathcal{D}(T) \to Y$ establishes a one-to-one correspondence between $C(T)$ and $\mathcal{R}(T)$. The generalized inverse of T is defined to be the linear extension of $\{T_1|C(T)\}^{-1}$ so as to have $\mathcal{R}(T) + \mathcal{R}(T)^{\perp}$ as its domain and $\mathcal{R}(T)^{\perp}$ as its null space. Clearly $C(T^\dagger) = \mathcal{R}(T)$, $\mathcal{R}(T^\dagger) = C(T)$, $\mathcal{D}(T^\dagger)$ is dense and $n(T^\dagger)$ is closed. This generalized inverse turns out to be the maximal generalized inverse with closed null space among those given by the following more encompassing definition An operator T^g is called a generalized inverse of T if (1) $\mathcal{R}(T) \subseteq \mathcal{D}(T^g)$, $\mathcal{R}(T^g) \subseteq \mathcal{D}(T)$; and (2) $T^g T = P_1$ on $\mathcal{D}(T)$; $TT^g = P_2$ on $\mathcal{D}(T^g)$, where P_1 and P_2 are the orthogonal projections of X, Y onto $\overline{\mathcal{R}(T^g)}$ and $\overline{\mathcal{R}(T)}$ respectively. Note that neither $\mathcal{D}(T)$ nor $\mathcal{D}(T^g)$ is necessarily dense. An operator T has generalized inverse in this sense if and only if $\mathcal{D}(T)$ is decomposable ([74], [23]). In this case T has a unique maximal generalized $T^\dagger = T^g_{max}$ with $\mathcal{D}(T^\dagger) = \mathcal{R}(T) \oplus \mathcal{R}(T)^\perp$ and $n(T^\dagger) = \mathcal{R}(T)^\perp$. However this $T^\dagger$ is not the only generalized inverse with a closed null space. To each subspace $S \subseteq \mathcal{R}(T)^\perp$ corresponds a unique generalized inverse T^g (for $T^\dagger$, $S = \mathcal{R}(T)^\perp$) with $\mathcal{D}(T^g) = \mathcal{R}(T) \oplus S$, $n(T^g) = S$, $\mathcal{R}(T^g) = \mathcal{R}(T^\dagger) = C(T)$. Taking for S any closed proper subspace of $\mathcal{R}(T)^\perp$, we see that there is an infinite number of generalized inverses with closed null space. There is also an infinite number of generalized inverses for which $\mathcal{D}(T^g)$ is not dense, or equivalently $\overline{n(T^g)} \neq \mathcal{R}(T)^\perp$, and an infinite number of generalized inverses with dense domain. For a generalized inverse with dense domain, $T^\dagger = T^g_{max}$ is the only generalized inverse with a closed null space.

Clearly if T is a closed operator, then $\mathcal{D}(T)$ is decomposable; consequently every closed linear operator T has a generalized inverse, whether or not $\mathcal{D}(T)$ is dense in X. When $\mathcal{D}(T)$ is dense and $\mathcal{D}(T^g)$ is required to be dense, the preceding generalized inverse reduces to Tseng's definition (see Section 4.1). If $n(T)$ is closed, then T is a closed operator if and only if $T^\dagger$ is a closed operator.

By suitably restricting the domain of T^g and/or the class of operators one can recover various definitions of generalized inverses from this setting. It is the author's feeling that the "best approximation solution" property and the projectional properties of $TT^\dagger$ and $TT^\dagger$ (or of suitable extensions of $TT^\dagger$ and $T^\dagger T$) should be part of any comprehensive theory of generalized inverses for arbitrary linear operators between Hilbert spaces. The preceding function-analytic approach with modifications given in [23] lends itself to such properties.

The notion of an inverse of a mapping is a function-theoretic notion and is not restricted to linear operators. It is natural therefore to seek notions of pseudo or generalized inverses of nonlinear mappings.(*) We conclude this section by indicating some possible directions, hoping to pursue these later.

At the outset it seems that a natural approach to generalized inverses of nonlinear operators would be via the theory of normal solvability. Using for simplicity the setting of Section 8 we call a continuously Fréchet differentiable operator F <u>normally solvable</u> if the operator equation $F(x) = y$ is strictly normally solvable for each $y \in Y$. Let $F: X \to Y$ be a normally solvable operator. We call an operator F^∂: $Y \to X$ a <u>semi-inverse</u> of F if for each fixed $y \in Y$, $F^\partial y$ satisfies the following conditions:

(i) $\inf\{\|F(x) - y\|: x \in X\} = \|F(F^\partial y) - y\|$

(ii) $y - F(F^\partial y) \in n([F'(F^\partial y)]^*)$ implies $F(F^\partial y) = y$.

In general there is a (nondenumerable) infinity of semi-inverses. In the case of a bounded linear operator on a Hilbert space, this definition reduces to the definition of a semi-inverse given on page .

Another possible approach is via the generalized inverse of the derivative of F, i.e. by defining the generalized inverse of a continuously differentiable operator as the primitive of the generalized inverse of the derivative. Such a definition however does not seem to have wide applications. Also there seems to be little hope of introducing a meaningful definition of a generalized inverse for general nonlinear operators from a range-null-space point of view since we do not have any relations such as $n(T^*)^\perp = \overline{\Re(T)}$ which played an essential role in the case of a linear operator. On the other hand, an approach using a range-null-space point of

---

(*) Professor L. C. Young has informed the author that generalized inversion for mappings on $\mathbb{R}^2$ is implicit in some work of Carathéodory on reciprocal relationships of boundaries of domains under conformal representations.

view may turn out fruitful at least for continuously differentiable operators which have "adjoints" (see Remark 13.4 in [36]), in view of the following relations which hold for such mappings:

$$\Re(F)^\perp \subset \mathfrak{n}(F^*)^\perp \quad \mathfrak{n}(F^*)^\perp \subset \overline{\Re(F)}$$
$$\Re(F^*)^\perp \subset \mathfrak{n}(F), \quad \mathfrak{n}(F)^\perp \subset \overline{\Re(F^*)} \ ,$$

where F^* denotes the "adjoint" of the nonlinear operator F. Another possible approach is via nonlinear projections along the lines of Section 3 for linear operators.

REFERENCES

[1] E. H. Moore, On the reciprocal of the general algebraic matrix, Abstract in Bull. Amer. Math. Soc. 26 (1920), 394-395.

[2] E. H. Moore, General Analysis, Part 1, Memoirs, Amer. Philos. Soc. 1 (1935), esp. pp. 147-209.

[3] R. Penrose, A generalized inverse for matrices, Proc. Cambridge Philos. Soc. 51 (1955), 406-413.

[4] R. Penrose, On best approximate solution of linear matrix equations, Proc. Cambridge Philos. Soc. 52 (1956), 17-19.

[5] Yu. Ya. Tseng, Sur les solutions des equations operatrices fonctionnelles entre les espaces unitaires. Solutions extremales. Solutions virtuelles, C. R. Acad. Sci. Paris, 228 (1949), 640-641.

[6] Yu. Ya. Tseng, Generalized inverses of unbounded operators between two unitary spaces, Dokl. Akad. Nauk SSSR (N.S.) 67 (1949), 431-434 (Math. Reviews 11 (1950), p. 115).

[7] Yu. Ya Tseng, Properties and classification of generalized inverses of closed operators, Dokl. Akad. Nauk. SSSR (N.S.) 67 (1949), 607-610 (Math. Reviews 11 (1950), p. 115).

[8] Yu. Ya. Tseng, Virtual solutions and generalized inversions, Uspehi Math. Nauk (N.S.) 11 (1956), 213-215 (Math. Reviews 18 (1957), p. 749).

[9] T. N. Greville, The pseudo inverse of a rectangular or singular matrix and its application to the solution of systems of linear equations, SIAM Reivew 1 (1959), 38-43.

[10] T. N. Greville, Some applications of the pseudo inverse of a matrix, SIAM Review 2 (1960), 15-22.

[11] T. O. Lewis, T. L. Boullion, and P. L. Odell, A bibliography on generalized matrix inverses, pp. 283-315 in [73].

[12] A. E. Taylor, *Introduction to Functional Analysis*, John Wiley, New York, 1958.

[13] K. Yosida, *Functional Analysis*, 2nd ed., Springer-Verlag, Berlin-New York, 1958.

[14] W. V. Petryshyn, On generalized inverses and uniform convergence of $(I - \beta K)^n$ with applications to iterative methods, J. Math. Anal. Appl. 18 (1967), 417-439.

[15] C. A. Desoer and B. H. Whalen, A note on pseudo-inverses, J. Soc. Indust. Appl. Math. 11 (1963), 442-447.

[16] F. J. Beutler, The operator theory of the pseudo-inverse I. Bounded Operators, II. Unbounded operators with arbitrary range, J. Math. Anal. Appl. 10 (1965), 451-470, 471-493.

[17] G. F. Votruba, Generalized Inverses and Singular Equations in Functional Analysis, Ph.D. dissertation, University of Michigan, Ann Arbor, 1963.

[18] S. G. Mikhlin, The Problem of the Minimum of a Quadratic Functional, Holden-Day, San Francisco, 1965.

[19] K. O. Friedrichs, Functional Analysis and Applications (Lecture given in 1949-50; notes by F. Ficken), Institute of Mathematical Sciences, New York University.

[20] H. Hamburger, Non-symmetric operators in Hilbert space, in Proceedings of the Symposium on Spectral Theory and Differential Problems, pp. 67-112, Oklahoma Agricultural and Mechanical College, Stillwater, Oklahoma. (See p. 81).

[21] R. D. Sheffield, On pseudo-inverses of linear transformations in Banach space, Oak Ridge Nat. Lab. Report #2133, 1956.

[22] N. Minamide and K. Nakamura, A restricted pseudo-inverse and its application to constrained minima, SIAM J. Appl. Math. 19 (1970), 167-177.

[23] M. Z. Nashed, A retrospective and prospective survey of generalized inverses of operators, MRC Report No. 1125, Mathematics Research Center, University of Wisconsin, Madison, 1971.

[24] W. T. Reid, Generalized inverses of differential and integral operators, pp. 1-25 in [73].

[25] O. Wyler, Green's operators, Ann. Mat. Pura Appl. 66 (1964), 251-264.

[26] M. R. Hestenes, Relative self-adjoint operators in Hilbert spaces, Pacific J. Math. 11 (1961), 1315-1357.

[27] F. E. Browder, Functional analysis and partial differential equations, I. Math. Annalen 138 (1959), 55-79.

[28] W. J. Kammerer and M. Z. Nashed, Iterative methods for best approximate solutions of linear integral equations of the first and second kinds, MRC Report No. 1117, Mathematics Research Center, University of Wisconsin, Madison, 1971.

[29] R. D. Milne, An oblique matrix pseudoinverse, SIAM J. Appl. Math. 10 (1968), 931-944.

[30] J. E. Scroggs and P. L. Odell, An alternate definition of a pseudoinverse of a matrix, SIAM J. Appl. Math. 14 (1966), 796-810.

[31] F. J. Murray, On complementary manifolds and projections in spaces L_p and ℓ_p, Trans. Amer. Math. Soc. 41 (1937), 138-152.

[32] A. Ben-Israel and A. Charnes, Contributions to the theory of generalized inverses, J. Soc. Indust. Appl. Math. 11 (1963), 667-699.

[33] I. Fredholm, Sur une classe d'equations fonctionelles, Acta Math. 27 (1903), 365-390.

[34] W. A. Hurwitz, On the pseudo-resolvent to the kernel of an integral equation, Trans. Amer. Math. Soc. 13 (1912), 405-418.

[35] D. Hilbert, Grundzüge einer allgemeinen Theorie der linearen Integralgleichungen, Teubner, Leipniz and Berlin, 1912; a reprint of six papers which appeared originally in the Göttingen Nachrichten (1904, pp. 49-51; 1905, pp. 307-338; 1906, pp. 157-227; 1906, pp. 439-480; 1910, pp. 355-417).

[36] M. Z. Nashed, Differentiability and Related Properties of Nonlinear Operators: Some Aspects of the Role of Differentials in Nonlinear Functional Analysis, this volume, pp. 103-309.

[37] W. W. Elliott, Generalized Green's functions for compatible differential systems, Amer. J. Math. 50 (1928), 243-258.

[38] W. T. Reid, Generalized Green's matrices for compatible systems of differential equations, Amer. J. Math. 53 (1931), 443-459.

[39] W. Greub and W. C. Rheinboldt, Non-self-adjoint boundary value problems in ordinary differential equations, J. Res. Nat. Bur. Standards Sect. B., 64B (1960), 83-90.

[40] O. Wyler, On two-point boundary problems, Ann. Mat. Pura Appl. 67 (1965), 127-142.

[41] J. S. Bradley, Adjoint quasi-differential operators of Euler type, Pacific J. Math. 16 (1966), 213-237.

[42] J. S. Bradley, Generalized Green's matrices for compatible differential systems, Michigan Math.J. 13 (1966), 97-108.

[43] W. S. Loud, Generalized inverses and generalized Green's functions, SIAM J. Appl. Math. 14 (1966), 342-369.

[44] W. S. Loud, Some examples of generalized Green's functions and generalized Green's matrices, SIAM Rev. 12 (1970), 194-210.

[45] W. T. Reid, Generalized Green's matrices for two-point boundary problems, SIAM J. Appl. Math. 15 (1967), 856-870.

[46] W. T. Reid, Principal solutions of non-oscillatory linear differential systems, J. Math. Anal. Appl. 9 (1964), 397-423.

[47] E. M. Landesman, Hilbert-space methods in elliptic partial differential equations, Pacific J. Math. 21(1967), 113-131.

[48] C. Kallina, A Green's function approach to perturbations of periodic solutions, Pacific J. Math. 29 (1969), 325-334.

[49] A. Halany and A. Moro, A boundary value problem and its adjoint, Ann. Mat. Pura Appl. 79 (1968), 399-411.

[50] L. V. Kantorovich, Functional analysis and applied mathematics, Uspehi Mat. Nauk (3) 6 (1948), 89-185.

[51] L. V. Kantorovich and G. P. Akilov, Functional Analysis in Normed Spaces, Pergamon Press, London-New York, 1964.

[52] M. Z. Nashed, Steepest descent for singular linear operator equations, SIAM J. Numer. Anal. 7 (1970), 358-362.

[53] M. R. Hestenes and E. Stiefel, Method of conjugate gradients for solving linear systems, J. Res. Nat. Bur. Standards Sect. B, 49 (1952), 409-436.

[54] M. R. Hestenes, The conjugate gradient method for solving linear systems, Proceedings of Symposia on Applied Mathematics, Vol. VI: Numerical Analysis, McGraw-Hill, New York, 1956, pp. 83-112.

[55] R. M. Hayes, Iterative methods for solving linear problems in Hilbert space, in Contributions to the Solutions of Systems of Linear Equations and Determination of Eigenvalues, O. Taussky, ed., Nat. Bur. of Standards, Appl. Math. Series, Vol. 39, 1954, pp. 55-69.

[56] H. A. Antosiewicz and W. C. Rheinboldt, Conjugate direction methods and the method of steepest descent, in A Survey of Numerical Analysis, J. Todd, ed., McGraw-Hill, New York, 1962, pp. 501-512.

[57] J. W. Daniel, The conjugate gradient method for linear and nonlinear operator equations, SIAM J. Numer. Anal. 4 (1967), 10-26. See also J. W. Daniel, A correction concerning the convergence rate of the conjugate gradient method, ibid. 7 (1970), 277-280.

[58] W. J. Kammerer and M. Z. Nashed, On the convergence of the conjugate gradient method for singular linear operator equations, SIAM J. Numer. Anal., to appear.

[59] W. J. Kammerer and M. Z. Nashed, Steepest descent for singular linear operators with nonclosed range, J. Applicable Analysis 1 (1971).

[60] E. S. Kennedy and W. R. Transue, A medieval iterative algorism, Amer. Math. Monthly 63 (1956), 80-83.

[61] D. Showalter, Representation and computation of the pseudoinverse, Proc. Amer. Math. Soc. 18 (1967), 584-586.

[62] A. T. Lonseth, Approximate solutions of Fredholm-type integral equations, Bull. Amer. Math. Soc. 60 (1954), 415-430.

[63] D. W. Showalter and A. Ben-Israel, Representation and computation of the generalized inverse of a bounded linear operator between two Hilbert spaces, Accad. Naz. Dei Lincei 48 (1970), 184-194.

[64] H. Bialy, Iterative Behandlung linearer Funcktionalgleichungen, Arch. Rational Mech. Anal. 4 (1959), 166-176.

[65] H. B. Keller, The solution of singular and semidefinite linear systems by iteration, SIAM J. Numer. Anal. 2 (1965), 281-290.

[66] J. Douglas, Jr., and C. M. Pearcy, On the convergence of the alternating direction procedures in the presence of singular operators, Numer. Math. 5 (1963), 175-184.

[67] R. B. Kellogg and J. Spanier, On optimal alternating direction parameters for singular matrices, Math. Comp. 19 (1965), 448-452.

[68] S. I. Pohožaev, Normal solvability of nonlinear operators, Dokl. Akad. Nauk SSSR 184 (1969), 40-43. Translated in Soviet Math. Dokl. 10 (1969), 35-38.

[69] S. I. Pohožaev, Nonlinear operators which have a weakly closed range of values and quasilinear elliptic equations, Mat. Sb. 78 (120) (1969), 237-259. Translated in Math. USSR Sb. 7 (1969).

[70] S. I. Pohožaev, Normal solvability of nonlinear mappings in uniformly convex Banach spaces, Funkcional Anal. i Priložen. 3 (1969), 80-84. Translated in Functional Anal. Appl. 3 (1969), 147-151.

[71] F. Hausdorff, Zur Theorie der Linearen Metrischen Raumme, J. Reine Angew. Math. 167 (1932), 294-311.

[72] F. E. Browder, On the Fredholm alternative for nonlinear operators, Bull. Amer. Math. Soc. 76 (1970), 993-998.

[73] T. L. Boullion and P. L. Odell, eds., Theory and Applications of Generalized Inverses of Matrices: Symposium Proceedings, Texas Technological College, Mathematics Series, No. 4, Lubbock, Texas, 1968.

[74] E. Arghiriade, Sur l'inverse généralisée d'un opérateur linéaire dans les espaces de Hilbert, Atti Acad. Naz. Lincei Rend. Cl. Sci. Fis. Mat. Natur. Ser. 8, 45(1968), 471-477.

[75] E. Arghiriade and A. Dragomir, Remarques sur quelques théoremes relatives l'inverse generalisée d'un opérateur linéaire dans les espaces de Hilbert, Atti Acad. Naz. Lincei Rend. Cl. Sci. Fis. Mat. Natur. Ser. 8, 46 (1969), 333-338.

[76] E. Arghiriade and E. Boros, L'inverse généralisée d'un opérateur linéaire dans un espace à produit intérieur, Atti. Acad. Naz. Lincei Rend. Cl. Sci. Fis. Mat. Natur. Ser. 8, 46 (1969), 646-649.

[77] S. Kurepa, Generalized inverses of an operator with closed range, Glasnik Mat. 23 (1968), 207-214.

[78] M. R. Hestenes, A ternary algebra with applications to matrices and linear transformations, Arch. Rational Mech. Anal. 11 (1962), 138-194.

On Polynomial Operators and Equations

PATRICIA M. PRENTER

Introduction

     Perhaps the simplest of all non-linear operators on a normed linear space are the so-called polynomials operators. Equations in such operators are the linear space analog of ordinary polynomials in one or several variables over the fields of real or complex numbers. Such equations encompass a broad spectrum of applied problems including all linear equations. Most often the polynomial nature of many non-linear problems goes unrecognized by researchers. This is most likely due to the fact that unlike the polynomials in a single variable which have proved a challenge to some of the greatest mathematicians, polynomial operators have received little attention. Whether this situation is due to an inherent valuelessness of these operators or to simple oversight remains to be seen. Hopefully, one should be able to exploit their semi-linear character to wrest more extensive results for these equations than one can obtain in the general non-linear setting.

     This article gives a brief survey of the research in this field over the last fifteen years as well as some recent unpublished results. Part one of the paper deals mainly with qualitiative results such as a quadratic formula in Banach Space, Weierstrass Theory in normed linear spaces, eigenfunction expansions for quadratic operators, and existence theory for solutions of polynomial equations. Part two treats some computational techniques for solving polynomial equations in normed linear spaces.

Part One. Qualitative Theory 361

1.   Polynomials in a linear space 362
2.   Quadratic formulas in Banach space. 368
3.   Interpolation and approximation by polynomial operators 371
4.   Eigenfunctions expansions for quadratic operators 378

Part Two. Computational Techniques 383

5.   Neumann expansions for quadratic expansions . . 384
6.   Solution by continued fractions 385
7.   Solution via a contraction mapping 387
8.   Other iterative techniques 393
9.   Concluding remarks 394

1. Polynomials in a linear space

Let X and Y be linear spaces over the field of real (or complex) numbers. For each integer $k \geq 1$ let X^k denote the direct product

$$\underbrace{X \times X \times X \ldots \times X}_{k \text{ times}}$$

A k-linear operator M, or X to Y, is a function in X^k to Y which is linear and homogeneous in each of its arguments separately. That is, for each $i = 1, 2, \ldots, k$

$$M(x_1, x_2, \ldots, x_i + y_i, \ldots, x_k) = M(x_1, x_2, \ldots, x_i, \ldots, x_k) + M(x_1, \ldots, y_i, \ldots, x_k)$$

and

$$M(x_1, x_2, \ldots, ax_i, \ldots, x_k) = aM(x_1, x_2, \ldots, x_i, \ldots, x_k).$$

A 0-<u>linear operator</u> L_0, on X, is a constant function on X into Y. That is, for some fixed $y \in Y$, $L_0 x = y$ for all $x \in X$. We identify a 0-linear operator L_0 with its range so that $L_0 x = L_0$ for all $x \in X$. Given a k-linear operator M on X ($k \geq 1$) and an $x \in X$, we set

$$Mx^k = \underbrace{M(x, x, \ldots, x)}_{k \text{ times}}.$$

For each $k = 0, 1, \ldots, n$, let L_k be a k-linear operator on X. Then, the operator P on X into Y, given by

(2.1) $$Px = L_n x^n + L_{n-1} x^{n-1} + \ldots + L_1 x + L_0$$

is called as n-th degree <u>polynomial operator</u> and equation (2.1) is called a <u>polynomial of degree n in X</u>. Clearly the family of all polynomial operators on X includes the family of all linear operators on X and with the exception of the linear operators, its members are non-linear. Solving the equation $Px = 0$ is equivalent to finding any roots of the polynomial P which lie in X.

Examples of polynomials abound. Many of the equations of elasticity theory are of this type, the Chandrasekhar equation of radiative transfer [2] is quadratic, and all of the examples worked on in the papers of the Cesari school [1, 6] are quadratics or cubics. We give several examples here and leave the rest to the readers experience and imagination.

<u>Example 1.</u> In the case of ordinary differential equations, the famous <u>Riccati equation</u>

(2.2) $$\frac{dy}{dt} + a(t)y + b(t)y^2 = c(t)$$

$$y(0) = c$$

is quadratic. Assuming the coefficient functions a, b, c to be continuous, the operator P defined by

$$p(y) = L_2 y^2 + L_1 y + L_0 .$$

Where $L_2(y, z)(t) = b(t) y(t) z(t)$, $L_1 y = a(t)y + y'$, and $L_0 = -c(t)$ may be regarded as a quadratic operator from the space $C'[0, 1]$ into the space $C[0, 1]$.
Clearly $P(y) = 0$ is the Riccati equation (2.2).

Example 2. The <u>Hammerstein Integral Equations</u> provide us with a second example. Let $K(s, t_1, t_2, t_3)$ be a square integrable function of four variables for which

$$\int_0^1 \int_0^1 \cdots \int_0^1 |k(s, t_1, t_2, t_3)|^2 \, dt_1 dt_2 dt_3 ds < \infty .$$

Then

$$C(x_1, x_2, x_3) = \int_0^1 \int_0^1 \int_0^1 k(s, t_1, t_2, t_3) x(t_1) x(t_2) x(t_3) \, dt_1 dt_2 dt_3$$

is a 3-linear operator on $(L^2[0, 1])^3$ to $L_2[0, 1]$. Let $y \in L_2[0, 1]$ and let λ be a scalar. The equation

(2.3) $$Cx^3 - \lambda x = y$$

is a <u>cubic Fredholm integral equation</u> on $L_2[0, 1]$ into $L_2[0, 1]$.

Example 3. The <u>Navier-Stokes equations</u> provide us with a polynomial partial differential equation which is quadratic. Here the problem is to find a velocity field $u = u(x, t)$ and a pressure field $p = p(x,t)$ which, for $x = (x_1, x_2, \ldots, x_n) \in A$ (A a bounded simply connected region in C^n) and for $t > 0$ satisfy the differential equations

(2.4)
$$u_t + u \cdot \text{grad } u = -\text{grad } p + \Delta u$$
$$\text{div } u = 0$$

where Δ is the Laplacian, div is the divergence, and $u(x,t)$ satisfies the initial conditions

$$u(x, 0) = u_0(x), \quad x \in A$$

and the boundary condition

$$u(x, t) = 0, \quad x \in \text{boundary of } A.$$

Letting $L_2(u, v) = u \cdot \text{grad } v$, $L_1 u = u_t - \Delta u$, it is apparent that (2.4) is a quadratic equation in u.

Example 4. <u>Fréchet Derivatives of Operators.</u>

Differentiation of operators in a Banach space is the topic of several papers in this proceeding. The papers of Tapia and Nashed treat differentiability of non-linear operators and several other papers exploit the properties of such derivatives (i.e. J. E. Dennis' paper on Newton's method). In general, let F be a function mapping an open subset V of a Banach space X into a subset W of a Banach space Y and let $x_0 \in V$. Let $\mathcal{L}_n(X, Y)$ denote the family of n-linear operators from X^n to Y. If there exists a continuous, linear operator $U \in \mathcal{L}_1(X, Y)$ such that

$$\| f(x_0 + \Delta x) - f(x_0) - U(\Delta x) \| = o(\| \Delta x \|)$$

then $U = f'(x_0)$ is called the <u>Fréchet derivative</u> of f <u>at</u> x_0. Equivalently

$$U(x) = \lim_{t \to 0} \frac{f(x_0 + tx) - f(x_0)}{t}$$

where convergence is uniform on the sphere $\{x: \|x\| = 1\}$. One then extends this definition to obtain higher derivatives. In particular if $f: X \to Y$ then the function $f': X \to \mathcal{L}_1(X, Y)$. Let $Z = \mathcal{L}_1(X, Y)$ and $g = f'$. If g' exists then $g': X \to \mathcal{L}_1(X, Z) = \mathcal{L}_1(X, \mathcal{L}_1(X, Y))$. But $\mathcal{L}_1(X, \mathcal{L}_1(X, Y))$ is isomorphic to $\mathcal{L}_2(X, Y)$. Thus $g'(x_0) = f''(x_0)$ is a bilinear operator on X^2 to Y. In general the nth <u>Fréchet derivative</u> $f^{(n)}(x_0)$ <u>at</u> x_0 <u>of a function</u> f, <u>mapping a normed linear space</u> X <u>into a normed linear space</u> Y <u>is an n-linear operator on</u> X^n <u>to</u> Y. For real-valued functions f of n real variables $(x_1, x_2, \ldots, x_n) = x$, $f'(x) = \text{grad } f(x)$ and $f''(x) = \left(\frac{\partial^2 f(x)}{\partial x_i \partial x_j}\right)$ is the Hessian of f at x.

Now suppose X and Y are normed linear spaces and for each integer $n \geq 1$, L_n is a n-linear operator. The operator L_n is said to be <u>bounded</u> provided there exists a real constant $M > 0$ for which

$$\|L_n(x_1, x_2, \ldots, x_n)\| \leq M \|x_1\| \cdot \|x_2\| \cdot \ldots \cdot \|x_n\|.$$

Analogous to the 1-linear case, it can be proved that an n-linear, operator L_n, $n \geq 1$, is bounded if and only if it is continuous. Continuity of L_n is defined in terms of the product topology on X^n where $L_n: X^n \to Y$. If we define

$$\|L_n\| = \inf\{M: \|L_n(x_1, x_2, \ldots, x_n)\| \leq M \|x_1\| \cdot \|x_2\| \cdot \ldots \cdot \|x_n\|\}$$

then

$$\|L_n\| = \sup\{\|L_n(x_1, x_2, \ldots, x_n)\|: \|x_i\| = 1, \; i = 1, 2, \ldots, n\}.$$

It is clear that a 0-linear operator is unbounded, but if L_n, $n \geq 1$, is a bounded n-linear operator then $\|L_n x^n\| \leq \|L_n\| \|x\|^n$. In this sense $\|L_0 x\| = \|L_0\| = \|L_0\| \|x\|^0$ is diagonally bounded. We say that a <u>polynomial operator of degree n is bounded on</u> X provided there exist constant M_i, $i = 0, 1, \ldots, n$ such that

$$\|Px\| = \|L_n x^n + L_{n-1} x^{n-1} + \ldots + L_0\|$$

$$\leq M_n \|x\|^n + M_{n-1} \|x\|^{n-1} + \ldots + \|L_0\|.$$

Thus a polynomial operator such as L_0 can be bounded even though its component k-linear operators L_k are not bounded.

Let $\mathcal{L}_n(X, Y)$, $n = 0, 1, \ldots$, denote the set of n-linear operators from X^n to Y. If $X = Y$, we shall simply write $\mathcal{L}_n(X)$. It is clear that for each $n \geq 1$ the set $B_n(X, Y) = \{L \in \mathcal{L}_n(X, Y) : L \text{ is bounded}\}$ is a Banach space whenever Y is a Banach space.

Now let L_2 be a bilinear operator belonging to $\mathcal{L}_2(X, Y)$. Then for each $x \in X$, $L_2 x \in \mathcal{L}_1(X, Y)$ where $L_2 x$ is that linear operator defined by

$$(L_2 x)(z) = L_2(x, z)$$

for each $z \in X$. In general if $L \in \mathcal{L}_n(X, Y)$, $n > 1$, then for each $x \in X$, $L(x) \in \mathcal{L}_{n-1}(X, Y)$ is that (n-1)-linear operator defined by

$$(L(x))(x_2, x_3, \ldots, x_n) = L(x, x_2, \ldots, x_n)$$

and for each $x_1, x_2, \ldots, x_k \in X$, $1 < k \leq n$, $L(x_1, x_2, \ldots, x_k) \in \mathcal{L}_{n-k}(X, Y)$ is that (n-k)-linear operator defined by

$$(L_1(x_1, x_2, \ldots, x_k))(z_{k+1}, \ldots, z_n) = L(x_1, \ldots, x_k, z_{k+1}, \ldots, z_n)$$

for each $z_{k+1}, z_{k+2}, \ldots, z_n$ in X.

It is also useful to note [3] that $\mathcal{L}_n(X, Y)$ is isomorphic to $\mathcal{L}_1(X, \mathcal{L}_{n-1}(X, Y))$ which is isomorphic to $\mathcal{L}_1(X, \mathcal{L}_1(X, \mathcal{L}_1(X, \ldots, \mathcal{L}_1(X, Y) \ldots)))$. The isomorphisms become isometries when the $\mathcal{L}_k$'s are replaced by B_k's.

2. Quadratic Formulas in Banach Space

The first to examine polynomial operators in any depth was L. B. Rall. In his initial paper [15] he confined himself to quadratic equations

(2.1) $\qquad Qx = Bxx + Lx + y = 0$

where X and Y are linear spaces, B is a bilinear operator in $B_2(X,Y)$, L is a linear operator in $B_1(X,Y)$, and $y = L_0$ belongs to Y. Multiplicity of solutions to (2.1) as well as the analog of the Neumann expansion for the quadratic equation

(2.2) $\qquad x = y + \lambda Bxx$

are studied. Additional existence theory is obtained via a quadratic formula in Banach space when the operator B has certain factorization properties. He initially proves

<u>Theorem 2.1.</u> <u>If</u> $Qx_0 = 0$ <u>the quadratic equation</u> $Qu = 0$ <u>has a solution</u> $u \neq x_0$ <u>if and only if the homogeneous quadratic equation</u> $Bxx + [Q'(x_0)]^{-1}x = 0$ <u>has a solution</u> $x \neq 0$,

and

<u>Theorem 2.2</u> (Rall). <u>If</u> $Qx_0 = 0$ <u>and</u> $[Q'(x_0)]^{-1}$ <u>exists, the solution</u> x_0 <u>of</u> $Qx = 0$ <u>is unique in the sphere</u>

$$\|x - x_0\| < \frac{1}{\|\bar{B}\| \, \|[Q'(x_0)]^{-1}\|} \, .$$

He then classifies quadratic equations into three kinds. An equation $Qu = 0$ is said to be <u>first kind</u> if $Q'(x_0) = 0$ for some $x_0 \in X$. If $Qu = 0$ is not first kind, $Qu = 0$ is said to be <u>second kind</u> if $[Q'(x_0)]^{-1}$ exists for some $x_0 \in X$. Equations which are neither of first nor second kind are said to be third kind. Every equation of first kind with $Q'(x_0) = 0$ can be put in <u>normal form</u>

$$\bar{B}xx = z$$

where $\bar{B}$ is the symmetric bilinear operator

$$\bar{B}xy = \frac{1}{2}(Bxy + Byx)$$

$u = u - x_0$, $\bar{B} = \frac{1}{2}Q''(x_0)$, and $z = -Q(x_0)$. Putting such an equation in normal form is the normed linear space analog of "completing the square" and the roots of such equations occur in pairs. Similarly every equation $Qu = 0$ of second kind with $Q'(x_0)$ non-singular can be put in <u>normal form</u>

(2.3) $$\bar{B}xx + Ix + z = 0$$

where $x = u - x_0$, $\bar{B} = \frac{1}{2}[Q'(x_0)]^{-1} Q''(x_0)$, and $z = [Q'(x_0)]^{-1} y$.

The validity of a natural generalization of the quadratic formula to equations of first and second kind depends on the possibility of expressing a square root of the linear operator $B(\bar{B}xx)$ as $\bar{B}x$. This can, of course, always be done if $\bar{B}$ and x are real (or complex) numbers. To this end one defines the <u>factor set</u> of a symmetric bilinear operator $\bar{B}$. A bilinear operator B is said to be <u>symmetric</u> provided

$$Bxz = Bzx$$

for all $x, z \in X$. Although it is certainly not true that all bilinear operators are symmetric one can always symmetrize such an operator. In fact, if we define $B^* \in \mathcal{L}_2(X, Y)$ by

$$B^* xz = Bzx$$

then the operator $\bar{B}$ defined by

$$\bar{B}xz = \frac{1}{2}(B^* xz + Bxz)$$

is <u>symmetric</u>. It is important to note that any solution to the quadratic equation $Qx = Bxx + Lx + y = 0$ is also a solution to the symmetrized equation $\bar{Q}x = \bar{B}xx + Lx + y = 0$ and conversely. Thus one may as well assume $\bar{B} = B$ is symmetric. The subset

$$F\{\bar{B}\} = \{x: (\bar{B}x)^2 = \bar{B}(\bar{B}xx)\}$$

is called the <u>factor set of</u> $\bar{B}$. If Q is an equation of first kind then

<u>Theorem 2.3</u> (Rall). <u>The equation of first kind</u> $\bar{B}xx = z$ <u>has a solution</u> $x \in F\{\bar{B}\}$ <u>if and only if</u> $(\bar{B}z)^{\frac{1}{2}}$ <u>exists such that</u> $\bar{B}x = (\bar{B}z)^{\frac{1}{2}}$ <u>and</u> $(\bar{B}z)^{\frac{1}{2}}x = z$. <u>The equation of second kind</u> $\bar{B}xx + Ix + z = 0$ <u>has a solution</u> $x \in F\{\bar{B}\}$ <u>if and only if</u> $(I - 4\bar{B}z)^{\frac{1}{2}}$ <u>exists such that</u> $\bar{B}x = \frac{1}{2}\{-I + (I - 4\bar{B}z)^{\frac{1}{2}}\}$ <u>and</u> $\frac{1}{2}[I + (I - 4\bar{B}z)^{\frac{1}{2}}]x + z = 0$.

Letting the <u>null space</u> $0(\bar{B}) = \{x: \bar{B}x = 0\}$ where θ is the zero operator $\theta x = 0$ for all $x \in X$,

<u>Theorem 2.4</u> (Rall). <u>If</u> $0\{\bar{B}\} = \{0\}$, <u>the quadratic equation of first kind</u> $\bar{B}xx = z$ <u>has at most one solution</u> $x \in F\{\bar{B}\}$ <u>for each distinct square root of</u> $\bar{B}z$.

If Q is an equation of second kind then

Theorem 2.5 (Rall). <u>The quadratic equation of second kind</u> $\bar{B}xx + Ix + z = 0$ <u>has at most one solution</u> $x \in F\{\bar{B}\}$ <u>for each distinct pair</u> $(I - 4\bar{B}z)^{\frac{1}{2}}, -(I - 4\bar{B}z)^{\frac{1}{2}}$ <u>of square roots of</u> $(I - 4\bar{B}z)$.

The reader should note that these results, while elegant, only give us information about solutions in the factor set of $\bar{B}$. It is not difficult to construct differential equations which are quadratic (Q is unbounded) for which the factor set is very small or even empty and for which there are infinitely many solutions (A forthcoming paper of Allgower and Prenter treats existence theory and multiplicity of solutions of boundary value problem for quadratic differential operators on a Sobolev space). Despite the limitations of those particular results of Rall they are important for the insights they lend into the structure of these operators and for their originality in introducing a whole new way of looking at a large class of non-linear problems.

3. Interpolation and Approximation by Polynomial Operators

Being the linear space analog of polynomials in one variable it is natural to suspect there may be a polynomial operator analog to the classical Weierstrass Theorem and to the Stone-Weierstrass Theorem. That such is indeed the case was proven by Prenter [11], and [14] and generalized to polynomial algebras by Wulbert [22]. The classical Weierstrass Theorem tells us the family of all polynomials P[a,b] defined on the closed interval [a,b] is dense in the family of all real valued continuous functions C[a,b] defined on [a,b] where C[a,b] carries the Tchebycheff or uniform norm

$$\|f\| = \sup_{t \in [a,b]} |f(t)|.$$

The Stone-Weierstrass Theorem affords a simple proof that the real-valued polynomials in n variables defined on a compact set K are dense in the family C(K) of real-valued, continuous functions on K in the uniform norm

$$\|f\| = \sup_{x \in K} |f(x)|$$

where $f \in C(K)$. The linear space analog to these theorems, then, would tell us that the family $P(K, Y)$ of all continuous polynomials mapping a compact subset K of a normed linear space X into a normed linear space Y is dense in the family $C(K, Y)$ of continuous functions mapping K to Y in the uniform norm topology

$$\|f\| = \sup_{x \in K} \|f(x)\|.$$

The Hilbert space form of this theorem is proved in [11].

Theorem 3.1. *Let X be a real, separable Hilbert space. The family of continuous polynomials on X into X, restricted to a compact subset K of X, is dense in the set C(K) of continuous functions on X into X restricted to K where C(K) carries the uniform norm topology.*

The initial vehicle for proving this theorem was to first prove it in n-dimensions by observing that every n-linear operator on such a space has a simple matrix representation and then to invoke the classical Weierstrass Theorem for n variables coordinate-wise. The proof for infinite dimensions is accomplished by using projections $P_n F P_n$ of $F \in C(K)$ onto finite dimensional subspaces and then taking a limit via a generalized Dini Theorem. The same techniques do not work in a normed linear space X since in general one is not guaranteed uniformly bounded projections P_n of X onto n-dimensional subspaces, $n = 1, 2, \ldots$ (see the papers of Sobczyk [17], and Goodner [5]). By adaptation of a theorem of Vainberg [21], Prenter [14] obtained a Weierstrass Theorem for normed linear spaces on compact sets with property M.

Definition 3.1. A compact set K in a real normed linear space X has property M, M > 0, if for each ε-net, $\varepsilon > 0$, $\{x_1, x_2, \ldots, x_p\} = A$ in K there exists a continuous projection S of X onto span A such that

$$\|S(x - x_i)\| \le M \|x - x_i\|$$

for each $x \in K$ and each $x_i \in A$.

Let K be compact in X and let X be a real normed linear space. Let $\tilde{C}(K, Y)$ be the set of continuous functions on X to Y restricted to K and let P(K, Y) be the family of continuous polynomial operators on K to Y. Then

Theorem 3.2 (A Weierstrass Theorem for Real Normed Linear Space). *Let* X *and* Y *be real normed linear spaces. Let* K *be a compact subset of* X *which has property* M, M > 0. *Then* P(K, Y) *is dense in* $\tilde{C}(K, Y)$ *where* $\tilde{C}(K, Y)$ *carries the uniform norm topology.*

Is greater generality possible? If X is a real separable Banach space, Locker has shown that Prenter's proof of Theorem 3.1 works verbatum in this broader setting. An even more startling result is due to Wulbert generalizing from [22] and from the Stone-Weierstrass Theorem [18]. Recall that a subset of the family C(K) of continuous real-valued functions on a compact Hausdorff space K is said to form an algebra provided A is a linear space closed under both a multiplication and a scalar multiplication obeying

$$\begin{aligned}
fg + fh &= f(g + h) \\
f(gh) &= (fg)h \\
(f + g)h &= fh + gh \\
\alpha(fg) &= (\alpha f)g = f(\alpha g) .
\end{aligned}$$

A set B in C(K) is said to be separating provided for each two distinct points x and y in K there is a function $f \in B$ such that $f(x) \ne f(y)$. An algebra A is a unitary algebra provided there exists an element $1 \in A$ such that

$$1 \cdot f = f \cdot 1 = f$$

for all $f \in A$. The Stone Approximation Theorem (Stone-Weierstrass Theorem) states:

Theorem 3.3 (Stone Approximation Theorem). <u>Every unitary, separating algebra in</u> $C(K)$ <u>is dense in</u> $C(K)$ <u>in the uniform norm topology</u>.

To generalize this to a normed linear space setting let K be a compact Hausdorff space and let Y be a normed linear space. Let $C(K, Y)$ denote the space of continuous Y-valued functions defined on K. Topologize $C(K, Y)$ by defining neighborhoods $U(0, V)$ of zero to be

$$\{f \in C(K, Y): f(k) \in V \text{ for all } k \in K\}$$

when V is an arbitrary neighborhood of zero in Y. Let $P(Y)$ be the family of all continuous polynomial operators on Y into Y.

Definition 3.2 (Wulbert). A subset A of $C(K, Y)$ is a <u>polynomial algebra</u> if

1. A is a linear space, and
2. $p \circ f$ is in A for all $f \in A$ and all $p \in P(Y)$ where $(p \circ f)(x) = p(f(x))$.

Theorem 3.4 (Wulbert). <u>Let</u> A <u>be a point separating polynomial algebra in</u> $C(K, Y)$ <u>and let</u> E <u>be a complemented subspace of</u> Y. <u>Then</u>

$$A' = \{f \in A: \text{range of } f \subseteq E\}$$

<u>is a point separating polynomial algebra in</u> $C(K, E)$.

Using Theorem 3.4 Wulbert proves

Theorem 3.5 (A Stone-Weierstrass Theorem for Normed Linear Spaces). *A point separating polynomial algebra which contains the constants is dense in* $C(K, Y)$.

Generalizations to polynomials on a normed linear space over a complex field can be obtained by defining conjugation operators to n-linear operators (see [11]).

In light of the preceding theorems it is not surprising that there is also a polynomial operator analog to the Lagrange and Hermite interpolating polynomials in one variable. If $\{y_1, y_2, \ldots, y_n\}$ are real numbers and $\{x_1, x_2, \ldots, x_n\}$ are distinct real numbers there exists exactly one polynomial $p(x)$ of degree less than or equal to $(n-1)$ solving the interpolation problem $p(x_i) = y_i$, $1 \leq i \leq n$. Furthermore, this polynomial, the Lagrange polynomial of degree $(n-1)$, is given by

$$p(x) = \sum_{i=1}^{n} \ell_i(x) y_i$$

where

$$\ell_i(x) = \frac{w(x)}{(x - x_i) w'(x_i)} = [w'(x_i)]^{-1} \frac{w(x)}{(x - x_i)}$$

where $w(x) = (x - x_1)(x - x_2) \cdots (x - x_n)$. Now suppose X is a Banach space and that $\{y_1, y_2, \ldots, y_n\}$ are points of X. Let $\{x_1, x_2, \ldots, x_n\}$ be distinct points of X. Then there exists a continuous polynomial operator p of degree $n-1$ or less solving the interpolation problem $p(x_i) = y_i$. If L is an n-linear operator on X^n let $\partial_i L$ denote the $(n-1)$-linear from X^{n-1} to $\mathcal{L}_1(X, Y)$ defined by

$$\partial_i L(x_1, x_2, \ldots, x_{i-1}, x_{i+1}, x_n)$$
$$= L(x_1, x_2, \ldots, x_{i-1}, \underline{\qquad}, x_{i+1}, \ldots, x_n)$$

375

where

$$(L(x_1, x_2, \ldots, x_{i-1}, \underline{}, x_{i+1}, \ldots, x_n))(x)$$

$$= L(x_1, x_2, \ldots, x_{i-1}, x, x_{i+1}, \ldots, x_n).$$

Letting $w(x) = L(x-x_1, x-x_2, \ldots, x-x_n)$, an alternate symbol for $\partial_i \omega(x)$ is

$$\partial_i \omega(x) = \frac{w(x)}{(x-x_i)}.$$

The following theorems are proved in [13] when $X = Y$.

Theorem 3.6. <u>If there exists an n-linear operator L such that $[w'(x_i)]^{-1}$ exists for each $i = 1, 2, \ldots, n$ where $w'(x_i)$ is the Fréchet derivative of w at x_i and</u>

$$w(x) = L(x-x_1, x-x_2, \ldots, x-x_n),$$

<u>then the Lagrange polynomial $p(x)$ of degree $(n-1)$ given by</u>

$$p(x) = \sum_{i=1}^{n} \ell_i(x) y_i$$

<u>where</u> $\ell_i(x) = [w'(x_i)]^{-1} \dfrac{w(x)}{(x-x_i)} = [w'(x_i)]^{-1} \partial_i w(x)$ <u>solves the interpolation problem</u> $p(x_i) = y_i$, $1 \leq i \leq n$.

And

Theorem 3.7. <u>Let $x_1, x_2, \ldots, x_n$ be distinct points of a Banach space X. Then for each $i = 1, 2, \ldots, n$ there exists an n-linear operator L_i for which $[w_i'(x_i)]^{-1}$ exists where</u>

$$w_i(x) = L_i(x - x_1, \ldots, x - x_n).$$

Furthermore, the L_i's can be chosen so that $w_i'(x_i) = I$ where I is the identity operator in $B_1(X, X)$.

As a direct result of these two theorems we have

Theorem 3.8. The interpolation problem $p(x_i) = y_i$ can always be solved by a polynomial $y(x)$ of degree $(n-1)$ having a Lagrange representation given by

$$y(x) = \sum_{i=1}^{n} \ell_i(x) y_i$$

where $\ell_i(x) = [w_i'(x_i)]^{-1} \dfrac{w_i(x)}{(x-x_i)}$ for appropriately chosen n-linear operators $L_1, L_2, \ldots, L_n$.

The Banach space analog to ordinary Hermite interpolating polynomials also exists where their representation in terms of Fréchet derivatives is almost identical to the canonical representation for Hermite interpolating polynomials in one variable. In the event X is a separable Hilbert space one can write down a simple equation for these polynomials in terms of inner product. For example, in the Lagrange case

$$(3.1) \qquad \ell_i(x) = \frac{\pi_i(x)}{\pi_i(x_i)}$$

where

$$\pi_i(x) = \prod_{\substack{j=1 \\ j \neq i}}^{n} (x - x_j, x_i - x_j)$$

where (,) denotes inner product. Thus one obtains a fairly simple algorithm for, say, a real valued polynomial in m variables of degree $n-1$ solving the interpolation problem $p(x_i) = y_i$ where $\{x_1, x_2, \ldots, x_n\}$ are distinct points of E^m

and $\{y_1, y_2, \ldots, y_n\}$ are real numbers. The points $\{x_1, x_2, \ldots, x_n\}$ can be arbitrarily located (i. e. they need not be grid points of a square or triangular array). As of yet, however, no error estimate is available for (3.1) when $m \geq 2$. Such an estimate would, of course, be highly desirable.

4 Eigenfunction Expansions for Quadratic Operators

The study of eigenfunction expansions for quadratic operators on a Hilbert space was initiated by J. R. Phillips [10]. Proceeding formally by analogy with the linear case, suppose you wish to solve the quadratic equation

$$(4.1) \qquad \lambda x - Bxx = y$$

by means of an eigenfunction expansion where B is a bilinear operator in $B_2(X, X)$, $y = L_0 \in X$, and X is a real, separable Hilbert space. If there exists a complete orthonormal sequence of eigenfunctions (ϕ_i) in X such that

$$(4.2) \qquad \begin{aligned} B\phi_i\phi_j &= \lambda_i \delta_{ij} \phi_i \\ \varphi_i \varphi_j &= \delta_{ij} \end{aligned}$$

where δ_{ij} is the Kronecker delta, then

$$(4.3) \qquad \begin{aligned} \lambda x - Bxx &= \sum_i \lambda(x, \phi_i)\phi_i - \sum_i \lambda_i (x, \phi_i)^2 \phi_i \\ &= \sum_i (y, \phi_i)\phi_i \ . \end{aligned}$$

Equating coefficients one finds that (x, ϕ_i) is a root of the quadratic equation

$$(4.4) \qquad \lambda_i(x, \phi_i)^2 - \lambda(x, \phi_i) + (y, \phi_i) = 0 \ .$$

The first condition of (4.2) is very strong and demands that the operator B have rather severe projective properties with respect to its invariant subspaces.

An operator F (linear or nonlinear) is said to be <u>compact</u> if and only if F maps bounded sets into precompact sets. That is, if K is a bounded subset of X, the closure $\overline{F(K)}$ is compact. Equivalently, every sequence (x_n) in a bounded set $K \subseteq X$ has a subsequence (x_{n_k}) such that the sequence (Fx_{n_k}) converges in X. A linear operator L whose domain is all of X is self-adjoint if and only if

$$(Ax, y) = (x, Ay)$$

for all $x, y \in X$. A bilinear operator B whose domain is all of $X \times X$ is said to be <u>self-adjoint</u> provided it is symmetric and $(Bxy, z) = (y, Bxz)$ for all x, y, and z in X. The real number λ is said to be an <u>eigenvalue</u> for B provided there exists a nonzero <u>eigenvector</u> $x \in X$ such that

$$Bxx = \lambda x .$$

Corresponding to each eigenvector x there is only one eigenvalue. On the other hand, to each eigenvalue there may correspond more than one eigenvector. However the analogy with the linear problem breaks down. For example, if $\lambda \neq 0$ is an eigenvalue, then every real number different from zero is an eigenvalue. To see this simply observe that if x is an eigenvector corresponding to $\lambda \neq 0$ and α is any real number different from zero

$$B(\alpha x, \alpha x) = \lambda \alpha^2 x = (\lambda \alpha)(\alpha x) .$$

Using compactness and self-adjointness of B together with the extremal properties of continuous functions on compact sets Phillips proves

<u>Theorem 4.1.</u> <u>If</u> B <u>is a nonzero, compact self-adjoint bilinear operator on</u> X, <u>there exists at least one nonzero eigenvalue</u> λ <u>and a corresponding eigenvector</u> x_0 ($\|x_0\| = 1$) <u>such that</u>

$$Bx_0 x_0 = \lambda x_0$$

$$\lambda = (Bx_0 x_0, x_0) = \sup_{x \in S} (Bxx, x)$$

where S is the unit sphere, $\{x \in X : \|x\| = 1\}$, in X.

His proof is analogous with the proof [4] in the linear case when X is real and finite dimensional, and by a limit approach when X is infinite dimensional.

To obtain an eigenfunction expansion for Bxx one needs a projective property such as (4.2). Let ϕ be an eigenfunction guaranteed by Theorem 4.1 corresponding to the eigenvalue $\lambda_1 \neq 0$. We seek a second eigenfunction ϕ_2 corresponding to an eigenvalue $\lambda_2 \neq 0$ such that $(\phi_1, \phi_2) = \delta_{12}$. Since $B\phi_1 \phi_1$ is in the span of ϕ_1, ϕ_2 must be orthogonal to $M_1 = \text{span}\{\phi_1\}$. Letting $H_2 = (M_1)^\perp$ it follows that $X = H_2 \oplus M_1$ and H_2 is itself a subspace of X. If $H_2 \times H_2$ is invariant under B so that $Byz \in H_2$ for all y, z in H_2 then we can invoke Theorem 4.1 again to obtain $\lambda_2 \neq 0$ and ϕ_2 such that $B\phi_2 \phi_2 = \lambda_2 \phi_2$ where $\phi_2 \in H_2$ with $\|\phi_2\| = 1$. To this end Phillips proves the following very important theorem:

Theorem 4.2. Let λ be an eigenvalue of Q corresponding to the eigenvector x. Let M denote the invariant subspace of X spanned by x and let $M^\perp = \{y \in X : (y, x) = 0\}$. Then $M^\perp$ is invariant under B if and only if

$$(Bx - \lambda P_M) z = 0$$

for all $z \in H$ where P_M is the orthogonal projection of H onto M.

This theorem leads to

Theorem 4.3. A nonzero compact, self-adjoint bilinear operator B possesses a sequence (λ_k) of eigenvalues and a corresponding orthonormal system of eigenelements (x_k) if and only if for each k

$$(Bx_k - \lambda_k P_k)z = 0 \text{ for all } z \in H_k$$

where $M_k = \text{span}\{x_k\}$, P_k is the orthogonal projection of X onto M_k, $H_k = (M_1 + M_2 + \ldots + M_{k-1})^\perp$ for $k \geq 2$, and $H_1 = H$.

Furthermore

Theorem 4.4 (Phillips). The sequence of eigenvalues determined by Theorem 4.3 has the property that for each k

$$|\lambda_{k+1}| \leq |\lambda_k|.$$

If (λ_k) is infinite then $\lim_{k \to \infty} \lambda_k = 0$.

Note that if the conditions of Theorem 4.3 are satisfied and ϕ_i and ϕ_j are distinct eigenvectors, $B\phi_i \phi_j = 0$ as required by (4.2).

However, one must not assume that all compact self-adjoint bilinear operators give rise to orthonormal systems of eigenvectors obeying (4.2).

Example. A two way matrix

$$B = \begin{pmatrix} a_{111} & a_{112} & | & a_{211} & a_{212} \\ a_{121} & a_{122} & | & a_{221} & a_{222} \end{pmatrix}$$

gives rise to a bilinear operator on $(E^2)^2$ to E^2 when the linear operator Bx is defined by

$$Bx = \begin{pmatrix} a_{111}x_1 + a_{112}x_2 & a_{121}x_1 + a_{122}x_2 \\ a_{211}x_1 + a_{212}x_2 & a_{221}x_1 + a_{222}x_2 \end{pmatrix}$$

where $x = (x_1, x_2)$. Letting

$$B = \begin{pmatrix} 1 & 0 & 0 & -1 \\ 0 & -1 & -1 & 0 \end{pmatrix}$$

it is easy to see that B is compact and self-adjoint, and that the following are eigenvalues and normalized eigenvectors of B:

$$\lambda_1 = -1 \quad x_1 = (\tfrac{1}{2}, \sqrt{3/2}\,)$$
$$\lambda_2 = 1 \quad x_2 = (-\tfrac{1}{2}, \sqrt{3/2}\,)$$
$$\lambda_3 = 1 \quad x_3 = (1, 0)\ .$$

However $A = \{x_1, x_2, x_3\}$ does not contain an orthogonal pair and $[\text{span}\{x_k\}]^\perp$ is not invariant under B for $k = 1, 2, 3$. Thus the construction of a complete orthonormal system of eigenvectors for a given bilinear operator B may prove very difficult even if it should exist.

No numerical applications are given in [10]. One suspects that bilinear operators possessing the required projective properties may be rather sparse. Furthermore, as the example illustrates, construction of such eigenvectors may be next to impossible. Nonetheless, the qualitative results of this paper are very revealing. If well-behaved invariant subspaces are difficult to come by for these simplest of all nonlinear operators, how even more difficult it must be in the general non-linear setting. Thus any hopes for a spectral decomposition of an arbitrary non-linear operator are rather doubtful.

Part Two: Computational Techniques for Solving Polynomial Equations

We have seen (Section 2) that Rall [15] divides quadratic equations

$$Qx = \bar{B}xx + Lx + y$$

into equations of first, second, are third kind. Every equation of first kind has a normal form

$$\bar{B}xx = z$$

while every equation of second kind has a normal form

$$\bar{B}xx + \bar{I}x = z \ .$$

This classification corresponds to the classification of linear Fredholm and Volterra integral equations into equations of first and second kind. The similarity goes even deeper. It is well known that both the existence theory and the numerical solution of linear integral equations of second kind is far more tractable than that for equations of first kind. This characteristic is also true of quadratic equations of second kind as well as of polynomial equations of second kind. We shall call a polynomial (equation) $Px = 0$ in a linear space X to Y an equation of <u>second kind</u> provided Px has the form

$$Px = L_n x^n + L_{n-1} x^{n-1} + \ldots + (L+I)x + L_0 = 0$$

where each L_n, $n \geq 2$, is an n-linear operator, L_0 is a constant, L is a linear operator, I is the identity map, and $L \neq -I$. The tractability of equations of second kind seems to arise from the fact that such equations can be written in the form

$$Fx = x$$

and thus lend themselves to standard fixed point theorems

via contraction mappings. It is thus not surprising that the only existence theorems that have been developed specifically for polynomial equations are for equations of second kind and all of these are iterative techniques. In this section we examine the iterative techniques of Marcus [7, 8], McFarland [9], and Rall [16, 15].

5. Neumann Expansions for Quadratic Equations

Let X be a normed linear space. If a given quadratic equation on X to X can be written in the form

(5.1) $$x = y + \lambda Bxx, \quad B \in B_2(X, Y) ,$$

with $y \neq 0$ and real $\lambda > 0$, it is natural to seek a solution to (5.1) in the form of a <u>Neumann type expansion</u>

(5.2) $$x = x_0 + \lambda x_1 + \lambda^2 x_2 + \ldots + \lambda^n x_n + \ldots .$$

This is the basis for the method suggested by Rall in [15]. Formal substitution of (5.2) in (5.1) and equation of like powers of λ gives

(5.3) $$\begin{aligned} x_0 &= y \\ x_1 &= Bx_0 x_0 \\ x_2 &= Bx_0 x_1 + Bx_1 x_0 \\ &\vdots \\ x_n &= \sum_{j=0}^{n-1} Bx_j \, x_{n-j-1} . \end{aligned}$$

Theorem 5.1 (Rall). <u>The series</u> (5.2) <u>with coefficients given by</u> (5.3) <u>converges to a solution of</u> (5.1) <u>provided</u>

$$0 < \lambda < \frac{1}{4\|B\| \cdot \|y\|} .$$

The solution x is unique in the sphere

$$\|u - x\| < \frac{\sqrt{1 - 4\lambda \|y\| \cdot \|B\|}}{2\lambda \|B\|} .$$

If $\xi = \dfrac{1 - \sqrt{1 - 4\lambda \|y\| \cdot \|B\|}}{2\lambda \|B\|}$, x is this solution to (5.1),

and $x^{(n)} = \sum_{j=0}^{n} \lambda^j x_j$, then

$$\|x - x^{(n)}\| \leq \xi - \|y\| \sum_{j=0}^{n} \frac{(2j)!}{j!(j+1)!} (\lambda \|y\| \cdot \|B\|)^j .$$

Rall applies this technique to Chandrasekhar's equation of radiative transfer [2]

$$x(s) = 1 + \lambda\, x(s) \int_0^1 \frac{s}{s+t} x(t)\, dt, \qquad 0 \leq s \leq 1 .$$

6. Solution by Continued Fractions

A second iterative technique, suggested by Rall and worked out by McFarland [9], is based on a continued fraction approach to the problem together with the fact that $Bx \in B_1(X, Y)$ whenever $B \in B_2(X, Y)$. He considers the quadratic equation

(6.1) $$Qx = Bxx + Lx - y = 0$$

where $B \in B_2(X, Y)$ and $L \in B_1(X, Y)$. Without loss of generality assume B is symmetric. If $(L + Bx_0)$ is nonsingular for some $x_0 \in X$, then

$$x_1 = (Bx_0 + L)^{-1} y$$

is defined. If $L + Bx_1$ is nonsingular, we can repeat the process to obtain $x_2 = (Bx_1 + L)^{-1} y$. In general, one would like the sequence (x_n) in X defined by

(6.2) $$x_{n+1} = (Bx_n + L)^{-1} y ,$$

to exist and to converge to a solution x of (6.1). McFarland has proved the following theorems.

Theorem 6.1. If L is nonsingular,

$$0 < \|L^{-1}\| \cdot \|B\| \cdot \|L^{-1}y\| \leq \tfrac{1}{4}$$

and

$$\frac{1 - [1 - 4\|L^{-1}\| \cdot \|B\| \cdot \|L^{-1}y\|]^{\frac{1}{2}}}{2} \leq \|L^{-1} Bx_0\| \leq \frac{1 + [1 - 4\|L^{-1}\| \cdot \|B\| \cdot \|L^{-1}y\|]^{\frac{1}{2}}}{2}$$

then the sequence (6.2) is defined. If x_n converges to x as $n \to \infty$ then x is a solution of (6.1).

Theorem 6.2 (McFarland). If L^{-1} exists and

$$0 < \|L^{-1}\| \cdot \|B\| \cdot \|L^{-1}y\| \leq \delta < \tfrac{1}{4}$$

and

POLYNOMIAL OPERATORS AND EQUATIONS

$$\frac{1-(1-4\delta)^{\frac{1}{2}}}{2} \leq \|L^{-1} B x_0\| < \frac{1}{2}$$

then the sequence (6.2) is defined and converges to a solution of (6.1). Furthermore

$$\|x - x_n\| \leq \frac{\beta^n}{1-\beta} \|(L + Bx_0)^{-1} y - x_0\|$$

where

$$\beta = \frac{\|L^{-1} Bx_0\|}{1 - \|L^{-1} Bx_0\|} < 1.$$

That this is indeed a continued fraction approach becomes apparent when one rewrites (6.2) sybolically to obtain

$$x_n = \cfrac{y}{L + \cfrac{By}{L + \cfrac{By}{L + \cfrac{By}{L + \ldots + \cfrac{By}{Bx_0 + L}}}}}.$$

Some numerical applications of this technique to the solution of nonlinear integral equations have been tried at Oregon State University.

7. Solution via Contraction Mappings

A study of sufficient conditions for the convergence of the direct iterative techniques $x^{n+1} = Px^n$ applied to both the infinite system

$$(7.1) \quad x_i = b_i + \sum_{k_1=1}^{\infty} c_{ik_1} x_{k_1} + \sum_{k_1=1}^{\infty} \sum_{k_2=1}^{\infty} c_{ik_1 k_2} x_{k_1} x_{k_2} + \cdots$$

$$\cdots + \sum_{k_1=1}^{\infty} \sum_{k_2=1}^{\infty} \cdots \sum_{k_n=1}^{\infty} c_{ik_1 k_2 \cdots k_n} x_{k_1} \cdots x_{k_n} + \cdots$$

of polynomial equations in an infinity of unknown complex numbers $x_1, x_2, \ldots$ and to its finite subsystems

$$(7.2) \quad x_i = b_i + \sum_{k_1=1}^{N} c_{ik_1} x_{k_1} + \cdots + \sum_{k_1=1}^{N} \cdots \sum_{k_M=1}^{N} c_{ik_1 k_2 \cdots k_M} x_{k_1} \cdots x_{k_M},$$

M and N positive integers and $i = 1, 2, \ldots$, is the subject of the papers [7] and [8] of Marcus. In both cases b_i, c_{ik_1}, $c_{ik_1 k_2}$ are given complex numbers. His conditions arise directly from application of the classical Contraction Mapping Theorem to these sets of equations and enable him to obtain error estimates for $\|x - x^n\|$ in terms of both the Tchebycheff and ℓ_1 norms where x^n is the nth iterate. He also obtains conditions under which solutions to system (7.2) can be extended to elements of ℓ_1 or ℓ_∞ converging to solutions of (7.1) along with error estimates for approximations of solutions of system (7.2) by these extended solutions. He is, however, unaware of the fact that the equations (7.2) are simply a polynomial equation

$$(7.3) \quad x = L_0 + L_1 x + \cdots + L_M x^M$$

of second kind in appropriate normed linear-spaces whereas his equations (7.1) are an <u>infinite series of second kind</u>

(7.4) $$x = L_0 + L_1 x + \ldots + L_n x^n + \ldots$$

in the same normed linear spaces. We take the liberty here of reformulating his entire theory in this more catholic setting to enable the reader to see it easily generalizes to arbitrary Hilbert spaces as well as to many Banach spaces.

To accomplish our reformulation let C denote the set of all infinite sequences (x_n) of complex numbers. Let

(7.5) $$\ell_1 = \{(x_n) \in C : \sum_{n=1}^{\infty} |x_n| < \infty \}$$

and let

(7.6) $$\ell_\infty = \{(x_n) \in C : \sup_{n \geq 1} |x_n| < \infty \} .$$

The spaces ℓ_1 and ℓ_∞ are both Banach spaces and as such are complete. Clearly the set $\{e_i : i = 1, 2, \ldots\} \subset C$ forms a basis for either of these spaces where

$$e_i = (\delta_{in} : n = 1, 2, \ldots)$$

and δ_{in} is the Kronecker delta. Now let L_M be an M-linear operator on the space X^M to X where $X = \ell_1$ or ℓ_∞. Then it is a simple matter to prove by analogy with the corresponding theory for linear operators on X (see [12]) that

<u>Theorem 7.1.</u> <u>Let L_M be a bounded M-linear operator on X^M to X where $X = \ell_1$. Then L_M has a matrix representation $(a_{i k_1 k_2 \ldots k_M})$, $j, i = 1, 2, \ldots, k_j = 1, 2, \ldots$ with respect to the basis $\{e_k : k = 1, 2, \ldots\}$. Furthermore</u>

(7.7) $$\sup_{k_1, k_2, \ldots, k_M \geq 1} \sum_{i=1}^{\infty} |a_{ik_1k_2\ldots k_M}| < \infty .$$

Conversely if the M-way matrix $(a_{ik_1k_2\ldots k_M})$ has property (7.7) then $(a_{ik_1k_2\ldots k_M})$ is the matrix representation of a bounded M-linear operator on X^M to X.

Similarly, for the space ℓ_∞

Theorem 7.2. Let $X = \ell_\infty$ and let the M-way matrix satisfy

(7.8) $$\sup_{i \geq 1} \sum_{k_1=1}^{\infty} \sum_{k_2=1}^{\infty} \cdots \sum_{k_M=1}^{\infty} |a_{ik_1k_2\ldots k_M}| < \infty .$$

Then $(a_{ik_1k_2\ldots k_M})$, $k_j = 1, 2, \ldots$, $i, j = 1, 2, \ldots$ is the matrix representation of a bounded M-linear operator on X^M to X.

The following theorem replaces Lemmas 1 and 2 of Marcus.

Theorem 7.3. Let L_M be a bounded M-linear operator on X^M to X where X is a normed linear space. Let

$$S_B(0) = \{x \in X : \|x\| \leq B\} .$$

Then MB^{M-1} is the smallest real number R satisfying

$$\|L_M x^M - L_M y^M\| \leq R \|x - y\|$$

for all $x, y \in S_B(0)$.

Proof: Let L_2 be a bounded bilinear operator on X^2 to X. Then for each pair $x, y \in S_B(0)$

$$\|L_2 x^2 - L_2 y^2\| = \|L_2 xx - L_2 xy + L_2 xy - L_2 yy\|$$

$$\leq \|L_2(x, (x-y)) + L_2((x-y), y)\|$$

$$\leq \|L_2\| \cdot \|x\| \cdot \|x-y\| + \|L_2\| \cdot \|y\| \cdot \|x-y\|$$

$$\leq 2B \cdot \|L_2\| \cdot \|x-y\| \quad .$$

Now simply proceed by induction observing that $B_M(X, Y)$ is isomorphic to $B_1(X, B_{M-1}(X, Y))$.

Recall the well known

<u>Contraction Mapping Theorem</u>. Let T be a contraction mapping of a closed subset U of a Banach space into U. That is, there exists an α, $0 < \alpha < 1$, such that

$$\|Tx - Ty\| \leq \alpha \|x - y\|$$

for all $x, y \in U$. Then the equation $Tx = x$ has exactly one solution in U. Furthermore the sequence of iterates

(A) $$x_{n+1} = Tx_n, \quad n = 0, 1, 2, \ldots, x_0 \in U$$

converges to this unique solution x and

(A') $$\|x - x_n\| < \frac{\alpha^n}{1-\alpha} \|x_0 - x_1\| \quad .$$

The following theorems are now simply proved.

Theorem 7.4. Let $X = \ell_\infty$ or ℓ_1 and let $\{L_n\}$ be a set of bounded n-linear operators, on X^n to X, $n = 1, 2, \ldots$. Let L_0 be the zero linear operator $L_0 = (b_i : i = 1, 2, \ldots)$. Then if

(7.9) $$\sum_{m=1}^{\infty} mB^{m-1} \|L_m\| < 1$$

and if

(7.10) $$\|L_0\| + \sum_{m=1}^{\infty} B^m \|L_m\| \leq B$$

the infinite series defined by

(7.11) $$Sx = L_0 + L_1 x + \ldots + L_m x^m + \ldots$$

is a contraction mapping of the closed sphere $S_B(0)$ into $S_B(0)$ and as such has a unique fixed point. Similarly if

(7.12) $$\sum_{m=1}^{N} m B^{m-1} \|L_m\| < 1$$

and if

(7.13) $$\|L_0\| + \sum_{m=1}^{N} B^m \|L_m\| \leq B$$

then the polynomial

$$P_N x = L_0 + L_1 x + \ldots + L_N x^N$$

is a contraction mapping of the closed sphere $S_B(0)$ into $S_B(0)$ and as such has a unique fixed point.

Proof: Simply apply Theorem 7.3.

Of course conditions (7.9) and (7.10) imply conditions (7.12) and (7.13). Combining Theorem 7.4 with Theorems 7.1 and 7.2 one arrives rather simply at the systems (7.1) and (7.2) together with the sufficient conditions of Marcus for the convergence of the iterative procedure (A) to the solution of such equations. One cannot directly compute solutions to equation (7.11) but one can compute solutions to (7.12) together with the error estimate (A') on the spaces ℓ_1^N and ℓ_∞^N

$$\ell_1^N = \{(x_1, x_2, \ldots, x_N): \|x\| = \sum_{i=1}^{N} |x_i|\}$$

$$\ell_\infty^N = \{(x_1, x_2, \ldots, x_N): \|x\| = \sup_{1 \leq i \leq N} |x_i|\}$$

Letting x^N be the solution to $P_N x = x$ in $S_B(0) \subseteq \ell_1^N(\ell_\infty^N)$ and $\bar{x}$ be the solution to $Sx = x$ in $S_B(0) \subseteq \ell_1(\ell_\infty)$ one can estimate $\|\bar{x} - \bar{x}^N\|$ where $\bar{x}^N = (\bar{x}_i : \bar{x}_i = x_i^N$, $1 \leq i \leq N$ and $\bar{x}_i = b_i$ for $i > N$) by imposing appropriate conditions on the ith, $i > N$, coordinates of Sx for all $x \in S_B(0)$ (see [7] and [8]). These methods are also applicable to ℓ^p spaces, $1 < p < \infty$.

8. Other Iterative Techniques

Rall [16] has also studied iterative techniques for solving polynomial equations of second kind by use of the Contraction Mapping Theorem. He observed that a polynomial

(8.1) $$Px = L_n x^n + L_{n-1} x^{n-1} + \ldots + Lx + L_0 = 0$$

on a Banach space X into a Banach space Y can be transformed to an equation of second kind

(8.2) $$h = F(h) = -(B_n h^n + \ldots + B_2 h^2 + B_0)$$

whenever there exists an $x_0 \in X$ such that $P'(x_0)$ is a nonsingular linear operator. Such a polynomial is said to be <u>regular at</u> x_0 and one obtains (8.2) from (8.1) through the transformations

$$B_k = (k!)^{-1} [P'(x_0)]^{-1} P^{(i)}(x_0)$$

$$h = x - x_0$$

$$B_0 = [P'(x_0)]^{-1} P(x_0)$$

where each B_k is a k-linear operator. Equations such as (8.2) lend themselves to the Banach-Cacciopoli contraction mapping theorem. He gives a penetrating analysis of the applicability of this theorem together with its intrinsic geometry through the use of the scalar majorant polynomial

$$f(B) = a_n B^n + a_{n-1} B^{n-1} + \ldots + a_2 B^2 + \|B_0\|$$

and its derivatives where $a_k = \|B_k\|$ and B is the radius of the sphere $S_B(0)$ invariant under F. He also gives conditions under which Newton's method and the modified Newton method are applicable together with a number of examples.

9. Concluding Remarks

As the reader can readily see, much work remains to be done on polynomials in a normed linear space both of a theoretical and computational nature. Existence theory is far from complete and what little there is is confined to local solutions on neighborhoods which are often of very small radius. With the exception of Rall's paper all published existence theorems depend in some way on contraction mapping theorems which are what force the local nature of the solutions. This situation is not surprising and is certainly not excluded simply to nonlinear polynomial operators. In fact, it permeates a large portion of nonlinear theory starting with the work of E. Schmidt and proceeding through the work of Hildebrandt and Graves, Cronin, Sather, Knightly, and Kantorovič to mention a few. Hopefully one should be able to prove some global theorems for polynomial equations. For example, one suspects it is possible to say something quite specific about the cardinality of zeroes of polynomials of nth degree. The previously mentioned paper of Allgower and Prenter studies polynomial differential operators both of first and second kind together with boundary-value problems for such equations. The null sets of these operators can be infinite dimensional and the boundary-value problems are quite intriguing. For example, one can prove that a large class of homogeneous cubic equations with three linearly independent boundary-values have exactly three solutions. Much work remains to be done on these problems.

Polynomials equations in Banach spaces also lend themselves, in a very natural way to a study of variational or projectional methods for their solutions. The reason they are especially adaptable to such methods is their semi-linear character. A forthcoming paper of the author studies existence theory for roots of polynomials equations in a Hilbert space H to a Hilbert space W as well as computational techniques for their solutions via such methods. The papers of Cesari [1], Locker [6], and Urabe [19, 20] are important to such a study and the theory is far from complete.

For those of a qualitative rather than computational frame of mind, it has been suggested that polynomial operators should carry a Galois theory. Such a theory, should it exist, may be very limited, but, nonetheless, interesting. The pessimistic note is prompted by the fact, as we have already seen [9], that a completely general spectral theory does not exist for these operators and that Rall's quadratic formula applies only to the factor set of the given quadratic operator.

REFERENCES

1. Cesari, L., Functional analysis and Galerkin's method, Michigan Math Journal, 11(1964), 336-384.

2. Chandrasekhar, S., Radiative Transfer, Dover, New York, 1960.

3. Gavurin, M. K., On K-ple operators in Banach spaces, Dokl. Akad. Nauk SSSR 22, no. 4(1939), 547-551.

4. Grueb, W. H., Linear Algebra, Springer-Verlag, New York, 1967.

5. Goodner, D. B., Projections in normed linear spaces, Transactions A. M. S. 69(1950), 89-108.

6. Locker, John, An existence analysis for nonlinear equations in Hilbert space, Transactions A. M. S., 13 (1967), 373-403.

7. Marcus, B., Solutions of infinite polynomial systems by iteration, Rendiconti Circulo Mathematico di Palermo, Serie II, vol. 11(1962), 5-24.

8. Marcus B., Error bounds for solutions of infinite polynomial systems by iteration, Rendiconti Circulo Mathematico di Palermo, Serie II, vol. 13(1964), 5-10.

9. McFarland, J. E., An iterative solution of the quadratic equation in Banach space, Proceedings of A. M. S., (1958), 824-830.

10. Phillips, John R., Eigenfunction expansions for self-adjoint bilinear operators in Hilbert space, Technical Report 27, Oregon State University, May, 1966.

11. Prenter, P. M., A Weierstrass Theorem for real, separable Hilbert spaces, MRC Report #868, April (1968) and Journal of Approximation Theory (to appear).

12. Prenter, P. M., Matrix representations of polynomail operators, MRC Report #929, August (1968).

13. Prenter, P. M., Lagrange and Hermite interpolation in Banach spaces, MRC Report #921, October (1968).

14. Prenter, P. M., A Weierstrass Theorem for real normed linear spaces, MRC Report #957, January (1969) and Bulletin of A. M. S., 75(1969), 860-862.

15. Rall, L. B., Quadratic equations in Banach spaces, Rendiconti Circulo Mathematico di Palermo, vol. 10 (1961), 314-332.

16. Rall, L. B., Solutions of abstract polynomial equations by iterative methods, MRC Report #892, August (1968), 1-35.

17. Sobczyk, A., On the extension of linear transformations, Transactions A. M. S., 55(1944), 153-169.

18. Stone, M. H., The generalized Weierstrass approximation theorem, Mathematics Magazine, 21(1948), 167-183, 237-254.

19. Urabe, Minoru, Galerkin's procedure for nonlinear periodic systems, Arch. Rational Mech. Anal., 20(1965), 120-152.

20. Urabe, Minoru, Galerkin's procedure for nonlinear periodic systems and its extension to multipoint boundary value problems for general nonlinear systems. Numerical solutions of nonlinear differential equations. Proceedings of an advanced symposium, M. R. C. Edited by Donald Greenspan, Wiley (1966).

21. Vainberg, M. M., Variational methods for the study of nonlinear operators, Moscow (1956), English translation by Holden-Day (1964), p. 319.

22. Wulbert, Daniel, University of Washington, private communication, November, 1969.

Department of Mathematics and Statistics
Colorado State University
Fort Collins, Colorado 80521

Applications and Methods for the Minimization of Functionals

JAMES W. DANIEL

Acknowledgment

The general outline of this material and occasional complete passages are taken from the author's forthcoming book, The approximate minimization of functionals, to be published in the Prentice-Hall Series in Automatic Computation; the author thanks the publishers for their permission to include this material.

1. Introduction and typical problems

Many problems of pure and applied mathematics either arise or can be formulated as variational problems, that is, as problems of locating a minimizing point for some (nonlinear) real-valued functional over a certain set. Such a setting is often beneficial from the analytic viewpoint of determining the existence and uniqueness of such points; however we shall primarily emphasize here the computational aspects of this approach, that is, how we can compute a solution to the problem by actually minimizing the appropriate functional.

The variety of minimization problems is immense; we consider some typical problems. A large class is that of optimal control problems [Hestenes (1966), Pontryagin et al (1962), Balakrishnan-Neustadt (1964)]. Here one seeks to minimize a cost functional

JAMES W. DANIEL

$$f(x,u) = \int_0^{t_F} c(t, x(t), u(t)) dt$$

over the set of points (functions) (x,u) satisfying constraints

$$\dot{x} \equiv \frac{dx}{dt} = s(t, x(t), u(t)), \quad x(0) \in X_I, \quad x(t_F) \in X_F,$$

$$x(t) \in X(t), \quad u(t) \in U(t) \quad \text{for} \quad t \in (0, t_F)$$

where X_I, X_F, $X(t)$, and $U(t)$ are certain specified sets and set functions.

As a special case with $s(t, x, u) \equiv u$ we have a basic problem of the <u>calculus of variations</u> [Courant-Hilbert (1953), Morrey (1966), Akhiezer (1962)], in which we seek to minimize

$$f(x) = \int_0^1 c(t, x(t), \dot{x}(t)) dt$$

subject to some boundary conditions on $x(t)$ such as

$$x(0) = x(1) = 0 .$$

Many problems of applied mathematics are of this latter form if we allow the variable t to represent a vector of dimension higher than one and $\dot{x}$ to represent the vector of first partial derivatives of x with respect to those variables.

A further special case is that in which $c(t, x, \dot{x}) = \frac{\dot{x}^2}{2} + G(t, x)$, in which case the minimization problem is equivalent to the two-point boundary value problem

$$\ddot{x} = g(t, x), \quad x(0) = x(1) = 0$$

where $g(t, x) = \frac{\partial G}{\partial x}(t, x)$. In the same way one can consider higher order differential equations of special forms, such as

$$\sum_{j=0}^{n} (-1)^{j+1} \frac{d^j}{dt^j} [q_j(t) \frac{d^j}{dt^j} x(t)] = g(t,x)$$

$$\frac{d^j}{dt^j} x(0) = \frac{d^j}{dt^j} x(1) = 0 \text{ for } 0 \le j \le n-1$$

as minimization problems.

A discrete analogue of the continuous optimal control problem described above is the <u>mathematical programming</u> problem [Abadie (1967), Fiacco-McCormick (1968), Hadley (1964), Mangasarian (1969), Zangwill (1969)] in which one seeks to minimize some function of a finite number of real variables subject to finitely many constraints. As we shall later see, such problems also often arise as discretized approximations to continuous optimal control problems. Many problems in <u>data approximation</u> [Lorentz (1966)] ultimately are of this form also and provide perhaps one of the most common forms of such problems. Here one has some data y_i depending on a parameter t measured at certain points t_i, i = 1, 2, ..., N; it is desired to approximate the formula generating the data by some expression

$$y(t) \sim g(t;\alpha)$$

where the choice of the parameter α determines the expression. We can do this by picking α so as to minimize some norm of the vector in E^N with components

$$y_i - g(t_i;\alpha) .$$

Essentially this technique has been applied lately in a somewhat novel fashion for the solution of such problems as differential or integral equations [Rosen-Meyer (1967), Rosen (1978), Rabinowitz (1968), Mikhlin-Smolitskiy(1967)]. For example, one seeks to solve the differential equation

Du = f in C, u = 0 on ∂C = boundary of C

where C is some domain and D is some differential operator. If $\varphi_1, \ldots, \varphi_N$ are some functions satisfying the boundary condition on ∂C, we try to choose numbers $\alpha_1, \ldots, \alpha_N$ to minimize some norm of the vector in E^M with components given by

$$[Du](t_i) - f(t_i), \quad i = 1, \ldots, M$$

where $\{t_i\}$ is some grid of g points over the domain C; $\alpha_1\varphi_1 + \ldots + \alpha_N\varphi_N$ is then taken as an approximate solution to the differential equation.

Since we are here interested primarily in the techniques of functional analysis, we shall not emphasize finite dimensional problems. From a computational viewpoint, however, it is clear that for our general problems we must devise numerical methods that use only finitely many numbers since computing machines are finite. Thus we must consider how to replace infinite dimensional problems by finite dimensional ones, and analyze the errors this causes. We have already seen one such method above for the problem $Du = f$. For the general optimal control or calculus of variations problem outlined above with $t_F = 1$ for simplicity, one might reasonably choose a positive integer n, set $k \equiv k_n \equiv \frac{1}{n}$, $t_i = ik$ for $0 \le i \le n$, and attempt to compute the finitely many numbers $x_n \equiv (x_{n,0}, \ldots, x_{n,n})$ and $u_n \equiv (u_{n,0}, \ldots, u_{n,n})$ which minimize $k \sum_{i=1}^{n} c(t_i, x_{n,i}, u_{n,i})$ over the set of vectors

satisfying $\dfrac{x_{n,i+1} - x_{n,i}}{k} = s(t_i, x_{n,i}, u_{n,i})$ for $0 \le i \le n-1$,

$x_{n,i} \in X(t_i)$ and $u_{n,i} \in U(t_i)$ for $0 \le i \le n$, and $x_{n,0} \in X_I$ and $x_{n,n} \in X_F$. This discretization method has proved useful in practice [Rosen (1966)].

A generally applicable discretization method, of special import for the two-point boundary value problems, is the <u>Rayleigh-Ritz method</u> in which one minimizes not the functional $f(x)$ with respect to x but rather $f(a_1 \varphi_1 + \ldots + a_n \varphi_n)$ with respect to the finitely many variables $a_1, \ldots, a_n$, where $\varphi_1, \ldots, \varphi_n$ are some fixed functions chosen propitiously. We wish to develop tools for analyzing these and other discretization methods. First however we must pause to remind the reader of some basic functional analysis necessary to understand abstract optimization problems.

2. Abstract analysis for optimization

We consider variational problems in a real Hilbert space for simplicity. Recall [Dunford-Schwartz (1962), Kantorovich (1948), Kantorovich-Akilov (1964), Taylor (1961)] that a real Hilbert space E is a vector space over the reals with a norm $\|\cdot\|$ determined by an inner product (or dot product) $<\cdot,\cdot>$, that is $\|x\|^2 = <x,x>$, such that E is complete in the topology (the <u>norm topology</u>) generated by $\|\cdot\|$. We denote norm convergence by $x_n \to x$. We can consider E as having another topology, the weak topology, for which a basis at the origin is given by all sets O of the form $O = \{x; |<y_1, x>| < a_1, \ldots, |<y_n, x>| < a_n\}$ for some real numbers $a_1, \ldots, a_n$ and points $y_1, \ldots, y_n \in E$. Convergence in this weak topology is denoted by $s_n \to x$; recall that $x_n \to x$ if and only if $|<y, x_n> - <y,x>|$ tends to zero for each fixed y in E. A bounded, weakly closed set S in E is <u>weakly sequentially compact</u>, that is, for any sequence $\{x_n\}$ from S there is a subsequence weakly converging to a point in S; in particular, a bounded, convex, norm-closed set is weakly sequentially compact.

Recall that a function f is called <u>sequentially lower semicontinuous</u>, for some topology, if the convergence of $\{x_n\}$ to x implies that $f(x) \le \liminf_{n \to \infty} f(x_n)$; this concept is important for minimization problems since every sequentially lower semicontinuous function defined on a sequentially compact set is bounded below there and attains its minimum. Hence a weakly sequentially lower semicontinuous functional achieves its minimum over each bounded, convex, norm-closed subset of Hilbert space, a fact often used to prove existence of solutions to constrained minimization problems. Theoretically one often proves the existence of solutions to unconstrained minimization problems by showing that the particular functional f tends to $+\infty$ with $\|x\|$, and hence the minimum over all of E must lie in the bounded, convex, norm-closed set $\{x; \|x\| \leq B\}$ for some B [Daniel (1971), Vainberg (1964)].

The concept of convexity for functions or sets, and various extensions of this concept, is fundamental in several aspects of minimization theory [Daniel (1971), Levitin-Poljak (1966a,b)]. Recall that a function f is <u>convex</u> (<u>strictly convex</u>) on a set C if and only if for each x_1, x_2 in C and λ in $(0,1)$ with $\lambda x_1 + (1-\lambda)x_2$ in C we have $f(\lambda x_1 + (1-\lambda)x_2) \leq (<) \lambda f(x_1) + (1-\lambda) f(x_2)$; a function is <u>quasi-convex</u> (<u>strongly quasi-convex</u>) on a set C if and only if for each x_1, x_2 in C and λ in $(0,1)$ with $\lambda x_1 + (1-\lambda)x_2$ in C we have $f(\lambda x_1 + (1-\lambda)x_2) \leq (<) \max\{f(x_1), f(x_2)\}$. Clearly a convex functional is quasi-convex and a strictly convex functional is strongly quasi-convex. The following statements indicate the importance of convexity [Daniel (1971)]: 1) A quasi-convex, norm-sequentially lower semicontinuous functional on a convex set is weakly sequentially lower semicontinuous there; 2) a strongly quasi-convex functional can achieve its minimum over a convex set at only one point; 3) every local minimum of a convex functional or a strongly quasi-convex functional on a convex set is in fact a global minimum.

3. Discretization of minimization problems

As we saw in Section 1, when we seek to minimize the weakly sequentially lower semicontinuous functional f over a weakly sequentially compact set C in a Hilbert space, computationally we are forced to treat only a finite dimensional approximating problem. Thus, to find an x^* in C solving the above MPC (minimization problem over C), we instead usually treat the $MPC_n - \varepsilon_n$ (ε_n-approximate minimization problem over C_n) to find an x_n^* in a subset C_n of a (usually finite dimensional) space E_n which satisfies

$f_n(x_n^*) \leq f_n(x_n) + \varepsilon_n$ for all x_n in C_n, where f_n is some approximation to f and $\varepsilon_n \geq 0$ tends to zero. Since x_n^* and x^* lie in different spaces, we need a technical device for comparing x^* and x_n^*, f and f_n, and C and C_n.

<u>Definition 3.1.</u> A <u>discretization</u> for the MPC is a family $\{E_n, f_n, p_n, r_n, C_n\}$ where f_n is a real valued functional on the subset C_n of the real normed space E_n and p_n maps E_n into E while r_n maps E into E_n.

We can now compare x^* to $p_n x_n^*$, $r_n x^*$ to x_n^*, $f(x)$ to $f_n(r_n x)$, and $f_n(x_n)$ to $f(p_n x_n)$; we wish to find conditions on the discretization allowing us to prove that $p_n x_n^*$ in some sense converges to x^*.

<u>Definition 3.2.</u> A discretization $\{E_n, f_n, p_n, r_n, C_n\}$ for the MPC is <u>consistent</u> if and only if

1) $\lim\sup_{n \to \infty} f_n(r_n x^*) \leq f(x^*)$ for some x^* solving the MPC and satisfying $r_n x^* \in C_n$ for all n,

2) $\lim\sup_{n \to \infty} [f(p_n x_n^*) - f_n(x_n^*)] \leq 0$ for all x_n^* solving the $MPC_n - \varepsilon_n$,

3) the sets $C^n \equiv p_n C_n \cup C$ are uniformly bounded and, if $z_{n_i} \in C^{n_i}$ with $z_{n_i} \to z$, then $z \in C$,

4) solutions x_n^* of the $MPC\text{-}\varepsilon_n$ exist for all n.

We remark that one might prove conditions 1) and 2) by proving them for all points in C or C_n, not just for solutions to the minimization problems; also condition 3) is trivial if $p_n C_n \subset C$ or if $e(C^n, C) \to 0$ where

$$e(C^n, C) \equiv \sup_{x \in C^n} d(x, C) \quad \text{and} \quad d(x, C) \equiv \inf_{y \in C} \|x - y\|.$$

With these tools in hand, it is straightforward to prove the following basic theorem on constrained optimization via discretization; details and further examples may be found in [Daniel (1968, 1971)].

Theorem 3.1. Let f be a weakly sequentially lower semicontinuous functional over a set containing C^n for large n and let $\{E_n, f_n, p_n, r_n, C_n\}$ be a consistent discretization of the MPC for a given weakly sequentially compact set C. Let x_n^* solve the MPC$_n$-ε_n for $\varepsilon_n \geq 0$ tending to zero. Then $\lim_{n\to\infty} f(p_n x_n^*) = \lim_{n\to\infty} f_n(x_n^*) = \inf_{x \in C} f(x)$, and all weak limit points of $\{p_n x_n^*\}$, at least one of which exists, lie in C and solve the MPC; in particular, if x^* is the unique solution to the MPC, then $p_n x_n^* \to x^*$.

Although the above theorem applies as well with some changes to unconstrained optimization with $C \equiv E$ and $C_n \equiv E_n$, it is not in general useful in the applications because of difficulties involved in proving the consistency of the discretization, primarily because x_n^* may be arbitrarily large. In many applications x_n^* cannot in fact become large, and one can develop a useful general theory for discretization of unconstrained optimization problems based on hypotheses of uniform growth for the functionals f_n; details and examples may be found in [Daniel (1968b, 1970)].

4. Practical discretizations: I

We now wish to indicate how the general theory can be used to analyze the numerical solution of some continuous optimization problems, including some we mentioned in Section 1.

A). Optimal control problems

In Section 1 we mentioned solving the general control problem by replacing the cost functional with a quadrature sum, replacing the differential equation constraints by difference equation constraints, and replacing the other continuous constraints by a finite subset. Because of increased computational ease that can be obtained in solving the finite dimensional problem, it has been suggested [Rosen (1966)] that one replace the difference equations by inequalities $x_{n,i+1} \leq x_{n,i} + k s(t_i, x_{n,i}, u_{n,i})$ and then approximately minimize $f_n(x_n, u_n) \equiv k \sum_{i=1}^{n} c(t_i, x_{n,i}, u_{n,i}) + \frac{1}{a_n} k \sum_{i=1}^{n} [s(t_{i-1}, x_{n,i-1}, u_{n,i-1}) - \frac{y_{n,i} - y_{n,i-1}}{k}]$ for a sequence of positive a_n tending to zero; note that f_n has been defined so as to penalize us severely if the difference inequalities are not in fact satisfied as equalities.

To apply our general theory, we need mappings p_n and r_n. We must apply p_n to points (x_n, u_n) satisfying $x_{n,i+1} = x_{n,i} + ks(t_i, x_{n,i}, u_{n,i}) - kb_{n,i}$, $b_{n,i} \geq 0$. If we define $w_n(t) \equiv p_n u_n$ as the step function constant on each $[t_i, t_{i+1})$ with value $u_{n,i}$ and $b_n(t)$ similarly from the $b_{n,i}$, then x_n looks like the numerical solution to $\dot{z} = s(t, z(t), w_n(t)) - b_n(t)$, $z(0) = x_{n,0}$ obtained by Euler's method; it therefore seems reasonable to let $p_n x_n$ be the solution z_n to this differential equation. The mapping r_n on the other hand must be applied to a solution (x^*, u^*) to the control problem. If u^* is piecewise continuous, then we can naturally define $(r_n u^*)_i = u^*(t_i)$ and $x_n \equiv r_n x^*$ via $x_{n,i+1} = x_{n,i} + ks(t_i, x_{n,i}, u^*(t_i))$, $x_{n,0} = x^*(0)$. Under some particular technical hypotheses [Daniel (1970, 1971)] it can be shown that these lead to a consistent discretization. We warn the reader however that some care must be taken to "expand" slightly the discretized constraints $x_{n,i} \in X(t_i)$, $u_{n,i} \in U(t_i)$, that is to replace these by $d(x_{n,i}, X(t_i)) \leq d_n$, $d(u_{n,i}, U(t_i)) \leq d_n$ for some positive d_n converging to zero; if this is not done, the discretized constraint set can be empty [Daniel (1970, 1971)]. Without proceeding with more technical details, let us merely say that the general discretization theory is well suited for analyzing the above numerical problem and leads to the result that, under appropriate hypotheses, $|x_{n,i}^* - x^*(t_i)|$ tends to zero uniformly in i, $\frac{x_{n,i+1}^* - x_{n,i}^*}{k}$ and u_n^* converge in a weaker sense to $(x^*)^{\cdot}(t_i)$ and u^* respectively [Daniel (1970), (1971)], and $f_n(x_n^*, u_n^*)$ converges to $f(x^*, u^*)$.

B). **Calculus of variations problems**

As we remarked in Section 1, the simplest problem in the calculus of variations is considerably simpler than the general control problem, of which it is a special case; correspondingly, the analysis of discretization methods is easier as well. Precisely, suppose we seek to minimize $f(x) \equiv \int_0^1 c(t,x,\dot{x})dt$, $x(0) = x(1) = 0$ by approximately minimizing $f_n(x_n) = k \sum_{i=0}^{n-1} c(t_i, x_{n,i}, \frac{x_{n,i+1} - x_{n,i}}{k})$, $x_{n,0} = x_{n,n} = 0$

where $k = k_n = \frac{1}{n}$ and $t_i = ik$ [Greenspan (1967)]. The mappings p_n and r_n can here be defined more simply than in the general control problem; we merely set $(r_n x)_i = x(t_i)$ and define $p_n x_n$ as the piecewise linear function interpolating the values $x_{n,i}$ at $t = t_i$. Under essentially the standard hypotheses used to prove existence of a solution to the original calculus of variations problem, one can use the unconstrained discretization theory to prove that $p_n x_n^*$ converges uniformly (and $(p_n x_n^*)^{\cdot}$ converges weakly in $L_2(0,1)$) to solutions x^* (and $(x^*)^{\cdot}$), and $f_n(x_n^*)$ converges to $f(x^*)$ [Daniel (1968b, 1971)].

C). **General Rayleigh-Ritz method**

In Section 1 we briefly mentioned the Rayleigh-Ritz method, in which one minimizes f over some finite dimensional subspace E_n spanned by elements $\{\varphi_1, \ldots, \varphi_n\}$ rather than over all of E; the simplest kind of convergence results for this approach can be obtained via the discretization viewpoint, although more detailed results require special analysis. We suppose that $\lim_{n \to \infty} d(x, E_n) = 0$ for each x in E, where $d(x, E_n) = \inf_{y \in E_n} \|x-y\|$. Let $f_n = f$, let p_n be the identity mapping, and let r_n be the best approximation mapping, that is, $\|x - r_n x\| = d(x, E_n)$. It is then simple to prove the following:

Theorem 4.1. Let f be weakly sequentially lower semicontinuous, $-f$ be norm sequentially lower semicontinuous, and $\lim_{\|x\| \to \infty} f(x) = \infty$. Let $\{E_n\}$ be a sequence of closed linear subspaces such that $\lim_{n \to \infty} d(x, E_n) = 0$ for all x in E. For each n let x_n^* satisfy $f(x_n^*) \le f(x_n) + \varepsilon_n$ for all x_n in E_n, where $\varepsilon_n \ge 0$ tends to zero. Then all weak limit points of x_n^*, at least one of which exists, minimize f over E, and $\lim_{n \to \infty} f(x_n^*) = \inf_{x \in E} f(x)$.

5. Convergence of minimizing sequences

In our basic discretization theory of Section 3 and the examples of Section 4, our fundamental result was always that $f(p_n x_n^*)$ converges to $f(x^*)$; the fact that $\{p_n x_n^*\}$ had weak limit points, all of which minimize f, follows directly from the "compactness" of C and the "continuity" of f. We wish now to demonstrate approaches which can be used in general to deduce stronger convergence results from the fact that $f(p_n x_n^*)$ converges to $f(x^*)$ [Levitin-Poljak (1966a,b), Poljak (1966), Daniel (1971)].

For greater generality as well as notational simplicity, we shall not require that we be considering a discretization process. Instead we shall now let $\{x_n\}$ denote a sequence taken from the space E, and consider what we can deduce from the assumption that $f(x_n)$ converges to $f(x^*)$; note that x_n is now playing the role of $p_n x_n^*$ in the preceding sections.

As we saw in our discretization analysis, the sequence $\{x_n\}$ need not always come from the set C over which we are minimizing f, but generally lies near C. Thus we introduce the following concept.

Definition 5.1. The sequence $\{x_n\}$ is an approximate minimizing sequence for f over C if $\lim_{n \to \infty} d(x_n, C) = 0$ and $\lim_{n \to \infty} f(x_n) = \inf_{x \in C} f(x)$.

The results which we have already mentioned for approximate minimizing sequences can be stated as follows:

Theorem 5.1. If f is a weakly sequentially lower semicontinuous functional on a norm neighborhood of the weakly sequentially compact set C and $\{x_n\}$ is an approximate minimizing sequence for f over C, then all weak limit points of $\{x_n\}$, at least one of which exists, minimize f over C.

A simple and very widely applicable concept which allows us to deduce stronger convergence results than that stated above is a uniform version of the generalization of convexity we mentioned in Section 2. We recall that a quasi-convex functional is called <u>uniformly quasi-convex</u> on some convex set C if and only if there is a real-valued, continuous, monotone increasing function $\delta(t)$ for $t \geq 0$ with $\delta(t) = 0$ if and only if $t = 0$, such that

$$f(\frac{x+y}{2}) \leq \max\{f(x), f(y)\} - \delta(\|x-y\|)$$

for all x,y in C.

Theorem 5.2. Let C be a norm-closed, bounded, convex subset of E, f a weakly sequentially lower semicontinuous functional on a norm neighborhood of C which is uniformly quasi-convex on C, and $\{x_n\}$ an approximate minimizing sequence for f over C. Then x_n norm converges to x^*, the unique point minimizing f over C.

Computationally one often desires bounds on $\|x_n - x^*\|$, not merely a statement of convergence. In a number of applied problems, one has that f is a twice differentiable functional whose second derivative satisfies $<h, f''_x h> \geq \delta(\|h\|)$ for all x in C, where δ is as in Theorem 5.2. This implies that f is uniformly quasi-convex and, under the remaining hypotheses of Theorem 5.1, allows us to deduce the error estimate $\delta(\|x_n - x^*\|) \leq f(x_n) - f(x^*)$. Further details may be found in [Daniel (1971)].

6. **Practical discretizations: II**

We wish to return now to another of the typical problems mentioned in Section 1, to which the discretization theory is applicable and for which much stronger convergence results are available by using essentially more sophisticated versions of the ideas of Section 5. We consider for simplicity the solution of

$$D^2 x(t) = g(t, x(t)), \quad t \in (0,1), \quad x(0) = x(1) = 0 \ ,$$

where D is the operator $\frac{d}{dt}$, although higher order problems can be treated just as easily [Ciarlet (1966), Ciarlet et al (1967, 1968a,b)]. We assume that $g(t,x)$ is continuous in (t,x) in $[0,1] \times (-\infty, \infty)$ and satisfies

1) $\dfrac{g(t,x) - g(t,y)}{x - y} \geq \gamma > -\pi^2$ if $x \neq y$

2) $\dfrac{g(t,x) - g(t,y)}{x - y} \leq M(c) < \infty$ if $|x| \leq c, \ |y| \leq c$.

Define the functional

$$f(x) \equiv \int_0^1 \{\tfrac{1}{2}(Dx(t))^2 + \int_0^{x(t)} g(t, z)dz\}dt \ .$$

It is easy to deduce that, if $x^*(t)$ is a classical solution to the differential equation then x^* minimizes f over E. Clearly also x^* is the unique minimizing point, since f is convex and in particular

$$f(x+y) \geq f(x) + \int_0^1 \{[Dx(t)][Dy(t)] + y(t)g(t,x(t))\}dt$$

$$+ (\gamma + \pi^2) \int_0^1 y^2(t)dt$$

which implies

$$f(x^* + y) \geq f(x^*) + (\gamma + \pi^2)\int_0^1 y^2(t)dt \ .$$

Moreover, if S_M is a subspace of dimension M spanned by the functions $\varphi_1, \ldots, \varphi_M$, then there exists a unique element ϕ_M in S_M minimizing f over S_M, $\phi_M = \sum_{i=1}^{M} a_i \varphi_i$, which is also the unique solution of

$$\frac{\partial f(\sum_{i=1}^{M} a_i \varphi_i)}{\partial a_i} = 0, \quad i = 1, \ldots, M ,$$

that is,

$$Ba + G(a) = 0$$

where $a = (a_1, \ldots, a_M)^T$, B is the matrix $B = ((B_{ij}))$,

$$B_{ij} \equiv <\varphi_i, \varphi_j> = \int_0^1 [D\varphi_i(t)][D\varphi_j(t)] dt, \quad G(a) = (G_1(a), \ldots, G_M(a))^T,$$

$$G_i(a) = \int_0^1 g(t, \sum_{j=1}^{M} a_j \varphi_j(t)) \varphi_i(t) dt .$$

Since this is just a Rayleigh-Ritz method, we know that ϕ_M converges weakly to x^*; by the methods of Section 5, we can prove it to be norm-convergence and give general error bounds. For various kinds of subspaces S_M, bounds on the error between ϕ_M and x^* have been computed; the basic argument for obtaining the bound is simple. If we write $\nabla f(x) = J(x)$, then since x^* minimizes f on E we have $J(x^*) = 0$, while $<J(\phi_M), \varphi_i> = 0$ since ϕ_M minimizes f over S_M. Thus $0 = <J(x^*) - J(\phi_M), \varphi_i> = <J'_{x_0}(x^* - \phi_M), \varphi_i>$ for some fixed x_0. Defining $[x, y] \equiv <J'_{x_0}(x), y>$, as a new inner product, we see that ϕ_M is the closest point to x^* in S_M in the sense of this inner product. Thus any

MINIMIZATION OF FUNCTIONALS

theorems about how well x^* can be approximated by elements of S_M can be used to lead to statements about the error $x^* - \phi_M$ in various norms.

Much of the theory has been developed for the case of S_M being various "piecewise polynomial" subspaces, making use of the well developed theory of spline and polynomial approximation. For example, let P denote the partitition $0 = t_0 < t_1 < \ldots < t_{N+1} = 1$ of $[0, 1]$. For $m \geq 1$, we define

$H_0^m(P) = \{\varphi(t);\ \varphi$ is in $C^{m-1}[0, 1],\ \varphi(0) = \varphi(1) = 0$, on $t \in [t_i, t_{i+1}]$ for $0 \leq i \leq N$ φ is a polynomial of degree at most $2m - 1\}$.

This space $H_0^m(P)$ is spanned by the $m(N+2)-2$ functions $S_{i,k}(t)$, $1 \leq i \leq N$, $0 \leq k\ m-1$ and $1 \leq k \leq m-1$, $i = 0, N+1$, where

$$D^\ell S_{i,k}(t_j) = \delta_{i,j} \delta_{k,\ell} \quad \text{for } 0 \leq \ell \leq m-1.$$

The functions $S_{i,k}(t)$ are zero except in $[t_{i-1}, t_{i+1}]$. For example, with $m = 1$, $H_0^1(P)$ is spanned by the N functions $S_{i,0}$, $1 \leq i \leq N$, where $S_{i,0}(t)$ is given by the roof function

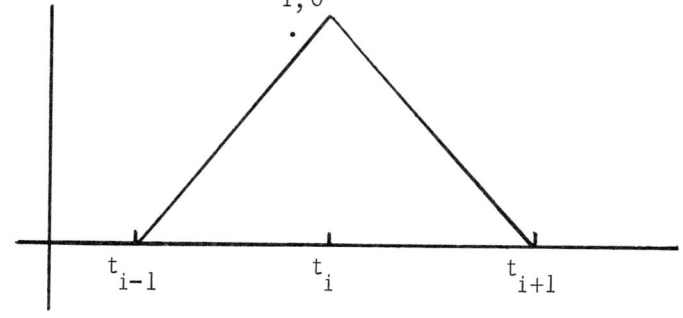

413

By using known results about approximation (in fact, interpolation) by elements of $H_0^m(P)$, we can give bounds for $\phi_M - x^*$, as described above, where $M = m(N+2) - 2$. For example [Ciarlet et al (1967)], if $|P| \equiv \max_{0 \le i \le N} |t_{i+1} - t_i|$, then if $x^* \in C^q[0, 1]$ with $q \ge 2m$, then there exists a constant K such that

$$\|D^k(\phi_M - x^*)\|_\infty \le K \|D^{2M} x^*\|_\infty |P|^{2M-1}, \quad k = 0, 1$$

where $\|u\|_\infty = \operatorname{ess\,sup}_{0 \le t \le 1} |u(t)|$ for $u \in L_\infty(0, 1)$. Thus, if $x^* \in C^2[0, 1]$, we can use H_0^1 as our subspace and find error bounds of order $|P|$. If $x^* \in C^4[0, 1]$ we can use H_0^2 and find bounds of order $|P|^3$. In fact, by more subtle arguments [Perrin-Price-Varga (1969)], one can show that the order of convergence for H_0^1 approximation is actually $|P|^2$.

In a practical sense however the above results are not meaningful unless one can compute ϕ_M, that is, solve

$$Ba + G(a) = 0 \ .$$

For the type of subspaces we are considering, the matrix B can be assumed to be known exactly since it is computed by integration using polynomials; using H_0^m spaces, the matrix in fact is a band matrix. The operator $G(a)$ however involves integration of $g(t, \sum_{i=1}^{M} a_i \varphi_i(t))$ which we cannot perform exactly; the use of a quadrature formula gives us a computable method [Herbold (1968), Herbold et al (1969)]. Suppose we use a quadrature formula

$$\int_{y_0}^{y_k} s(y) dy \doteq \sum_{i=0}^{k} \alpha_i s(y_i)$$

MINIMIZATION OF FUNCTIONALS

with error given by $K_1(\frac{y_k - y_0}{k})^{k_0+1} s^{(k_0)}(\xi)$ as usual. Given

the partition $P: 0 = t_0 < t_1 < \ldots < t_{N+1} = 1$, if we write

$$\int_0^1 s(t) \varphi_i(t) dt = \sum_{i=0}^{N} \int_{t_i}^{t_{i+1}} s(t) \varphi_i(t) dt$$

and apply the quadrature formula to each subinterval we obtain a quadrature formula

$$\int_0^1 s(t) \varphi_i(t) dt \sim \sum_{j=0}^{M_0} \beta_j s(t'_j) \varphi_i(t'_j)$$

where $M_0 = k(N+1)$ and the β_j and t'_j are obtainable from the α_j and t_i. If we use this sum to approximate the operator $G(a)$ we get $\bar{G}_k(a) = (\bar{G}_{k,1}(a), \ldots, \bar{G}_{k,M}(a))^T$,

$$\bar{G}_{k,i}(a) = \sum_{j=0}^{k(N+1)} \beta_j g(t'_j, \sum_{\ell=1}^{M} a_\ell \varphi_\ell(t'_j)) \varphi_i(t'_j) ,$$

and we now can solve

$$Ba + \bar{G}_k(a) = 0$$

numerically. This however is the gradient equation for the functional

$$f_k(a) = \int_0^1 \frac{1}{2} \{D[\sum_{j=1}^{M} a_j \varphi_j(t)]\}^2 dt + \sum_{j=0}^{k(N+1)} \beta_j \int^{\sum_{i=1}^{M} a_i \varphi_i(t'_j)} g(t'_j, \eta) d\eta .$$

Thus what we have now is a discretization scheme; rather than minimize $f(x)$ over E we now minimize $f_k(a_1, \ldots, a_M)$. The various parameters M, N, k, and the particular quadrature rules clearly must be related in some way to maintain our earlier error bounds. We merely state a sample of the type of result that can be obtained.

Theorem 6.1 [Herbold (1968)]. Suppose we use the subspaces $H_0^m(P_N)$ for a partition $P_N: 0 = t_0 < t_1 < \ldots < t_{N+1} = 1$, a subspace of dimension $m(N+2) - 2$, where the partition P_N satisfies $|P_N| \leq L \min_{0 \leq i \leq N} |t_{i+1} - t_i|$ for all N. Suppose $g(t, x)$ is so smooth that for any $\varphi \in H_0^m(P_N)$, $D^s g(t, \varphi(t))$ is continuous on each interval of the partition for $0 \leq s \leq k_0$, where k_0 describes the accuracy of the quadrature formula to be used and satisfies $k_0 \geq 4m - 1$. Suppose that the weights $\alpha_i \geq 0$, $\sum_{i=0}^{k} \alpha_i = y_k - y_0$. Suppose x^* solving the original problem is in $C^{2m}[0,1]$. Then $\phi_M = \phi_{m(N+2)-2}$ minimizing f over H_0^m and $\phi_{M,k}^*$ minimizing f_k are related by

$$\|D^\ell(\phi_M - \phi_{M,k}^*)\|_\infty = O(|P_N|^{2m}), \quad \ell = 0, 1,$$

and

$$\|D^\ell(x^* - \phi_{M,k}^*)\|_\infty = O(|P_N|^{2m-1}), \quad \ell = 0, 1.$$

To be more specific, consider use of $H_0^1(P_N)$; for a quadrature scheme one could use a two-point Gaussian scheme, but we consider

$$\int_{y_0}^{y} s(t) dt \sim \frac{y_2 - y_0}{6} [s(y_0) + 4s(y_1) + s(y_2)], \quad k_0 = 4.$$

Since $k_0 \geq m-1$, we deduce $\|D^\ell(\phi_M - \phi_{M,2}^*)\|_\infty = O(|P_N|^2)$, $\ell = 0, 1$; for the special case of H_0^1 we know that the error in $\phi_M - x^*$ is also $O(|P_N|^2)$, so we conclude, if $x^* \in C^2[0,1]$,

$$\|D^\ell(x^* - \phi_{M,2}^*)\|_\infty = O(|P_N|^2), \quad \ell = 0, 1, N \to \infty.$$

7. Minimizing functions of finitely many variables

Since we have been demonstrating the use of functional analysis, our emphasis in the preceding sections has been on analyzing the relationships between the discrete and continuous problems. From the really practical, computational viewpoint, however, the problem of how to compute the approximate solution to a finite dimensional minimization problem is of tremendous importance; thus we must now make some mention of computational algorithms.

Since many research papers, complete books, and large sections of other books have been written on this subject [Abadie (1967), Cea (1969), Daniel (1971), Fiacco-McCormick (1968), Fletcher (1969), Kowalik-Osborne (1968), Zangwill (1969), Zoutendijk (1969)], we can in our limited space only outline the basic concepts involved. The analysis can conveniently be stated in Hilbert spaces (in fact, in more general Banach spaces [Daniel (1971)]) since the concepts involved such as convergence, continuity, norm, et cetera, do not really depend on finite dimensionality; the reader may, if he wishes, assume everything to be finite dimensional in what follows. We remark that, in the case of finite dimensions, more careful analysis can lead to somewhat sharper results in some cases [Ostrowski (1966), Daniel (1971)].

We shall consider only gradient-like minimization algorithms, that is, algorithms which move from one approximate solution x_n to another x_{n+1} along a direction p_n in which the function is decreasing; that is, we have

$x_{n+1} = x_n + t_n p_n$ for a scalar t_n, where $\frac{d}{dt} f(x_n + tp_n)|_{t=0} = \langle \nabla f(x_n), p_n \rangle < 0$. We do not in general distinguish between constrained and unconstrained problems, except to require for constrained problems that p_n be a <u>feasible direction</u> for the constraint set C, that is, $x_n + tp_n \in C$ for $0 \le t \le 1$ if p_n is properly scaled. Given any such direction sequence and a reasonable way of choosing t_n (to which we return in a moment), one concludes that $\langle \nabla f(x_n), p_n \rangle$ converges to zero [Daniel (1971)]. For unconstrained problems one usually chooses p_n by some method that bounds the angle between p_n and $\nabla f(x_n)$ away from $90°$ and has $\|p_n\| = 1$, that is so that $-\langle \nabla f(x_n), p_n \rangle \ge \varepsilon \|\nabla f(x_n)\|$ for a fixed $\varepsilon > 0$; thus $\langle \nabla f(x_n), p_n \rangle \to 0$ implies $\nabla f(x_n) \to 0$ which usually implies that limit points of $\{x_n\}$ are critical points of f, that is, points where ∇f vanishes. If f is, for example, convex, this then implies that limit points of $\{x_n\}$ minimize f. For most good constrained minimization algorithms such as Frank-Wolfe, variable metric projected gradient, Newton, et cetera, one can also show that $\langle \nabla f(x_n), p_n \rangle \to 0$ is a useful condition [Topkis-Veinott (1967), Zangwill (1969), Daniel (1971)].

Intuitively, the most reasonable choice for t_n with $x_{n+1} = x_n + t_n p_n$ is as that value of t which minimizes $f(x_n + tp_n)$ with respect to t; while this approach and slight modifications of it can be proved to yield $\langle \nabla f(x_n), p_n \rangle \to 0$, the computational problems involved in this exact minimization usually cause people to use other choices of t_n. One possible idea is to evaluate $f(x_n + tp_n)$ for t at some values $t_{n,1}, t_{n,2}, t_{n,3}, \ldots, t_{n,k_n}$, and then let t_n be that t-value yielding the smallest f-value; computationally simple algorithms have been given [Cea (1969), Daniel (1971)] to implement this and can be proved to lead to $\langle \nabla f(x_n), p_n \rangle \to 0$ fairly generally. Other computationally oriented schemes have been proposed and analyzed, based on the idea of forcing a significant decrease in f from x_n to x_{n+1} [Goldstein

(1964), 1965, 1966, 1967), Elkin (1968, Armijo (1966), Daniel (1971)]; such computationally oriented schemes for decreasing rather than minimizing f each time seem very useful and generally are to be preferred.

While a large variety of direction choices may lead to convergent algorithms, the rate of convergence and the computational efficiency can vary drastically. For example, the most obvious choice of p_n is $\frac{-\nabla f(x_n)}{\|\nabla f(x_n)\|}$, the direction in which f is decreasing most rapidly at x_n; this choice in general yields very slow (linear) convergence. Other methods, primarily the conjugate direction or conjugate gradient or variable metric or Newton-like algorithms, of which the Davidon method is perhaps the best known [see Daniel (1971) and the many references therein], usually converge much more rapidly (often superlinearly), to some extent because they solve finite dimensional quadratic minimization problems in finitely many steps. A great assortment of these algorithms is available, depending on the nature of the problem, for example on whether or not one has access to just f, f and ∇f, or even f, ∇f, and f"; under any of these conditions, it appears that the conjugate direction-type algorithm tailored for that class of problems will be one of the best available for that class.

REFERENCES

1. Abadie, J. (1967), editor, <u>Methods of nonlinear programming</u>, North Holland, Amsterdam.

2. Akhiezer, N. (1962), <u>The calculus of variations</u>, Blaisdell, Waltham.

3. Armijo, L. (1966), "Minimization of functions having Lipschitz continuous first partial derivatives," Pacific J. Math., vol. 16, 1-3.

4. Balakrishnan, A. V., Neustadt, L. W. (1964), Computing methods in optimization problems, Academic Press, New York.

5. Cea, J. (1969), Lecture notes on optimization, l'Ecole d'Ete Analyse Numerique, France (French).

6. Ciarlet, P. G. (1966), "Variational methods for nonlinear boundary value problems," Dissertation, Case Inst. Tech.

7. Ciarlet, P. G., Schultz, M. H., Varga, R. S. (1967), "Numerical methods of high order accuracy for nonlinear boundary value problems. I. One dimensional problems," Numer. Math., vol. 9, 394-430.

8. Ciarlet, P. G., Schultz, M. H., Varga, R. S. (1968a), "Numerical methods of high order accuracy for nonlinear boundary value problems. II. Nonlinear boundary conditions," Numer. Math., vol. 11, 331-345.

9. Ciarlet, P. G., Schultz, M. H., Varga, R. S. (1968b), "Numerical methods of high order accuracy for nonlinear boundary value problems. IV. Periodic boundary conditions," Numer. Math., vol. 12, 266-279.

10. Courant, R., Hilbert, D. (1953), Methods of mathematical physics, vol. I. Interscience, New York.

11. Daniel, J. W. (1968), "On the approximate minimization of functionals," Computer Sci. Report #42, U. of Wisc., Madison. Also Math. Comp., 19

12. Daniel, J. W. (1970), "On the convergence of a numerical method for optimal control problems," JOTA, vol. 4, 330-342.

13. Daniel, J. W. (1971), The approximate minimization of functionals, Prentice-Hall, Englewood Cliffs, New Jersey

14. Dunford, N., Schwartz, J. (1962), *Linear operators, I.: General theory*, Interscience, New York.

15. Elkin, R. M. (1968), "Convergence theorems for Gauss-Seidel and other minimization algorithms," Computer Sci. Report #68-59, U. of Maryland, College Park.

16. Fiacco, A., McCormick, G. (1968), *Nonlinear programming: sequential unconstrained minimization techniques*, Wiley, New York.

17. Fletcher, R. (1969), editor, *Optimization*, Academic Press, London.

18. Goldstein, A. A. (1964), "Minimizing functionals on Hilbert space," 159-166 in *Computing methods in optimization problems*, ed. by Balakrishnan and Neustadt, Academic Press, New York.

19. Goldstein, A. A. (1965), "On steepest descent," J. SIAM Control, vol. 3, 147-151.

20. Goldstein, A. A. (1966), "Minimizing functionals on normed linear spaces," J. SIAM Control, vol. 4, 81-89.

21. Goldstein, A. A. (1967), *Constructive real analysis*, Harper and Row, New York.

22. Greenspan, D. (1967), "On approximating extremals of functionals, II," Int. J. Eng. Sci., vol. 5, 571-588.

23. Hadley, G. (1964), *Nonlinear and dynamic programming*, Addison-Wesley, Reading.

24. Herbold, R. J. (1968), "Consistent quadrature schemes for the numerical solution of boundary value problems by variational techniques," Dissertation, Case Western Reserve, Cleveland.

25. Herbold, R. J., Schultz, M. H., Varga, R. S. (1969), "Quadrature schemes for the numerical solution of boundary value problems by variational techniques," Aeq. Math., vol. 3, 96-119.

26. Hestenes, M. R. (1966), Calculus of variations and optimal control theory, Wiley, New York.

27. Kantorovich, L. V. (1948), "Functional analysis and applied mathematics, (Russian), Uspekhi Mat. Nauk, vol. 3, 89-185. Also translated as National Bureau of Standards Report #1509, in 1952.

28. Kantorovich, L. V., Akilov, G. P. (1964), Functional analysis in normed linear spaces, MacMillan, New York.

29. Kowalik, J., Osborne, M. (1968), Methods for unconstrained optimization problems, American Elsevier, New York.

30. Levitin, E. S., Poljak, B. T. (1966a), "Constrained minimization methods," (Russian), Zh. vych. Mat. mat. Fiz., vol. 6, 787-823. Also translated in USSR Comput. Math. Phys., vol. 6, 1-50.

31. Levitin, E. S., Poljak, B. T. (1966b), "Convergence of minimizing sequences in conditional extremum problems," Soviet Math. Dokl., vol. 7, 764-767.

32. Lorentz, G. G. (1966), Approximation of functions, Holt, Rinehart, and Winston, New York.

33. Mangasarian, O. L. (1969), Nonlinear programming, McGraw-Hill, New York.

34. Mikhlin, S. G., Smolitskiy, K. L. (1967), Approximate methods for solution of differential and integral equations, American Elsevier, New York.

35. Morrey, C. B., Jr. (1966), Multiple integrals in the calculus of variations, Springer-Verlag, New York.

36. Ostrowski, A. M. (1966), "Contributions to the theory of the method of steepest descent I," U.S. Army Math. Res. Cntr. Report #615, Madison, Wisconsin

37. Perrin, F. M., Price, H. S., Varga, R. S. (1969), "On higher order accuracies for numerical methods arising from variational methods for nonlinear two-point boundary value problems," Numer. Math., vol. 13, 180-198.

38. Poljak, B. T. (1966), "Existence theorems and convergence of minimizing sequences in extremum problems with restrictions," Soviet Math. Dokl., vol. 7, 72-75.

39. Pontryagin, L. S., Boltyanskii, V. G., Gamkrilidze, R. V., Miscenko, E. F. (1962), The mathematical theory of optimal processes, Wiley, New York.

40. Rabinowitz, P. (1968), "Applications of linear programming to numerical analysis," SIAM Rev., vol. 10, 121-160.

41. Rosen, J. B. (1966), "Iterative solution of nonlinear optimal control problems, J. SIAM Control, vol. 4, 223-244.

42. Rosen, J. B. (1968), "Approximate solution and error bounds for quasi-linear elliptic boundary value problems," Computer Sci. Report #30, U. of Wisc., Madison.

43. Rosen, J. B., Meyer, R. (1967), "Solution of nonlinear two point boundary value problems by linear programming," Computer Sci. Report #1, U. of Wisc., Madison.

44. Taylor, A. E. (1961), Introduction to functional analysis, Wiley, New York.

45. Topkis, D. M., Veinott, A. F., Jr. (1967), "On the convergence of some feasible direction algorithms for nonlinear programming," SIAM J. Control, vol. 5, 268-279.

46. Vainberg, M. M. (1964), <u>Variational methods for the study of nonlinear operators</u>, Holden-Day, San Francisco.

47. Zangwill, W. I. (1967), "Minimizing a function without calculating derivatives," Computer J., vol. 10, 293-296.

48. Zoutendijk, G. (1960), <u>Methods of feasible directions</u>, Elsevier, Amsterdam.

Presently visiting at the Mathematics Department, University of Texas, Austin, Texas

Toward a Unified Convergence Theory for Newton-Like Methods

J. E. DENNIS, JR.

1. Introduction

Let F be a nonlinear operator mapping some subset D of a real Banach space X into a subset of a real Banach space Y. Our concern here is with a certain technique often called "majorization," for analysis of iterative methods for generating approximate solutions to the equation

(1.1) $\qquad F(x) = 0, \quad x \in X, \; 0 \in Y \; .$

The personal bias of the author is toward the special problem $X = E^N = Y$ where E^N denotes the space of real N-tuples and the norms on the two spaces are determined by convenience. This problem will be denoted by

(1.2) $\qquad F(x) = \begin{pmatrix} f_1(x) \\ \vdots \\ f_i(x) \\ \vdots \\ f_N(x) \end{pmatrix} = \begin{pmatrix} 0 \\ \vdots \\ 0 \\ \vdots \\ 0 \end{pmatrix} = 0, \quad x \in E^N \; .$

Such a bias on the part of a numerical analyst requires no

explanation but deserves mention because of its influence on the development below. All the work given here is applicable to (1.1) but it owes much of its specific form to the desire to analyze practical computer algorithms for (1.2).

Since our main concern is with majorization as a technique for analysis of algorithms rather than with the algorithms themselves, we will not set out to survey the algorithms available for the problem. This would be a futile undertaking at the present time when new and very promising methods are being discovered frequently. We will mention some of our special favorites, either because they work well in practice or because they serve to illustrate the flexibility of the theory.

We are concerned here with a class of methods which we will call Newton-like. A Newton-like method can be defined as any iterative method of the form

(1.3) $$x_{n+1} = x_n - A_n^{-1} F(x_n) \quad n = 0, 1, \ldots; \; x_0 \text{ prechosen,}$$

for generating approximate solutions to (1.1). Here $\{A_n\}$ denotes a sequence of invertible linear operators. This is plainly too general and what is really implicit in the title is that A_n should be a conscious approximation to $F'(x_n)$, since when $A_n = F'(x_n)$, the method is the obvious generalization of the classical Newton or Newton-Raphsen method.

L. V. Kantorovich is generally credited with being the first to use the so-called 'method of majorants' in the convergence analysis of iterative schemes for solving nonlinear equations in abstract spaces. The most convenient reference is [19] though the original Russian reference is probably [12]. His approach was to study Newton's method for (1.1) by showing that the vector Newton correction was dominated, under certain hypotheses, by a null sequence of scalar Newton corrections for a related real function f. Kantorovich made no comment about the function f and its dependence on F or on the iterative method. Also, f was first exhibited under conditions which made its Newton sequence converge. This further obscured the relationship between f, F and the iteration scheme.

L. Collatz and J. Schroder [6], [34], [35], studied the relationship between a general vector iteration function and a majorizing real iteration function. This still gave no clue as to how to build the scalar iteration function given the vector iteration scheme.

W. C. Rheinboldt [32] in his paper on the convergence of these methods, was able to obtain very general convergence theorems by abandoning the F, f relationship and studying the iterative corrections in terms of nonlinear difference equations. J. M. Ortega [24] obtained a very short proof of the Kantorovich theorem by the same techniques. These ideas are collected in Chapter 12 of [25].

Although the F, f relationship was not overtly mentioned, one who is interested in the problem gains insight from Rheinboldt's work. We have given up some of Rheinboldt's generality to concentrate on developing this branch of the theory since it is more applicable to methods actually in use.

This work was begun in [12] and the results there are of the following type. Given an operator F and a rule A which assigns to each x an approximate Jacobian A(x), a simple technique is given to build a scalar function f and a scalar rule a(t) such that if the corresponding scalar Newton-like method for f converges then so does the vector method for F. In Section 2 we show that these techniques work in the much more useful case when the approximate Jacobian depends on factors other than the current iterate. In Section 3 we show that a well known theorem about steepest descent, the Kantorovich theorem, an improvement of Schmidt's theorem on the generalized secant method and some other theorems about more computationally useful methods are easily unified by these ideas.

In Section 4 we present an error analysis based on the majorant theory. A lower bound on the error is given which is method independent and computationally free. The upper error bound given is free for some methods and we propose and test a technique for its estimation in the case of Broyden's method. The results are very promising.

Section 5 is a discussion of the relationship between our "semilocal" analysis and local and global convergence theorems. Section 6 mentions some other attacks on the solution of (1.1).

2. Some Results on Majorization

Suppose we have a problem (1.1) and an initial point x_0 for a specific iterative method of the form (1.3). We want to know if (1.1) has a solution x^*, where x^* is in relation to x_0 and what will happen if we carry out the iteration from x_0. The idea that leads us to a solution is very simple. Think of carrying out the iteration and connecting successive iterates x_n, x_{n+1} by directed line segments. We now have a polygonal arc in X which becomes fuzzy after finitely many segments. Now we take x_0 between our left thumb and forefinger and smooth the arc out with our right hand along a convenient real line. Relabel the real nodes $t_0, t_1, \ldots,$ then notice that we have transformed the problem of what happens to the vector iteration sequence into a problem about a monotone nondecreasing real sequence. Elementary analysis tells us that such a real sequence is a Cauchy sequence if and only if it is bounded. Obviously the sequence $\{x_k\}$ is a Cauchy sequence if and only if $\{t_k\}$ is, and since X and E^1 are complete spaces, $\{x_k\}$ converges if $\{t_k\}$ converges. Suppose then that $\{t_k\}$ converges to t^* so that $\{x_k\}$ converges to some x^*. The form of the vector iteration must tell us if $F(x^*) = 0$ — we'll assume enough so it is — but we do know that x^* is no further from any x_n than t^* is from the corresponding t_n. Thus the scalar sequence can tell us almost everything we want to know about the vector iteration. Our first concern here will be with the construction of $\{t_k\}$ as a scalar iteration sequence generated by applying a scalar analog of (1.3) to a scalar analog of (1.1). In carrying out this construction it will be necessary to make certain assumptions which allow intervals between successive scalar iterates to sometimes be longer than the corresponding intervals between vector iterates. Hence we will not know that $\{t_k\}$ converges whenever $\{x_k\}$ does, but we will know the converse. In order to obtain a convergence theorem it is thus sufficient to make hypotheses, phrased in terms of the vector iteration, which make any limit of the vector iteration be a solution of (1.1) and force the scalar iteration to exist and converge.

The following lemma is a formal statement of our remarks concerning the polygonal arc and the real arc.

Lemma 2.1 [24]

Let $\{x_k\}$ be a sequence in X and $\{t_k\}$ a sequence of non-negative real numbers such that

(2.1) $\quad \|x_{k+1} - x_k\| \leq t_{k+1} - t_k, \quad k = 0, 1, \ldots$

and $t_k \to t^* > \infty$. Under these conditions, there exists a point $x^* \in X$ such that $x_k \to x^*$ and

(2.2) $\quad \|x^* - x_k\| \leq t^* - t_k, \quad k = 0, 1, \ldots$.

Definition 2.2 [19, page 269]

Let t_0 and t' be non-negative real numbers and let g be a continuously differentiable real function on $[t_0, t_0 + t']$ and G be a continuously differentiable operator on $\bar{N}(x_0, t') \subset X$ into X. Then the equation $t = g(t)$ will be said to majorize the equation $x = G(x)$ on $N(x_0, t')$ if

(2.3) $\quad \|G(x_0) - x_0\| \leq g(t_0) - t_0$

(2.4) $\quad \|G'(x)\| \leq g'(t),$ when $\|x - x_0\| \leq t - t_0 < t'$.

We will usually say g majorizes G instead of referring to the equations.

The following theorem hints that this concept will be useful in proving the existence and uniqueness of solutions.

Theorem 2.3 [19, page 697]

If g majorizes G on $\bar{N}(x_0, t')$ and g has a fixed point in $[t_0, t_0 + t']$ then G has a fixed point x^* in $\bar{N}(x_0, t')$. Furthermore $x_{n+1} = G(x_n)$ and $t_{k+1} = g(t_k)$, $k = 0, 1, \ldots$, converge to x^* and t^* respectively with the real sequence majorizing the vector sequence.
t^* is the smallest fixed point of g in $[t_0, t' + t_0]$. If in addition t^* is the only fixed point of g in $[t_0, t' + t_0]$ and $g(t' + t_0) \le t' + t_0$ then x^* is unique in $\bar{N}(x_0, t')$ and $x'_{n+1} = G(x'_n)$ $n = 0, 1, \ldots$ converges to x^* from any $x'_0 \in \bar{N}(x_0, t')$.

The next theorem shows how to construct a scalar iteration which majorizes the vector iteration (1.3).

Theorem 2.4

Let $F' \in \text{Lip}_K D_0$ (*), where D_0 is the closure of an open convex set and $D_0 \subset D$. Assume that for every n with $x_0, \ldots, x_n$ defined by (1.3) all in D_0, there is an invertible $A_n \in L(X, Y)$ and a positive real number a_n such that:

(2.5) $$\|A_n^{-1}\| \le a_n^{-1}$$

and for $\sigma \ge 1, \Delta > 0$, both independent of n,

(*) $F' \text{Lip}_K D_0 \equiv \forall x, y \in D_0, \|F'(x) - F'(y)\| \le K\|x - y\|$, where the derivative is the Fréchet derivative.

(2.6) $\quad a_n + \sigma K \sum_{j=1}^{n} \|x_j - x_{j-1}\| - \Delta \geq \|F'(x_n) - A_n\|$ (†).

Under these conditions, set

$$f(t) = \frac{1}{2} \sigma K t^2 - \Delta t + a_0 \|A_0^{-1} F(x_0)\| \quad \text{and}$$

$$t_{n+1} = t_n + f(t_n)/a_n; \quad t_0 = 0$$

majorizes

$$x_{n+1} = x_n - A_n^{-1} F(x_n); \quad x_0 \in D_0$$

as long as $\{x_n\}$ remains in D_0.

<u>Proof:</u> $t_1 - t_0 = a_0 \|A_0^{-1} F(x_0)\|/a_0 = \|x_1 - x_0\|$.

Assume by way of induction that $\{x_0, \ldots, x_n\} \subset D_0$ and $\|x_{j-1} - x_j\| \leq t_j - t_{j-1}$ for $j = 1, \ldots, n$. To complete the induction, notice that x_{n+1} is defined since $F(x_n)$ is defined and A_n^{-1} exists. Notice also that

$\sum_{j=1}^{n} \|x_j - x_{j-1}\| \leq t_n$. Hence

(†) The convention $\sum_{j=1}^{0} = 0$ is understood.

431

$$\|x_{n+1} - x_n\| \le \|A_n^{-1}\| \cdot \|F(x_n)\|$$

$$\le a_n^{-1}[\|F(x_{n-1} - A_{n-1}^{-1} F(x_{n-1})) - F(x_{n-1}) - F'(x_{n-1})(x_n - x_{n-1})\|$$

$$+ \|A_{n-1} - F'(x_{n-1})\| \cdot \|x_n - x_{n-1}\|]$$

$$\le a_n^{-1}[\tfrac{1}{2} K\|x_n - x_{n-1}\|^2 + (a_{n-1} + \sigma K t_{n-1} - \Delta)\|x_n - x_{n-1}\|]$$

$$\le a_n^{-1}[\tfrac{1}{2} \sigma K(t_n - t_{n-1})^2 + f'(t_{n-1})(t_n - t_{n-1}) + a_{n-1}(t_n - t_{n-1})]$$

$$\le a_n^{-1}[\tfrac{1}{2} f''(t_{n-1})(t_n - t_{n-1})^2 + f'(t_{n-1})(t_n - t_{n-1}) + f(t_{n-1})]$$

$$\le a_n^{-1} f(t_n) = t_{n+1} - t_n$$

since f is a quadratic and the induction is complete.

We think of (2.6) as expressing the fact that the sequence of Jacobian approximations is of "bounded deteriation". That is, although $A_n \not\to F'(x^*)$ as $x_n \to x^*$, the divergence is proportional to the distance the algorithm has moved from its starting point. (2.6) also says $f'(t_n) - (-a_n) \ge \|A_n - F'(x_n)\|$.

At this point we want some conditions which ensure that $\{t_n\}$ converges or equivalently, that $\{t_n\}$ is bounded. If $\{t_n\}$ converges to t^*, then t^* is the least upper bound for $\sum_{j=1}^{n} \|x_j - x_{j-1}\|$, independent of n. Hence if we then assume that $N(x_0, t^*) \subset D_0$, we will have immediately that $\{x_n\}$ exists and converges under the hypotheses of the previous theorem. This will be the topic of Theorem 2.6, but first we need the following theorem about existence of roots.

Theorem 2.5

Let $F' \in \text{Lip}_K D_0$ and let (2.5) hold for $n = 0$. In addition assume that

$$1 > \delta' \equiv \|F'(x_0) - A_0\|/a_0 \tag{2.7}$$

$$\frac{1}{2} \geq h' \equiv \frac{K\|A_0^{-1} F(x_0)\|}{a_0(1-\delta')^2} \tag{2.8}$$

$$N(x_0, r_0') \subset D_0, \text{ where } r_0' \equiv \frac{1-\sqrt{1-2h'}}{K}(1-\delta')a_0 . \tag{2.9}$$

Under these conditions, F has a unique root $x^* \in \overline{N}(x_0, r_0')$. If $h' < \frac{1}{2}$, it is also unique in $D_0 \cap N(x_0, r_1')$,
$r_1' \equiv \frac{1+\sqrt{1-2h'}}{K}(1-\delta')a_0$.

$$x_{n+1}' = x_n' - A_0^{-1} F(x_n') ; \tag{2.10}$$

$\|x_0' - x_0\| < r_2$ converges to x^* where $r_2 \leq r_1'$ and $N(x_0, r_2) \subset D_0$. $\|x_n' - x^*\| \leq |r_0' - t_n'|$, where $\{t_n'\}$ is generated by

$$t_{n+1}' = (2a_0)^{-1} K(t_n')^2 + \delta' t_n' + \|A_0^{-1} F_0\|; \ \|x_0 - x_0'\| \leq t_0' < r_2 \leq r_1'. \tag{2.11}$$

Proof: Set $g(t) = t + \bar{f}(t)/a_0$, where $\bar{f}(t) \equiv \frac{1}{2} Kt^2 - a_0(1-\delta')t + a_0\|A_0^{-1} F(x_0)\|$ and $G(x) = x - A_0^{-1} F(x)$. We will accomplish the proof by showing that if $t' \in [r_0', r_1')$, then $g(t)$

majorizes $G(x)$ on $\overline{N}(x_0, t') \cap D_0$.

First note that $\|G(x_0) - x_0\| = \|A_0^{-1} F(x_0)\| = g(0) - 0$.
Now let $x \in \overline{N}(x_0, t') \cap D_0$ and let t have the property that $\|x - x_0\| \le t < t'$.

$$g'(t) = 1 + \frac{\bar{f}'(t)}{a_0} = 1 + \frac{Kt}{a_0} - (1 - \delta') = (Kt + a_0 \delta')/a_0$$

$$\ge \|A_0^{-1}\| [K\|x - x_0\| + \|F'(x_0) - A_0\|]$$

$$\ge \|A_0^{-1}\| [\|F'(x) - F'(x_0)\| + \|F'(x_0) - A_0\|]$$

$$\ge \|I - A_0^{-1} F'(x)\| = \|G'(x)\|.$$

Clearly r_0' is the only fixed point of $g(t)$ in $[0, t']$ and $g(t') \le t'$ with equality if and only if $t' = r_0$.

It is often the case that one does not wish to recalculate the approximate derivative at every iteration but will instead use A_n in place of $A_{n+1}, \ldots, A_{n+p}$ and then calculate A_{n+p+1} and use it for q iterations. In the next theorem we give convergence conditions which show that these computationally convenient modifications are allowable so long as (2.5) and (2.6) hold for $n + p + 1$. The author studied this modification of Newton's method, $A_n = F'(x_n)$, in [11] and found the following kind of notation to be useful. Let $\{\alpha_n\}$ be a nondecreasing sequence of nonnegative real numbers such that $\alpha_0 = 0$ and $\alpha_n = \alpha_{n-1}$ or $\alpha_n = n$. Then the general iteration referred to above can be written as

(2.12) $$x_{n+1} = x_n - A_{\alpha_n}^{-1} F(x_n).$$

434

Theorem 2.6

Let the hypotheses of Theorem 2.4 hold and assume in addition that $\{a_n\}$ is uniformly bounded above and

$$(2.13) \qquad \frac{1}{2} \geq h \equiv \frac{\sigma K \|A_0^{-1} F(x_0)\| a_0}{\Delta^2}$$

$$(2.14) \qquad N(x_0, r_0) \subset D_0, \quad \text{where} \quad r_0 = \frac{1 - \sqrt{1-2h}}{\sigma K} \Delta .$$

Under these conditions F has a root x^* to which (2.12) converges according to

$$(2.15) \qquad \|x_{n+1} - x^*\| \leq r_0 - t_n - a_n^{-1} (\frac{1}{2} \sigma K t_n^2 - \Delta t_n$$
$$+ a_0 \|A_0^{-1} F(x_0)\|); \quad t_0 = 0 .$$

Let h' and r_1' be defined by Theorem 2.5 and x^* is the only root of F in $N(x_0, r_1') \cap D_0$ if $h' < \frac{1}{2}$.

Proof: Define $B_n = \frac{A_n}{\alpha_n}$ and $b_n = \frac{a_n}{\alpha_n}$, $n = 0, \ldots$. We will show that every x_n is in $N(x_0, r_0) \subset D_0$ and the majorizing relations of Theorem 2.4 hold. We will also show that t_n converges to r_0 and so $\{x_n\}$ must converge. We will then show that $\lim x_n$ is a root of F and that $h' \leq h$ and $r_0' \leq r_0$ so the unicity results of Theorem 2.5 apply.

Assume by way of induction that $\sum_{j=1}^{n} \|x_j - x_{n-1}\| < r_0$.

Now (2.5) obviously holds for B_n and b_n. Furthermore, so does (2.6), since $\sigma \geq 1$ and $\alpha_n \leq n$ and so

$$\|B_n - F'(x_n)\| \equiv \|A_{\alpha_n} - F'(x_n)\|$$

$$\leq \|A_{\alpha_n} - F'(x_{\alpha_n})\| + \|F'(x_{\alpha_n}) - F'(x_n)\|$$

$$\leq a_{\alpha_n} + \sigma K \sum_{j=1}^{\alpha_n} \|x_j - x_{j-1}\| - \Delta + K\|x_{\alpha_n} - x_n\|$$

$$\leq a_{\alpha_n} + \sigma K \sum_{j=1}^{n} \|x_j - x_{j-1}\| - \Delta \equiv b_n +$$

$$+ \sigma K \sum_{j=1}^{n} \|x_j - x_{j-1}\| - \Delta .$$

If we let f be defined by Theorem 2.4, r_0 is the smallest zero of f. Define $g_n(t) = t + f(t)/b_n$. We know from Theorem 2.4 that $\|x_{n+1} - x_n\| \leq g_n(t_n) - t_n$ and we will now show that $t_{n+1} < r_0$ and so $\{t_n\}$ is bounded and convergent. But then $\sum_{j=1}^{n+1} \|x_{j-1} - x_j\| \leq \sum_{j=1}^{n+1} t_j - t_{j-1} = t_{n+1} < r_0$ and so $\{x_n\}$ is defined by induction and convergent by Lemma 2.1.

By the mean value theorem there is some $\xi_n \in (t_n, r_0)$ such that

$$r_0 - t_{n+1} = g_n(r_0) - g_n(t_n) = g'_n(\xi_n)(r_0 - t_n)$$

$$= [1 + (\sigma K \xi_n - \Delta)/b_n](r_0 - t_n)$$

$$= b_n^{-1}[b_n + \sigma K \xi_n - \Delta](r_0 - t_n) .$$

We can use the bounds on a_n as well as (2.6) and (2.14) to give $0 \leq b_n^{-1}[b_n + \sigma K t_n - \Delta](r_0 - t_n) < r_0 - t_{n+1} < b_n^{-1}[b_n + \sigma K r_0 - \Delta](r_0 - t_n) \leq r_0 - t_n$. $0 < r_0 - t_{n+1} < r_0 - t_n$ and $\{t_n\}$ converges to some $t^* \leq r_0$. $0 = \lim_n t_{n+1} - t_n = \lim_n \dfrac{f(t_n)}{b_n} \geq \lim_n \dfrac{f(t_n)}{a} = \dfrac{f(t^*)}{a}$ where a denotes the uniform upper bound on $\{a_n\}$ and hence on $\{b_n\}$. But this means $f(t^*) = 0$ so $t^* = r_0$ and (2.15) holds.

We must now show that $x^* = \lim_n x_n$ is a root of F. $\|F(x_n)\| \leq \|B_n(x_{n+1} - x_n)\|$ and $\|x_{n+1} - x_n\| \to 0$ so it suffices to show that $\|B_n\|$ is uniformly bounded. This follows readily from (2.6) since

$$\|B_n\| \leq \|F'(x_n)\| + b_n - \Delta + \sigma K \sum_{j=1}^{n} \|x_j - x_{n-1}\|$$

$$\leq \|F'(x_0)\| + K\|x_0 - x_n\| + b_n - \Delta + \sigma K r_0$$

$$\leq \|F'(x_0)\| + (\sigma + 1) K r_0 + a - \Delta .$$

All that is left is to show that (2.7), (2.8) and (2.9) are satisfied so that the unicity assertions hold. We can see by setting $n = 0$ in (2.6) that $0 < \Delta \leq a_0 - \|F'(x_0) - A_0\|$. Hence, $0 < \Delta/a_0 \leq 1 - \delta'$ so (2.7) holds but also

$$\Delta^2/a_0^2 \leq (1 - \delta')^2 \text{ so } a_0/\Delta^2 \geq 1/a_0(1 - \delta')^2 .$$

This last inequality together with $\sigma \geq 1$ and (2.13) imply (2.8). It is now straightforward that $r_0' \leq r_0 \leq r_1'$.

3. Applications of the Majorant Theory

In this section we will give an indication of the flexibility of our version of the majorant theory. As we mentioned in the introduction, the specific methods mentioned here were mostly chosen to the aforementioned end. The purpose of the theory is to provide tools to analyze specific methods and not to prove general convergence theorems. However, we will prove one general theorem, Theorem 3.2. It is included for a variety of reasons. First it seems to reduce to a minimum the assumptions necessary to apply the majorant technique. Second, it gives an easy semilocal convergence theorem for the class of methods which use what Ortega and Rheinboldt call "consistent" derivative approximations. This result along with the very interesting Theorem 11.3.3, page 371, of the same reference, gives a unified result for most of the generalized secant methods, a thing which they seem to despair of majorization being able to do (see Section NR12.6[25]). In addition, two of the Davidon methods are readily analyzed by this theorem.

We give a separate result for the Schmidt [33] secant hypothesis in order to illustrate our opinion that most methods should be studied separately rather than having Theorem 3.2 applied to them. The theorem, Corollary 3.8, on Schmidt's method includes the method of Laarsonen [20]. We will also prove a version of the Contractive Mapping Theorem (3.9) and Goldstein's theorem (3.10) about descent using step lengths determined by the spectrum of the Hessian.

All our theorems will be about the iteration (2.12) and so the following definition will be useful.

Definition 3.1. $\{\alpha_n\}$ is a <u>recalculation sequence</u> if for every nonnegative integer n, α_n is a real number and $\alpha_n = n$ or $\alpha_n = \alpha_{n-1}$, $\alpha_0 = 0$.

Theorem 3.2

Let $F' \in \text{Lip}_K D_0$ and let $x_0 \in D_0$, A_0 be an invertible element of $L(X, Y)$ with $\|A_0^{-1} F(x_0)\| \leq \eta$, $\|A_0^{-1}\| < \beta$. Assume that there are nonnegative real numbers δ and γ such that for every n for which $x_0, \ldots, x_n$ as defined by (2.12), are in D_0,

(3.1) $$\|A_n - F'(x_n)\| \leq \delta_n + \gamma \sum_{j=1}^{n} \|x_j - x_{j-1}\| ;$$

$$\delta_n \leq \delta \text{ for } n > 0.$$

If, in addition

(3.2) $$1 > \beta\delta_0 + 2\beta\delta$$

(3.3) $$\frac{1}{2} \geq h \equiv \frac{(2\gamma + K)\beta\eta}{(1 - 2\beta\delta - \beta\delta_0)^2}$$

and

(3.4) $N(x_0, r_0) \subset D_0$, where $r_0 = \dfrac{1 - \sqrt{1 - 2h}}{\beta(2\gamma + K)} (1 - 2\beta\delta - \beta\delta_0)$

then $\{x_n\}$ generated by (2.12) exists in $N(x_0, r_0)$ and converges to x^*. x^* is the unique root of F in $\overline{N}(x_0, \dfrac{1 - \sqrt{1 - 2h'}}{\beta K}(1 - \beta\delta_0))$, where $h' \equiv \dfrac{\beta K \eta}{(1 - \beta\delta)^2}$ and if $h' < \dfrac{1}{2}$, then x^* is unique in $D_0 \cap N(x_0, \dfrac{1 + \sqrt{1 - 2h'}}{\beta K}(1 - \beta\delta_0))$.

Furthermore, set $t_1 = \eta$ and for $n \geq 1$

$$(3.5) \quad \|x_{n+1} - x^*\| \le r_0 - t_n - (\frac{1}{2}\beta(K+2\gamma)t_n^2$$

$$- (1-\beta(\delta_0+2\delta)t_n + \eta)/(1-\beta(\delta_{\alpha_n} + \delta_0) - \beta(K+\gamma)t_n)$$

Proof: The proof will be by Theorem 2.6. a_0 is of course $1/\beta$ but we must now find the other terms of $\{a_n\}$. Assume that $x_0, \ldots, x_n$ are all elements of $N(x_0, r_0)$ and

$$r_0 > \sum_{j=1}^{n} \|x_j - x_{j-1}\|.$$

$$\|A_n - A_0\| \le \|A_n - F'(x_n)\| + \|F'(x_n) - F'(x_0)\| + \|F'(x_0) - A_0\|$$

$$\le \delta_n + \gamma \sum_{j=1}^{n} \|x_j - x_{j-1}\| + K\|x_n - x_0\| + \delta_0$$

$$\le \delta_0 + \delta_n + (K+\gamma) \sum_{j=1}^{n} \|x_j - x_{j-1}\| < \delta_0 + \delta + (K+\gamma)r_0$$

$$\le \delta_0 + \delta + (1-\sqrt{1-2h})(1-2\beta\delta - \beta\delta_0)/\beta \le 1/\beta.$$

Hence $\|A_0^{-1}A_n - I\| \le \beta(\delta_0 + \delta_n) + \beta(K+\gamma) \sum_{j=1}^{n} \|x_j - x_{n-1}\|$ and by the Banach Lemma [ ], A_n^{-1} exists and is bounded in norm by $\beta(1-\beta\delta_n - \beta\delta_0 - \beta(K+\gamma) \sum_{j=1}^{n} \|x_j - x_{j-1}\|)^{-1} \equiv a_n^{-1}$, $n > 0$.
Note that $\{a_n\}$ is uniformly bounded by a_0. Now we need to find σ and Δ such that

$$(1-\beta\delta_n - \beta\delta_0 - \beta(K+\gamma)\sum_{j=1}^{n}\|x_j - x_{j-1}\|)\beta^{-1} + \sigma K \sum_{j=1}^{n}\|x_j - x_{j-1}\| - \Delta$$

$$\geq \delta_n + \gamma \sum_{j=1}^{n}\|x_j - x_{j-1}\| \ .$$

First let $n = 0$ and take $\Delta \equiv (1 - 2\beta\delta - \beta\delta_0)/\beta > 0$ by (3.2). $\sigma \equiv (K + 2\gamma)/K$ will then work. (3.3) is now (2.13) and (3.4) is (2.14).

We could have defined h' earlier in the statement of the theorem and given the less restrictive existence statement based on Theorem 2.5. This can be a very useful type of result (see [5] for an application) but we wanted to leave our main concern, convergence of (2.12), as uncluttered as possible.

Theorem 3.2 gives a wide generalization of a theorem of Rheinboldt [32] or, [25] page 425. Rheinboldt assumes that

(i) $A(x)$ is defined for each $x \in D_0$ and that

(ii) $\|A(x) - A(x_0)\| \leq \mu \|x - x_0\|$ as well as

(iii) $\|A(x) - F(x)\| \leq \delta_0 + \gamma \|x - x_0\|$.

The second is the most severe requirement since it would require that if x_n gets close to x_0 for any n then A_n must get close to A_0. The Davidon, or quasi-Newton, methods form a class of very useful methods which do not satisfy this requirement as indeed do the whole class of multipoint methods. In [13], [3], [17] one can find examples of the use of Theorem 3.2 in the analysis of Davidon or Davidon-related methods. In [12], the author kept (i) and (iii) but used (ii') $\|A(x) - A(x_0)\| \leq \mu_0 + \mu_1 \|x - x_0\|$; see that reference for the corresponding theorem.

We think of (3.1) as expressing the idea that A_n is obtained by some approximation rule which is of "bounded deterioration". Theorem 3.2 gives a semilocal theorem for any method based on a derivative approximation of bounded deterioration. Ortega and Rheinboldt [25] define what they call a <u>consistent derivative approximation.</u> Basically this says that the approximation rule depends on an outside vector parameter h_n and that given any $\delta > 0$, one can choose an $\varepsilon > 0$ such that as long as one restricts $\|h_n\| < \varepsilon$, (3.1) will hold with $\gamma = 0$. They define a <u>strongly consistent approximation</u> as one in which there is a constant c such that δ_n can be taken as $c\|h_n\|$. In the very elegant development mentioned at the beginning of the section, they show that in many of the well-known methods utilize consistent approximations. It is therefore worthwhile that we present the following corollary of Theorem 3.2, which also seems to bear on NR11.2-6, page 365 of [25].

<u>Corollary 3.3</u>

Let $F' \in \text{Lip}_K D_0$ and let $x_0 \in D_0$, A_0 be an invertible element of $L(X,Y)$ with $\|A_0^{-1} F(x_0)\| \leq \eta$, $\|A_0^{-1}\| \leq \beta$. Assume that for every $n > 0$ such that $x_0, \ldots, x_n$ in D_0 are defined by (2.12) and for every $h \in H \subset Z$, a normed space, there is an element $A_n(h) \in L(X,Y)$ such that, for some $c \geq 0$, independent of n and h

$$\|A_n(h) - F'(x_n)\| \leq c\|h\|.$$

If, in addition, $\|A_0 - F'(x_0)\| \leq \delta_0$ and $\|h_n\| \leq \varepsilon$, $\frac{1}{2} \geq \frac{\beta K \eta}{(1-2\beta c-\beta \delta_0)^2} \equiv h$ and $N(x_0, r_0) \subset D_0$ for $r_0 = \frac{1-\sqrt{1-2h}}{\beta K}(1 - 2\beta c \varepsilon - \beta \delta_0)$, then, for any recalculation sequence $\{\alpha_n\}$,

$$x_{n+1} = x_n - A_{\alpha_n}(h_{\alpha_n})^{-1} F(x_n)$$

converges to x^*. x^* is the unique root of F in $\overline{N}(x_0, \frac{1-\sqrt{1-2h'}}{K}(1-\beta\delta_0))$, where $h' \equiv \frac{\beta K \eta}{(1-\beta\delta_0)^2}$ and if $h' < \frac{1}{2}$,

then x^* is unique in $D_0 \cap N(x_0, \frac{1+\sqrt{1-2h'}}{K}(1-\beta\delta_0))$.

Furthermore, set $t_1 = \eta$ and for $n \geq 1$

$$\|x^* - x_{n+1}\| \leq r_0 - t_n$$

$$-(\frac{1}{2}\beta K t_n^2 - (1-\beta(\delta_0 + 2c\varepsilon))t_n + \eta)/(1-\beta(c\|h_n\| + \delta_0) - \beta K t_{\alpha_n}).$$

Proof: Set $\delta_n = c\|h_n\| \leq c\varepsilon \equiv \delta$ and $\gamma = 0$ in Theorem 3.2.

This theorem is structured exactly like the theorem on the discretized Newton's method given by the author in [12]. There h_n is a constant times $F(x_n)$.

Notice that the scalar majorizing iteration in (3.5) is Newton's method only if $\delta = 0 = \gamma$ i.e., the vector iteration is. This case is given now.

Corollary 3.4 (The Kantorovich Theorem)

Let $F \in \text{Lip}_K D_0$ and for some $x_0 \in D_0$ let $[F'(x_0)]^{-1}$ exist with $\|F'(x_0)^{-1}\| \leq \beta$ and $\|F'(x_0)^{-1} F(x_0)\| \leq \eta$. If, in addition,

(3.6) $$\frac{1}{2} \geq h \equiv \beta K \eta$$

and

(3.7) $$N(x_0, r_0) \subset D_0, \text{ where } r_0 = \frac{1-\sqrt{1-2h}}{\beta K}$$

and $\{\alpha_n\}$ is an arbitrary recalculation sequence, then Newton's method with recalculation sequence $\{\alpha_n\}$ converges from x_0 to x^*, the unique root of F in $\overline{N(x_0, r_0)}$. If $h < \frac{1}{2}$, then x^* is also unique in $D_0 \cap N(x_0, r_1)$, where

$$r_1 = \frac{1 + \sqrt{1-2h}}{\beta K} \ .$$

The sequence $\{x_n\}$ generated by Newton's method with any recalculation sequence $\{\alpha_n\}$ satisfies

$$\|x_{n+1} - x^*\| \leq r_0 - t_n - (\frac{1}{2} Kt_n^2 - t_n + \eta)/(1 - Kt_{\alpha_n}); \ t_0 = 0 \ .$$

Proof: Set $\gamma = 0 = \delta_n = \delta$, $n = 0, 1, \ldots$ in Theorem 3.2.

In [11], [37] and [31] one can find more elegant error bounds for the Kantorovich Theorem. Since the ones given here are furnished by Newton's method for $\{t_n\}$, they will reflect quadratic convergence.

At this point we will apply Theorem 2.6 to Schmidt [37] and Laarsonen's [20] versions of the generalized secant method. These methods are not particularly practical but for some special problems they can be quite useful. This theorem is instructive because, although Corollary 3.3 would apply we get much better results by going back to Theorem 2.6.

We will assume, with Schmidt that for any $x, y \in D_0$, there is a divided difference operator $\delta F(x, y) \in L(X, Y)$ such that

(3.8) $\qquad \delta F(x, y)(x - y) = F(x) - F(y)$

and if $u \in D_0$,

(3.9) $\qquad \|\delta F(x, y) - \delta F(y, u)\| \leq a \|x - u\| + b \|x - y\| + b \|y - u\|$,

where $a \geq 0$ and $b \geq 0$ are independent of x, y, u.

Lemma 3.5$^{(\dagger)}$: Conditions (3.8) and (3.9) imply that $\delta F(x,x)$ is the Frechet derivative $F'(x)$ and $F' \in \text{Lip}_{2(a+b)} D_0$.

Proof: Let $x \in D_0$, $\varepsilon > 0$ and if $a + b \neq 0$, pick $\delta \leq \dfrac{\varepsilon}{a+b}$ such that $N(x, \delta) \subset D_0$. Then for $\|\Delta x\| < \delta$,

$$\| F(x + \Delta x) - F(x) - +F(x,x)(\Delta x) \| = \| +F(x + \Delta x, x)\Delta x - +F(x,x)\Delta x \|$$

$$\leq \| +F(x + \Delta x, x) - +F(x,x) \| \cdot \| \Delta x \|$$

$$\leq (a+b) \|\Delta x\| \cdot \|\Delta x\| \leq \varepsilon \cdot \|\Delta x\| .$$

If $a = b = 0$, then by (3.9) there is an $A \in L(X, Y)$ such that $\delta F(x,y) \equiv A$ for every $x, y \in D_0$. Hence, from (3.8) we can take δ arbitrary in the above proof and $F'(x) \equiv A$.

To complete the proof, let $x, y \in D_0$ and by (3.9)

$$\| F'(x) - F'(y) \| \leq \| \delta F(x,x) - \delta F(x,y) \| + \| \delta F(x,y) - \delta F(y,y) \|$$

$$\leq a\|x-y\| + b\|x-y\| + a\|x-y\| + b\|x-y\|$$

$$= 2(a+b) \|x-y\| .$$

Schmidt's iteration is, as one would suspect,

$$x_{n+1} = x_n - \delta F(x_n, x_{n-1})^{-1} F(x_n); \quad x_0, x_{-1}$$

given. We will call this the <u>backward secant method</u> for, since Schmidt doesn't assume $\delta F(x_n, x_{n-1}) = \delta F(x_{n-1}, x_n)$, another generalization of the secant method would be

$$x_{n+1} = x_n - \delta F(x_{n-1}, x_n)^{-1} F(x_n); \quad x_0, x_{-1} \text{ given} .$$

($\dagger$) The author has since found this lemma in [25].

We will call this the <u>forward secant method</u>.

Laarsonen [20] makes the following assumption. For some $M \geq 0$,

(3.10) $\quad \|\delta F(x', x'') - \delta F(y', y'')\| \leq M[\|x' - y'\| + \|x'' - y''\|]$

for every x', x'', y', y'' in D_0.

<u>Lemma 3.6</u>: Condition (3.10) implies condition (3.9) with $a = 0$ and $b = M$.

<u>Proof</u>: Let $x, y, u \in D_0$, then by (3.10)

$$\|\delta F(x, y) - \delta F(y, u)\| \leq M[\|x-y\| + \|y-u\|] + 0 \cdot \|x-u\|.$$

<u>Lemma 3.7</u>: Conditions (3.8) and (3.10) imply that $\delta F(x, y)$ is the Frechet derivative $F'(x)$ and $F' \in \text{Lip}_{2M} D_0$.

<u>Proof</u>: By Lemma 3.6 we can set $a = 0$, $b = M$ and apply Lemma 3.5.

Laarsonen considers the following iteration. Given $y_0, \bar{y}_0$ he forms $\delta F(y_0, \bar{y}_0)$ and proceeds by

$$y_{n+1} = y_n - \delta F(y_n, \bar{y}_n)^{-1} F(y_n)$$

$$\bar{y}_{n+1} = y_{n+1} - \delta F(y_n, \bar{y}_n)^{-1} F(y_{n+1}).$$

Notice that by (3.8), $\delta F(y_n, \bar{y}_n)^{-1} F(y_n) = y_n - \bar{y}_n + \delta F(y_n, \bar{y}_n)^{-1} F(\bar{y}_n)$. Hence, if we set $y_0 = x_{-1}$, $\bar{y}_0 = x_0$, then

CONVERGENCE OF NEWTON-LIKE METHODS

$$x_1 = x_0 - \delta F(x_{-1}, x_0)^{-1} F(x_0) \qquad (= y_1)$$

$$x_2 = x_1 - \delta F(x_{-1}, x_0)^{-1} F(x_1) \qquad (= \bar{y}_1)$$

$$x_3 = x_2 - \delta F(x_1, x_2)^{-1} F(x_2) \qquad (= y_2)$$

$$\vdots \qquad\qquad\qquad \vdots$$

$$x_{2n+1} = x_{2n} - \delta F(x_{2n-1}, x_{2n})^{-1} F(x_{2n}) \qquad (= y_{n+1})$$

$$x_{2n+2} = x_{2n+1} - \delta F(x_{2n-1}, x_{2n})^{-1} F(x_{2n+1}) \; (= \bar{y}_{n+1})$$

is the same iteration. Note that this is the forward secant method with recalculation sequence $\alpha_{2n+1} = 2n+1$ and $\alpha_{2n+2} = 2n+1$, if we adopt the convention that the divided difference is indexed by the highest subscript of its two arguments. This observation makes Laarsonen's proof that the $\{y_n\}$ and $\{\bar{y}_n\}$ converge faster than quadratically somewhat meaningless.

We want to prove a theorem now about the most general possible mixture of the forward and backward secant methods, i.e., (2.12) with A_n either $\delta F(x_n, x_{n-1})$ or $\delta F(x_{n-1}, x_n)$ and with a different choice allowed for a different n.

In view of Lemma 3.6, we will assume (3.8) and (3.9) since the theorem under conditions (3.8) and (3.10) can be gotten immediately by setting $a = 0$, $b = M$.

Corollary 3.8

Let (3.8) and (3.9) hold and let $x_{-1}, x_0 \in D_0$ have the following properties: for A_0 either $\delta F(x_0, x_{-1})$ or $\delta F(x_{-1}, x_0)$, $\|A_0^{-1}\| \leq \beta$; $\|x_{-1} - x_0\| \leq \eta_{-1}$; $\|A_0^{-1} F(x_0)\| \leq \eta$ and

(3.11) $$\beta(a+b)\eta_{-1} < 1 ,$$

(3.12) $$\frac{1}{4} \geq \frac{(a+b)\beta\eta}{(1-\beta(a+b)\eta_{-1})^2} \equiv h$$

and

(3.13) $N(x_0, r_0) \subset D_0$ where $r_0 = \dfrac{1-\sqrt{1-4h}}{2\beta(a+b)}(1-\beta(a+b)\eta_{-1})$.

Then, for an arbitrary recalculation sequence $\{\alpha_n\}$

$$x_{n+1} = x_n - A_{\alpha_n}^{-1} F(x_n); \quad n = 0, 1, \ldots ;$$

$A_{\alpha_n} \in \{\delta F(x_{\alpha_n}, x_{\alpha_n - 1}), \delta F(x_{\alpha_n - 1}, x_{\alpha_n})\}$ converges to x^*, a root of F, according to

$$\|x^* - x_{n+1}\| \leq r_0 - t_n - (1 - \beta(a+b)(t_{\alpha_n} + t_{\alpha_n - 1} + \eta_{-1}))^{-1}$$
$$\cdot (\beta(a+b)t_n^2 - (1-\beta(a+b)\eta_{-1})t_n + \eta) ;$$

$$t_{-1} = -\eta_{-1}, \quad t_0 = 0 .$$

Proof: The proof will be accomplished by finding the analog of (2.6) with $A_n = \delta F(x_n, x_{n-1})$ or $\delta F(x_{n-1}, x_n)$. K is $2(a+b)$ by Lemma 3.4. First we need to satisfy (2.5). We can dispense with the ambiguity of A_n by noting that for either choice, $\|A_n - F(x_{n-1})\| \leq (a+b)\|x_n - x_{n-1}\|$ by Lemma 3.4 and (3.9). Hence,

$$\|A_n - A_0\| \leq \|A_n - F'(x_{n-1})\| + \|F'(x_{n-1}) - F'(x_0)\| + \|F'(x_0) - A_0\|$$
$$\leq (a+b)\|x_n - x_{n-1}\| + 2(a+b)\|x_{n-1} - x_0\| + (a+b)\eta_{-1} .$$

Hence, as long as $\sum_{j=1}^{n} \|x_j - x_{j-1}\| < r_0$

$$\|A_0^{-1} A_n - I\| \leq \beta(a+b)(2 \sum_{j=1}^{n-1} \|x_j - x_{j-1}\| + \|x_n - x_{n-1}\| + \eta_{-1})$$

$$< 2\beta(a+b)r_0 + \beta(a+b)\eta_{-1}$$

$$= (1 - \sqrt{1-4h})(1 - \beta(a+b)\eta_{-1}) + \beta(a+b)\eta_{-1}$$

$$\leq 1 ,$$

and so as before we can take

$$a_n = [1 - \beta(a+b)(2 \sum_{j=1}^{n-1} \|x_j - x_{j-1}\| + \|x_n - x_{n-1}\|)\| + \eta_{-1}]\beta^{-1} \leq \beta^{-1}.$$

Now (2.6) holds with $\Delta = \beta^{-1} - (a+b)\eta_{-1}$ ($\Delta > 0$ by (3.11)) $\sigma = 1$ since $\|F'(x_n) - A_n\| \leq (a+b)\|x_n - x_{n-1}\|$ is an easy consequence of Lemma 3.4 and (3.9). (3.12) is (2.12) and (3.13) is (2.13) so the result follows.

<u>Remark 3.6.</u> It is reassuring to note that the majorizing $\{t_n\}$ sequence is generated by the secant method for f with recalculation sequence $\{\alpha_n\}$. If we use Schmidt's iteration, i.e., $\alpha_n = n$, this of course means that the error bounds will reflect the order of convergence of the secant method. Another very interesting point is that if $x_{-1} = x_0$, i.e., one starts the iteration with Newton's method, then the hypotheses of Corollary 3.8 are exactly those of Corollary 3.4, the Kantorovich theorem.

The following shows how this version of the majorant theory relates to the contractive mapping. It is another example of a better theorem through direct analysis than by application of Theorem 3.2.

Theorem 3.9

Let P be a nonlinear mapping of D_0 into itself such that $P' \in \text{Lip}_K D_0$. Assume that for some $x_0 \in D_0$, the following conditions are satisfied. $\|P'(x_0)\| \leq 1$, $\|P(x_0)\| \leq \eta$, $\frac{1}{2} \geq h \equiv \frac{K\eta}{(1-\delta)^2}$ and $N(x_0, r_0) \subset D_0$, where $r_0 = \frac{1-\sqrt{1-2h}}{K}(1-\delta)$. Under these conditions P has a unique fixes point x^* in $\overline{N}(x_0, r_0)$ and if $h < \frac{1}{2}$, x^* is unique in $D_0 \cap N(x_0, r_1)$, $r_1 = \frac{1+\sqrt{1-2h}}{K}(1-\delta)$. Furthermore, x^* can be reached by successive approximations from any $x_0' \in N(x_0, r_2) \subset D_0 \cap N(x_0, r_1)$ and $x_{n+1}' = P(x_n')$ satisfies

$$\|x^* - x_{n+1}'\| \leq |r_0 - \frac{1}{2}K t_n'^2 - \delta t_n' - \eta|, \quad t_0' \geq \|x_0' - x_0\|.$$

Proof: Apply Theorem 2.5 to $F = I - P$ and $A_0 = I$.

This theorem is closely related to a theorem of the author [9], where $P = I - A^{-1}F$ but F'' was assumed to exist. Rheinboldt later gave that theorem under the present continuity conditions, [32], but only for $x_0' = x_0$, $t_0' = 0$. See [9] also for comparison with an earlier theorem of Ghinea [15].

The following theorem is included to show a relationship between gradient methods and Newton-like methods and also to show a use of Theorem 2.6 in which $\{a_n\}$ is not determined by the Banach Lemma.

Theorem 3.10

Let f be a nonlinear functional defined on $D \subset X$, where X is a real Hilbert space and let $f'' \in \text{Lip}_K(D_0)$ for D_0 the closure of an open convex subset of D. Let $x_0 \in D_0$ and $\delta > 0 < \mu < \lambda$ where μ and λ bound the spectrum of $f''(x_0)$ and $\delta < 1/\lambda$.

If $\{a_n\}$ is a sequence in $[\delta, 2/\lambda - \delta]$ and if $\frac{1}{2} \geq h \equiv \dfrac{K\|f'(x_0)\|}{\mu^2}$ and $N(x_0, r_0) \subset D_0$ where $r_0 = \dfrac{1 - \sqrt{1-2h}}{K}\mu$ then f has a unique minimizing point $x^* \in \overline{N}(x_0, r_0)$. If $h < \frac{1}{2}$ then x^* is unique in $D_0 \cap N(x_0, \frac{1+\sqrt{1-2h}}{K}\mu)$. Furthermore x^* is the limit of

$$x_{n+1} = x_n - a_n^{-1} f'(x_n)$$

and

$$\|x^* - x_{n+1}\| \leq r_0 - t_n - a_n^{-1}(\tfrac{1}{2} K t_n^2 - \mu t_n + \|f'(x_0)\|); \quad t_0 = 0.$$

Proof: The proof is quite simple. We will let $A_n = a_n I$ and $F = f'$ and apply Theorem 2.6 to ensure convergence to x^*, a zero of f' in $\overline{N}(x_0, r_0)$. We will show that $f''(x)$ is positive semidefinite at every point in $\overline{N}(x_0, r_0)$ and hence f is convex on that closed convex set and the theorem on page 123 of [16] ensures that $f(x^*)$ is the minimum of f on $\overline{N}(x_0, r_0)$ and is attained only at x^*.

First let us consider the hypotheses of Theorem 2.4. $0 < \delta < 1/\lambda < 1/\mu$ so $0 < \mu\delta < \lambda\delta < 1$. Also, $2/\lambda - \delta > 1/\lambda > \delta$ so the positive sequence $\{a_n\}$ can be chosen. Clearly (2.5) is satisfied so let us consider (2.6). Let $x_0, \ldots, x_n \in N(x_0, r_0)$ and by the lemma and corollary, pages 129 and 131 of [16]:

$$\|A_n - F'(x_n)\| = \|a_n I - f''(x_n)\| \leq \|a_n I - f''(x_0)\| + K\|x_n - x_0\|$$

$$\leq a_n^{-1} \|I - a_n^{-1} f''(x_0)\| + K\|x_n - x_0\|$$

$$\leq a_n^{-1} \max\{|1 - a_n^{-1}\lambda|, |1 - a_n^{-1}\mu|\} + K\|x_n - x_0\|$$

$$\leq a_n - \mu + K\|x_n - x_0\|.$$

Hence (2.6) is satisfied with $\Delta = \mu$ and $\sigma = 1$ and the conditions on h and r are exactly (2.13) and (2.14) so the iteration converges to x^*, the only root of f' in the region specified.

We now complete the proof by showing that $f''(x)$ is positive semidefinite on $\bar{N}(x_0, r_0)$. First note that $f''(x_0)$ is invertible and its inverse is bounded in norm by μ^{-1}. Thus, if $\|x - x_0\| < r_0$

$$\|I - \mu^{-1} f''(x)\| \le \mu^{-1} K \|x - x_0\| < 1 - \sqrt{1-2h} \le 1$$

and so by the Banach lemma, $f''(x)$ is invertible and

$$\|f''(x)^{-1}\| \le \frac{\mu^{-1}}{1 - \mu^{-1} K \|x-x_0\|} = \frac{1}{\mu - K\|x-x_0\|} \;.$$

Hence the spectrum of $f''(x)$ is bounded below by $\mu - K\|x - x_0\| > 0$. By continuity we can conclude that $f''(x)$ is positive semidefinite on $\bar{N}(x_0, r_0)$. This completes the proof but notice that if $a_n = Kt_n - \mu$ then the sequence of error bounds is quadratically convergent.

4. Error Analysis

Theorems of the type given in the previous section are usually dismissed by people who are involved in programming various methods for (1.2) as being too pessimistic for most problems, although for trumped up problems they may be sharp. These objections are often justified as the numerical examples given here show. Furthermore, such bounds are dependent on an estimate of K. For some special problems this presents no difficulty but for most it is virtually impossible.

We will give here a method independent lower bound on the error which requires no additional computation. The upper bound we furnish is somewhat less desirable but it lends itself to a feasible computational procedure. These upper and lower bounds also give the traditional order of convergence results for the methods discussed.

Theorem 4.1

Let x_0, $\{A_n\}$, $\{a_n\}$ satisfy the conditions of Theorem 2.6, then for any $n \geq 0$,

$$\frac{1}{2} \|x_{n+1} - x_n\| \leq \|x^* - x_n\| \leq 2 \frac{a_{\alpha_n}}{\Delta - \sigma Kt_n} \|A_n^{-1} F(x_n)\|$$

(4.1)
$$\|x^* - x_{n+1}\| \leq (2 \frac{a_{\alpha_n}}{\Delta - \sigma Kt_n} - 1) \|A_n^{-1} F(x_n)\|$$

$$\|x_{n+1} - x^*\| \leq [\|I - A_{\alpha_n}^{-1} F'(x_{\alpha_n})\| + a_{\alpha_n}^{-1} K(\|x_{\alpha_n} - x_n\|$$

$$+ \|x_n - x^*\|)] \|x_n - x^*\| .$$

Proof: First we show that the lower bound on $\|x^* - x_n\|$ holds. This simple proof generalizes the bound for Newton's method given by the author [9]

$$\|x_{n+1} - x_n\| =$$

$$\|\int_{x^*}^{x_n} A_{\alpha_n}^{-1} F'(x) dx\| = \|\int_{x^*}^{x_n} (A_{\alpha_n}^{-1} F'(x) - I) dx + x_n - x^*\|$$

$$\leq (\sup_{(x_n, x^*)} \|A_{\alpha_n}^{-1} F'(x) - I\| + 1) \|x_n - x^*\|$$

$$\leq [\,\|A_{\alpha_n}^{-1} F'(x_{\alpha_n}) - I\| + \|A_{\alpha_n}^{-1}(F'(x_n) - F'(x_{\alpha_n}))\|$$

$$+ \sup_{(x_n, x^*)} \|A_{\alpha_n}^{-1}(F'(x) - F'(x_n))\| + 1]\|x_n - x^*\|$$

$$\leq [a_{\alpha_n}^{-1}(a_{\alpha_n} + \sigma K t_{\alpha_n} - \Delta + K(t_n - t_{\alpha_n}) + K\|x_n - x^*\|)$$

$$+ 1]\|x_n - x^*\|$$

$$\leq [a_{\alpha_n}^{-1}(a_{\alpha_n} + \sigma K(\|x^* - x_n\| + t_n) - \Delta) + 1]\|x_n - x^*\|$$

$$\leq [a_{\alpha_n}^{-1}(a_{\alpha_n} - \Delta\sqrt{1-2h}) + 1]\|x_n - x^*\|$$

$$\leq (2 - a_{\alpha_n}^{-1} \Delta\sqrt{1-2h})\|x_n - x^*\| \leq 2\|x_n - x^*\|\,.$$

Now set $y_0 = x_n$, $B_m = A_{\alpha_{n+m}}$ and $b_m = a_{\alpha_{n+m}}$. Also let $\Delta_n = \Delta - \sigma K t_n$. Note that $\|B_m^{-1}\| \leq b_m^{-1}$ and $y_m = x_{n+m}$, $m \geq 0$ so

$$b_m + \sigma K \sum_{j=1}^{m} \|y_j - y_{j-1}\| - \Delta_n = b_m + \sigma K \sum_{j=n+1} \|x_j - x_{j-1}\| + \sigma K t_n - \Delta$$

$$\geq a_{\alpha_{n+m}} + \sigma K \sum_{j=1}^{n+m} \|x_j - x_{j-1}\| - \Delta$$

$$\geq \|A_{\alpha_{n+m}} - F'(x_{n+m})\| = \|B_m - F'(y_m)\|\,.$$

Also, $h_n \equiv \dfrac{\sigma K \|B_0^{-1} F'(y_0)\| b_0}{\Delta_n^2} \equiv \dfrac{\sigma K \|A_{\alpha_n}^{-1} F'(x_n)\| a_{\alpha_{n+m}}}{(\Delta - \sigma Kt_n)^2}$

$\leq \dfrac{\sigma K(t_{n+1} - t_n) a_{\alpha_{n+m}}}{(\Delta - \sigma Kt_n)^2} = \sigma Kf(t_n)/(\Delta - \sigma Kt_n)^2$.

Hence $h_n \leq \dfrac{1}{2}$ is equivalent to

$$\Delta^2 - 2\Delta\sigma Kt_n + (\sigma Kt_n)^2 \geq 2\sigma K(\dfrac{1}{2}\sigma Kt_n^2 - \Delta t_n + a_0 \|A_0^{-1} F(x_0)\|)$$

or

$$\Delta^2 \geq 2\sigma K a_0 \|A_0^{-1} F(x_0)\| \quad \text{i.e., } h \leq \dfrac{1}{2} \quad .$$

Now we need $N(y_0, r_n) \subset N(x_0, r_0)$, where $r_n = \dfrac{1 - \sqrt{1-2h_n}}{\sigma K} \Delta_n$.
(Ignore the inconsistency of "r_1".) If $x \in N(y_0, r_n)$, then
$\|x - x_0\| \leq \|x - y_0\| + \|y_0 - x_0\| < r_n + t_n$. $\sigma K(r_n + t_n) = \Delta_n$
$- \Delta_n \sqrt{1-2h_n} + \sigma Kt_n = \Delta - \Delta_n \sqrt{1-2h_n}$ and

$\Delta_n \sqrt{1-2h_n} = \sqrt{\Delta_n^2 - 2h_n \Delta_n^2} \geq \sqrt{\Delta^2 - 2\Delta\sigma Kt_n + (\sigma Kt_n)^2 - 2\sigma Kf(t_n)}$

$\geq \sqrt{\Delta^2 - 2\sigma K a_0 \|A_0^{-1} F(x_0)\|} \geq \Delta\sqrt{1-2h}$, so

$$r_n + t_n \leq r_0 \quad .$$

Hence, from Theorem 2.6,

$$\|x^* - x_n\| \le r_n \le \frac{1 - (1-2h_n)^{\frac{1}{2}}}{\sigma K} \Delta_n$$

$$= 2 \frac{a_{\alpha_n}}{\Delta_n} \|A_{\alpha_n}^{-1} F(x_n)\| = 2 \frac{a_{\alpha_n}}{\Delta_n} \|x_{n+1} - x_n\|.$$

Also $\|x^* - x_{n+1}\| \le r_n - t_{n+1} = r_n - t_n + t_n - t_{n+1} < (2 \frac{a_{\alpha_n}}{\Delta_n} - 1)$

$\|x_{n+1} - x_n\|$ and $\|x^* - x_{n+1}\| \le a_{\alpha_n}^{-1} (\|A_{\alpha_n} - F'(x)\|) \|x^* - x_n\|$,

for some $x \in (x_n, x^*)$. Hence

$$\|x^* - x_{n+1}\| \le a_{\alpha_n}^{-1} (\|A_{\alpha_n} - F'(x_{\alpha_n})\| + \|F'(x_{\alpha_n}) - F'(x_n)\|$$

$$+ \|F'(x_n) - F'(x)\|) \|x_n - x^*\|$$

$$\le a_{\alpha_n}^{-1} (a_{\alpha_n} + \sigma K \sum_{j=1}^{\alpha_n} \|x_j - x_{j-1}\| - \Delta$$

$$+ K \|x_n - x_{\alpha_n}\| + K \|x_n - x^*\|) \|x_n - x^*\|.$$

This completes the proof but the presence of the quotient $q_n \equiv \frac{a_{\alpha_n}}{\Delta_n} = - \frac{a_{\alpha_n}}{f'(t_n)}$ makes for an unsatisfying upper bound. Sometimes this can be dealt with theoretically and sometimes it must be estimated. Let $e_n \equiv \|x_n - x^*\|$.

Case 4.1. Newton's method

When $\alpha_n = n$, $q_n = 1$ by Corollary 3.4 and so (4.1) is simply

$$\frac{1}{2} \|x_{n+1} - x_n\| \le e_n \le 2 \|x_{n+1} - x_n\| \ .$$

Note also that $e_{n+1} \le \dfrac{\beta K e_n^2}{1-\beta K t_n} \le \dfrac{\beta K}{\sqrt{1-2h}} e_n^2$ and so the convergence is quadratic if $h < \dfrac{1}{2}$. For general $\{\alpha_n\}$,

$$e_{n+1} \le (\tfrac{1}{\beta} - Kt_{\alpha_n})^{-1} (K \|x_n - x_{\alpha_n}\| + Ke_n)e_n$$

$$\le (\tfrac{1}{\beta} - Kt_{\alpha_n})^{-1} K(2e_n + e_{\alpha_n})e_n \ .$$

Case 4.2. The secant method

We see from Corollary 3.8 that $\Delta = \dfrac{1}{\beta} - (a+b)\eta_{-1}$ and $a_n = \dfrac{1}{\beta} - (a+b)(t_n + t_{n-1} + \eta_{-1}) = \Delta - (a+b)(t_n + t_{n-1})$. Thus $q_n = a_n/(\Delta - Kt_n) = [\Delta - (a+b)(t_n + t_{n-1})]/[\Delta - 2(a+b)t_n]$, and so $q_n - 1 = (a+b)(t_n - t_{n-1})/[\Delta - 2(a+b)t_n]$ and for $h < \dfrac{1}{2}$ q_n converges to 1 as the sequence converges. Also

$$e_{n+1} \le [\tfrac{1}{\beta} - (a+b)(t_{\alpha_n} + t_{\alpha_n -1}) + \eta_{-1}]^{-1} [(a+b) \|x_{\alpha_n} - x_{\alpha_n -1}\|$$

$$+ 2(a+b)(\|x_n - x_{\alpha_n}\| + e_n)]e_n$$

$$\le a_{\alpha_n}^{-1} (a+b)(e_{\alpha_n -1} + 3e_{\alpha_n} + 4e_n)e_n \ .$$

If $\alpha_n = n$ and $h < \dfrac{1}{2}$, we get $e_{n+1} = O(e_n^2 + e_{n-1}e_n)$ which is characteristic of the secant method.

Case 4.3. The consistent methods

We have immediately from Theorem 4.1 that

$$e_{n+1} \leq a_{\alpha_n}^{-1}(c\|h_{\alpha_n}\| + K\|x_{\alpha_n} - x_n\| + Ke_n)e_n$$

$$\leq a_{\alpha_n}^{-1}(c\|h_{\alpha_n}\| + Ke_{\alpha_n} + 2Ke_n)e_n .$$

This easily gives the usual order of convergence theorems for these methods. [25]

Case 4.4. Broyden's method. [4]

In [13] the author showed that Theorem 3.2 applies to Broyden's method with $\|\cdot\| = \|\cdot\|_2$, $\sigma = 3K/2$ and $\delta_n = \delta_0 = \|A_0 - F'(x_0)\|$. Thus

$$q_n = \frac{a_n}{\Delta - (K+2\gamma)t_n} = \frac{\Delta + \delta - 2.5Kt_n}{\Delta - 4Kt_n} .$$ q_n is easily shown to be an increasing function of t_n and hence of n. We have

$$q_n \leq q \equiv \frac{\Delta + \delta - 2.5Kr_0}{\Delta - rKr_0}$$ and so

$$q = \frac{\Delta + \delta - \frac{5}{8}(1 - \sqrt{1-2h})\Delta}{\Delta - (1 - \sqrt{1-2h})} = \frac{3\Delta + 8\delta + 5\sqrt{1-2h}\Delta}{\sqrt{1-2h}\ \Delta} .$$

Now suppose we monitor $\|x_{n+1} - x_n\|$ until we have a suitable lower bound on the error. Then get a fresh Jacobian approximation so that if we think of the iteration beginning again, the first step is Newton's method and $\frac{1}{2}\|x_1 - x_0\| \leq e_0 \leq 2\|x_1 - x_0\|$, $e_1 \leq \|x_1 - x_0\|$. If our stopping criteria are satisfied then x_1 is the approximation, if not, $\delta = 0$

and so $q = 11$ seems a reasonable guess. $q = 11$ is obtained by setting $\sqrt{1-2h} = \frac{1}{2}$, the middle of its range. Another strategy is to note that $q_n = \frac{a_n}{\Delta_n} \leq \frac{a_n}{\Delta \sqrt{1-2h}}$. Now if we restart as above with Newton's method, then $\Delta_0 = a_0$. Estimate $\sqrt{1-2h} = \frac{1}{2}$ and $q_n = 2a_n/a_0 = (2/a_0)(1/1/a_n)$. a_n^{-1} can easily be estimated, say by $\|A_n^{-1}\|_p$ for $p = 1, 2, \infty$, since A_n^{-1} is always at hand in Broyden's method.

The numerical results reported below give a promising indication of the possible utility of these results. They also show that the majorizing t sequence does not furnish very good error bounds. In the Broyden's method examples we always begin with the exact Jacobian.

Let $\eta_i = \|x_i - x_{i+1}\|_2$, $e_i = \|x_i - x^*\|_2$.

Example 4.1. $x^2 + y^2 - 20x + 32$

$y^2 - 2x$. $\qquad x^* = (2,2)$

The first table gives information for Newton's method with $x_0 = (2.3, 1.85)$ and the hypotheses of Corollary 3.4 fulfilled. We do not include the $i = 0$ line to save space.

The line $i = 4$ has probably already been contaminated by roundoff error. The poor behavior of the scalar error estimate persisted for the other examples so we will not report it any longer.

The next table is constructed for the same equations using Broyden's method, $x_0 = (2.95, .88)$. $q = 2\|A_0^{-1}\|/\|A_n^{-1}\|$ and the subscript indicates the norm used.

i	η_i	e_i	$r_0 - t_i$	predicted interval
1	.7146492025D-02	.7149479456D-02	.8320446694D-01	.3573246013D-02, 1.4292840 50D-02
2	.3221297266D-05	.3221298534D-05	.4636409663D-02	.1610648633D-05, .6442594532D-05
3	.1391720662D-11	.1391436949D-11	~ D-03	.0695860331D-11, .2783441324D-11
4	.1307876228D-15	.2204460049D-15	~ D-09	.0653938114D-15, .2615752456D-15
5	same	same	~ D-14	same

I	η_i	e_i	q_1	q_2	q_∞	predicted interval $(\eta/2, 2q\eta)$
1	.9539606731D-00	.6323845961D-00	2.926	3.110	3.094	.4769 8033D-00, 1.9079 2132D-00
2	.2899355043D-00	.3255479135D-00	3.991	4.341	4.396	.1449 6775D-00, .5798 7100D-00
3	.3689028235D-01	.3663861108D-01	3.691	3.912	3.888	.1844 5141D-01, .7378 0564D-01
4	.1415088357D-02	.1213431650D-02	3.564	3.844	3.854	.0707 5441D-02, .2830 1764D-02
5	.4287557874D-03	.4331892147D-03	4.227	4.562	4.747	.2143 7789D-03, .8575 1156D-03
6	.4597521897D-05	.4588756299D-05	4.176	4.517	4.703	.2298 7609D-05, .9195 0436D-05
7	.8768674246D-08	.8767377337D-08	4.184	4.524	4.712	.4384 371D-08, 1.1537 3484D-08
8	.1297091838D-11	.1297401933D-11	4.185	4.424	4.711	.0648 5459D-11, .2894 1836D-11

CONVERGENCE OF NEWTON-LIKE METHODS

At the eighth iteration we found

$$F'(x_8) - A_8 = \begin{pmatrix} .3567738148D-01 & -.2738939435D-01 \\ -.1215904942D-01 & .9334298441D-02 \end{pmatrix}$$

although A_0 was taken equal to $F'(x_0)$. This behavior was standard and seems to indicate that the Broyden derivative approximation is indeed not "consistent".

Example 4.2. $xy^3 - 2\sin(x+1) + 1$
$$e^{-y^2+1} + x^2y - 2 \qquad x^* = (-1,1)$$

Newton's method with $x_0 = (-2.5, 2.650)$.

i	η_i	e_i	predicated interval $(\frac{1}{2\eta}, 2\eta)$	
1	.6016594846D-00	.1296201928D+01	.30082974D-00,	1.20331896D-00
2	.3883058540D-00	.7039972865D-00	.19415292D-00,	.77661168D-00
3	.2095972966D-00	.3202191845D-00	.1047964 D-00,	.41919456D-00
4	.9167067968D-01	.1135580322D-00	.45835339D-01,	1.83341356D-01
5	.2094700818D-01	.2200378637D-01	.10473504D-01,	.41894016D-01
6	.1054821076D-02	.1057406436D-02	.0527415 D-02,	.21096416D-02
7	.2585464788D-05	.2585480542D-05	.12927323D-05,	.51709292D-05
8	.1575410507D-10	.1575410508D-10	.07877092D-10,	.31808208D-10

This is a very interesting table because we began just outside the region $h < \frac{1}{2}$ and e_1 is not in the predicted interval. e_2 is in the predicted interval, but barely, i.e., the $2\eta_2$ upper bound is needed.

This next table represents the results for Broyden's method with $x_0 = (-1.25, .85)$ — a very good intial guess.

461

i	η_i	e_i	q_1	q_2	q_∞	predicted interval
1	.7233820824D-02	.8126656004D-02	1.968	1.955	1.954	.36169104D-02, 1.446764 6D-02
2	.1093438913D-02	.1045041261D-02	1.772	1.718	1.730	.05467144D-02, .21868576D-02
3	.5810618988D-04	.5601539103D-04	1.869	1.783	1.778	.29053094D-04, 1.16212376D-04
4	.2086623106D-05	.2094463025D-05	1.932	1.815	1.801	.1043315 D-05, .41732400D-05
5	.7838068550D-08	.7839974329D-08	1.923	1.812	1.799	.39190342D-08, 1.56761368D-08
6	.1904736056D-11	.1905857198D-11	1.925	1.811	1.799	.09523680D-11, .38094720D-11
7	.1121356170D-14	.1121356170D-14	1.924	1.811	1.798	.05006780D-14, .22427120D-14

We have in hand several other examples including a quite nonlinear 3 × 3 case. The behavior is consistently the same. Once one gets close to the root, so that error estimates become important, the predicted interval furnishes quite good information. It also seems to be the case that q_1 or q_∞ (or perhaps $.5(q_1 + q_\infty)$) is quite as useful as the more costly q_2. All three seem better than the estimate $q = 11$.

All the numerical results as well as some stimulating discussions were provided by Mr. Garth Baker of the Cornell University Computer Science Department. The computations were performed on the Cornell University's IBM Model 65.

5. The Relationship of the Majorant Theory to Local and Global Analysis

The theory given here yields theorems which are local in nature but they differ from local theorems in some important ways. Ortega and Rheinboldt [25] call such theorems "semilocal" and the Kantorovich theorem is the prototype. We wish to discuss here our view as to the particular place of these semilocal results with respect to "local" and "global" convergence theorems.

A local convergence theorem for a method is usually the extension to general nonlinear systems of a global convergence theorem for linear systems. Most often the idea of the proof is that one assumes the starting point for the iteration is _close enough_ to the solution to ensure that the linear terms of the Taylor series are a _sufficiently good_ approximation to the function for one iteration of the linear version of the algorithm on the linearized problem to give a closer point to the solution of the nonlinear problem. Some disadvantages are obvious, one is put off by the italicized phrases as well as by the assumption that a solution exists. These flaws are absent from the semilocal analysis. There is a parameter which balances the local nonlinearity of the function against the size of the residuals at the initial point. If the parameter is less than a prescribed value then one has a root and the iteration will converge to it. Unfortunately, the local nonlinearity of the function is usually measured by a Lipschitz constant on the Fréchet derivative and, unless the iteration in question is Newton's method [36], there is currently no standard computational technique for estimating this number.

Local and semilocal convergence theorems would seem to have as their primary disadvantage the fact that one must have an initial point satisfying certain conditions which make it perhaps already a good approximate solution. Global convergence theorems certainly do not require the iteration to start from a good point. This is their single but very important advantage.

When one makes the transition from a local to a global analysis for a given method one is asking for the "right thing", convergence, to happen at many more points. Clearly, one must make some assumptions to counter the compounded effect of Murphy's Law: "If it is possible for a wrong thing to happen, it will." These assumptions usually take the form of either modifications to the algorithm or additional assumptions on F or both. Obviously a true global analysis would require at least enough more of the function to make the iteration possible from every point in X. At this point, however, we wish to distinguish between two extreme types of global convergence theorems on the basis of these additional function requirements.

The first and by far the most common kind of global analysis is based on a change in the algorithm and usually requires only a little more of the function than that the local assumptions hold globally and perhaps some extra continuity conditions. The responsibility for ensuring convergence is given to the algorithm and one often modifies it in very basic ways. The prototype for this kind of theorem is the global convergence theorem for Newton's method given by Goldstein, [16, p. 46]. Here the algorithm is changed by taking a step equal to a scalar multiple of the Newton step. The scalar multiple is chosen to decrease the norm of the residuals until the iterates get within the local convergence region of the basic algorithm, to which one then reverts. We will call this "globalizing the algorithm" since it is not really a global theorem about the basic algorithm.

A second technique for global analysis could be called "particularizing the problem". Here one keeps to the basic algorithm and simply restricts the function F to a reasonable class for which the algorithm is globally convergent. In analogy with summability theory, one seeks the werkfeld of the method. Baluev [25], Slugin [25], Greenspan and Parter [25], Ortega and Rheinboldt [25], and Vandergraft [38] have contributed to this problem for the case of Newton's method. The work was motivated by Chapygin's method for differential equations and the class of function is rather severely restricted by order properties.

This latter form of analysis is still in its infancy but it is obviously an outgrowth of the folk art of finding an algorithm to work for ones own special class of problems. An interesting case in point is the nonlinear Gauss-Seidel iteration in all its various forms. The important developments in the convergence analysis for these algorithms is traced in Ortega and Rheinboldt [25] up to the point of recent work by Rheinboldt and Moré. The algorithms arose because of their applicability to systems arising from the discretization of boundary value problems for elliptic partial differential equations. The class of problems for which they actually converge is still being extended. This form of global analysis is often quite elegant but it is probably going to be some time before it will make much impact on the general problem. Its effect will increase as more algorithms are analyzed in this way to eventually form a basis for deciding which algorithm to use for a particular problem.

The technique of globalizing the algorithm has reached the point where theorems of real computational significance are immerging, particularly in the work of Powell [27], [28]. Powell provides such a detailed account of how to globalize, i.e., proceed from a bad point, that his convergence theorems are practically certifications of associated computer programs. This is getting very close to what this author (and probably most everyone else) feels should be the ideal analysis. It should provide for convergence subject to computable error bounds of a computationally implementable algorithm under conveniently verifiable hypotheses.

We have now discussed local, semilocal, and two kinds of global analyses and stated what we feel should be the goal of the analysis of an algorithm. Where does the semilocal point of view fit? There is little doubt as to its place with respect to local analyses. The author's rule of thumb is that a semilocal theorem is preferred unless it takes more than one page to state, in which case the reader would be sufficiently confused that the advantage of greater definiteness is lost anyway.

The relationship between global and semilocal analyses is more interesting. Certainly a comparison of semilocal theorems for various methods is an indication of their relative robustness and should serve as a guide to which methods are worthy of one of the two types of global analysis.

An ideal semilocal analysis will form the heart of any ideal analysis based on globalizing the algorithm since it would tell when to shift to the basic algorithm and would provide error bounds from that point in the iteration. This is probably as soon as error bounds are of much interest anyway. The process of globalizing can thus be seen as the solution to finding the good initial guess that the basic algorithm requires and the role of the embedded semilocal analysis is to recognize a good initial guess and furnish error bounds. The form of the semilocal theory presented here is probably the proper one for this use. It has yet to reach the ideal stage where the hypotheses are conveniently verifiable except in cases where the Lipschitz constant mentioned above can conveniently be found. We have shown that it is applicable to a wide class of basic iterative methods and that good error bounds can be obtained.

The form of the semilocal theory given here is probably too tied to the idea of a local linear approximation to the function to be much help in the type of global analysis we have called "particularizing the problem". On the other hand, the form of that theory we have described is not the most useful one. What is really wanted is not quite so much restriction on the function, and convergence from almost every, instead of every, initial guess. Theorems of this type for

Newton's method are to be found in the references given above. This "almost everywhere" convergence analysis by particularizing the problem is a very promising technique as the following simple example will illustrate.

Example 5.1. Let $X = Y = E^1$ and $f(t) = (t-1)(t-3)$. Clearly Newton's method converges from any point except $t = 2$ and hence no true global theorem based on particularizing the problem can apply. The Kantorovich theorem, Corollary 3.4 ensures convergence from any point not in $(2 - \sqrt{.5}, 2 + \sqrt{.5})$ and the corresponding "almost everywhere" convergence theorem, Theorem 13.3.4 of [25] applied to f and then -f, ensures convergence from every point except $t = 2$.

6. Related Work

A real survey of all recent work on the solution of nonlinear operator equations is really unnecessary due to the recent appearance of the Ortega-Rheinboldt book. This happily allows us to mention topics at our whim and refer the reader to the aforementioned book for a real survey.

R. H. Moore [22] takes the dual of our point of view. Instead of having a sequence A_n approximate $F'(x_n)$ he has a sequence F_n approximate F. This is a useful point of view, for example when $F(x) = 0$ is an integral equation and F_n is the analog of the equation which uses an n-point quadrature rule in place of the integral operator.

Lohr and Rall [29] study the problem of generating an efficient recalculation sequence for Newton's method as a part of applying the method to a particular problem.

The use of a sequence $\{A_n\}$ in which each A_n has a left and not a right inverse seems tied to a local analysis but results in a method which generally has the same order of convergence as the two-sided inverse case [8]. Right inverses on the other hand, allow a semilocal analysis but the error bounds, not the error sequence, have the expected order properties [10]. Ben-Israel has done a semilocal analysis for the case when A_n is the generalized inverse of $F'(x_n)$ by making assumptions about the Taylor series of F

which are automatically satisfied if A_n is invertible. This work is discussed in the accompanying article by Nashed.

Recent work by Boggs [1], shows real promise for computational solutions to (1.2). It has long been known that (1.2) can be written in various ways as an initial value problem for a system of differential equations. Newton's method corresponds to Euler's method with steplength one and if the differential equation problem is set on a finite interval one applies say, a Runga-Kutta method [2], [23] to obtain the approximate value of the solution at the right end point. Now the true value of the equation at the right end point is x^* and so perhaps Newton's method will converge from the approximate value. Certainly this will be the case if one uses a sufficiently small step size and a convergent integration method on a problem with the requisite continuity.

On the other hand, suppose one considers the initial value problem

(6.1)
$$x'(t) = -(F'(x))^{-1} F(x), \qquad t \in [0, \infty)$$
$$\text{and} \quad x(0) = x_0 .$$

Under the conditions imposed here on F, $x(t) = x^*$ is an assymptotically stable solution and so if $x(t)$ is any solution to (6.1), $x(t) \to x^*$ as $t \to \infty$. Now the key point is that one is totally uninterested in $x(t)$ and so there is no reason to use an accurate, i.e., high order, method for the numerical integration of (6.1). One is interested in a method which will find the asymptote and do it with a large step size and hence less work. This is exactly the property of A-stability [7], and so there is good reason [7], [14] to prefer the trapezoidal rule. Boggs develops these ideas in [1] and finds that Broyden's method works extremely well as predictor and/or corrector.

REFERENCES

[1] Boggs, P. T., The solution of nonlinear operator equations by A-stable integration techniques, Cornell University, Department of Computer Science Technical Report 70-72, 1970.

[2] Bosarge, W. E., Infinite dimensional iterative methods and applications, IBM Publication 320-2347 (Houston, Texas).

[3] Brown, K. M., and J. E. Dennis, A new algorithm for the nonlinear least squares problem, Cornell University, Department of Computer Science Technical Report 70-57, 1970.

[4] Broyden, C. G., A class of method for solving non-linear simultaneous equations, Math. Comp. 19 (1965), pp. 577-593.

[5] Bryan, C. A., Approximate solutions to nonlinear integral equations, SIAM J. Numer. Anal. 5 (1968), pp. 151-155.

[6] Collatz, L., Functional Analysis and Numerical Mathematics, Academic Press, 1966.

[7] Dahlquist, G., A special stability problem for linear multi-step methods, BIT 3 (1963), pp. 27-43.

[8] Daniel, J. W., On Newton-like methods, University of Wisconsin, Department of Computer Science Technical Report 21, 1968.

[9] Dennis, J. E., A stationary Newton method for nonlinear functional equations, SIAM J. Numer. Anal. 4 (1967), pp. 222-232.

[10] _____, On Newton-like methods, Numer. Math. 11 (1968), pp. 324-330.

[11] _____, On the Kantorovich hypothesis for Newton's method, SIAM J. Numer. Anal. 6(1969), pp. 493-507.

[12] _____, On the convergence of Newton-like methods, in Numerical Methods for Nonlinear Algebraic Equations, Edited by P. Rabinowitz, Gordon and Breach, 1970.

[13] _____, On the convergence of Broyden's method for nonlinear systems of equations, Cornell University, Department of Computer Science Technical Report 69-48, 1968.

[14] _____, and R. A. Sweet, Some minimum properties of the trapezoidal rule, Cornell University, Department of Computer Science Technical Report 70-61, 1970.

[15] Ghinea, M., Sur la résolution des équations opérationnelles dans les expaces de Banach, Rev. Francoise Traitement Information Chiffres 8 (1965), pp. 3-22.

[16] Goldstein, A. A., Constructive Real Analysis, Harper-Row, 1967.

[17] Goldfarb, D., and J. Greenstadt, working paper.

[18] Kantorovich, L. V., The majorant principle and Newton's method, Dokl. Akad. Nauk. 76 (1951), pp. 17-20.

[19] _____, and G. P. Akilov, Functional Analysis in Normed Spaces, Pergammon Press, 1964.

[20] Laarsonen, P., Ein Übergradratisch Konvergenter Iterativer Algorithmus, Ann. Acad. Sci. Fenn, Ser. AI 405 (1969), 9 pages.

[21] Moore, R. H., Newton's method and variations, Nonlinear Integral Equations, Edited by P. Anselone, University of Wisconsin Press, 1964, pp. 65-98.

[22] _____, Approximations to nonlinear operator equations and Newton's method, Numer. Math. 12 (1968), pp. 23-34.

[23] Meyer, G., On solving nonlinear equations with a one-parameter operator imbedding, SIAM J. Numer. Anal. 5 (1968), pp. 739-752.

[24] Ortega, J. M., The Newton-Kantorovich theorem, Amer. Math. Monthly 75 (1968), pp. 658-660.

[25] _____ and W. C. Rheinboldt, Iterative Solution of Nonlinear Equations in Several Variables, Academic Press, 1970.

[26] Powell, M. J. D., A survey of numerical methods for unconstrained optimization, SIAM Rev. 12 (1970), pp. 73-78.

[27] _____, A hybrid method for nonlinear equations, in Numerical Methods for Nonlinear Algebraic Equations, Edited by P. Rabinowitz, Gordon and Breach, 1970.

[28] _____, A new algorithm for unconstrained optimization, to appear.

[29] Rall, L. B., and L. R. Lohr, Efficient use of Newton's method, ICC Bulletin 6 (1967), pp. 99-103.

[30] _____, Computational Solution of Nonlinear Operator Equations, Wiley, 1969.

[31] _____ and R. A. Tapia, The Kantorovich theorem and error estimates for Newton's method, University of Wisconsin, M.R.C. Technical Report #1043, 1970.

[32] Rheinboldt, W. C., A unified convergence theory for a class of iterative processes, SIAM J. Numer. Anal. 5 (1968), pp. 42-63.

[33] Schmidt, J. W., Ein Übertragung der Regula Falsi auf Gleichungen in Banachraumen, ZAMM 34 (1963), Part I, pp. 1-8, Part II, pp. 97-110.

[34] Schroder, J., Nichtlineare Majoranten beim Verfahren der Schritweisen Naherung, Ach. Math. 7 (1956), pp. 471-484.

[35] _____, Uber das Newtonsche Verfahren, Arch. Rat. Mech. Anal. 1 (1957), pp. 154-180.

[36] Roberts, S. M., and J. S. Shipman, The Kantorovich theorem and two-point boundary value problems, IBM J. Res. Develop. (1966), pp. 402-406.

[37] Tapia, R. A., The Kantorovich theorem, University of Wisconsin, M.R.C. Technical Report 1007, 1969.

[38] Vandergraft, J., Newton's method for convex operators in partially ordered spaces, SIAM J. Numer. Anal. 4 (1967), pp. 406-432.

Research supported in part by National Science Grant No. GJ-844.

Department of Computer Sciences
Cornell University
Ithaca, New York 14850

Operator Solutions of Nonlinear Equations in Optimal Control Problems

DAVID L. RUSSELL

1. The Stabilization Problem

In this article we shall be concerned with a <u>linear control system</u> in Hilbert space:

(1.1) $$\dot{x} = Ax + Bu .$$

The <u>state vector</u> x and the <u>control vector</u> u lie in Hilbert spaces H_1 and H_2 with inner products $(\ ,\ )_1$ and $(\ ,\ )_2$ respectively. A is a linear operator on H_1 and B is a linear transformation from H_2 into H_1. To facilitate this expository treatment we will assume A and B are both bounded relative to the norms $\|\ \|_1$ and $\|\ \|_2$ induced on H_1 and H_2 by $(\ ,\ )_1$ and $(\ ,\ )_2$. In the paper [A] by D. L. Lukes and the author it is shown that many of the results described here can be extended to cover cases where A is an unbounded operator. As long as A and B are bounded operators there is no difficulty in speaking of solutions of the differential equation (1.1). Given the initial time (w.l.o.g. $= 0$), initial state x_0, and measurable control function $u: [0, \infty) \to H_2$ with

$$\int_0^T \|u(t)\|_2 \, dt < \infty$$

for $T > 0$, there is exactly one vector function $x: [0, \infty) \to H_1$ such that x and u together satisfy (1.1) and

(1.2) $$x(0) = x_0 \ .$$

The derivative $\dot{x}(t)$ exists in the sense

(1.3) $$\lim_{\tau \to t} \left\| \frac{x(\tau) - x(t)}{\tau - t} - \dot{x}(t) \right\|_1 = 0 \ .$$

Moreover, the solution x is given explicitly by the variation of parameters formula

$$x(t) = e^{At} x_0 + \int_0^t e^{A(t-s)} Bu(s) ds \ .$$

Here e^{At} is the fundamental operator solution of the uncontrolled system

(1.4) $$\dot{x} = Ax$$

which corresponds to the control $u \equiv 0$. Its vector solutions, which have the form $x(t) = e^{At} x_0$, may have certain undesirable properties: these solutions may be highly oscillatory or we may have $\lim_{t \to \infty} \|x(t)\|_1 = \infty$. Such behavior may represent unsatisfactory, or dangerous, operation of the physical plant modelled by (1.4).

The purpose of the control term Bu is to enable us to prevent undesirable behavior as described above. Usually this means that u should be chosen so that the corresponding solution x of (1.1) satisfies $\lim_{t \to \infty} \|x(t)\|_1 = 0$. There are many ways in which this can be done but, because of simplicity of implementation, the one favored in practice is a <u>linear feedback control policy</u>

(1.5) $$u(t) \equiv K\,x(t)\,,$$

wherein $u(t)$ is a fixed linear function of the instantaneous state $x(t)$ given by the <u>feedback operator</u> K. When (1.5) is substituted in (1.1) we obtain the <u>closed loop system</u>

(1.6) $$\dot{x} = (A + BK)x\,.$$

A fortuitous choice of K may result in a system (1.6) all of whose solutions x have the property $\lim_{t \to \infty} \|x(t)\|_1 = 0$. If so, we say that the <u>linear feedback policy</u> (1.5) <u>stabilizes</u> the system (1.1).

Until fairly recently, stabilizing feedback matrices K for finite dimensional systems (1.1) were found via more or less experimental techniques such as the root locus method. For systems of modest dimension one could, given a little luck, find a satisfactory K. But the method becomes hopelessly impractical for systems of large dimension and infinite dimensional systems are out of the question altogether.

Even if one could obtain a stabilizing K via experimental methods there always remained the gnawing question: Is this the <u>best</u> choice of K? Will some other $\hat{K}$ yield a closed loop system $\dot{x} = (A + B\hat{K})x$ whose performance, measured in some appropriate manner, is superior to the performance of (1.6)? It is noteworthy that in attempting to answer this <u>optimization</u> question one is led to a nonlinear equation involving operators in H_1. A particular solution of this equation lends to an <u>algorithm</u> enabling one to find, approximately, a stabilizing $\hat{K}$ whenever one exists. Moreover, $\hat{K}$ is optimal with respect to a certain quadratic performance criterion.

2. <u>The Quadratic Criterion</u>

Let W and U be bounded self-adjoint operators on H_1 and H_2 respectively. Thus

$$(x,\,Wy)_1 = (Wx,\,y)_1,\quad x, y \in H_1$$

$$(u, Uv)_2 = (Uu, v)_2, \quad u, v \in H_2.$$

We assume W is positive semidefinite and U is positive definite. Together with the boundedness of W and U this implies that there are positive numbers ω, μ_1 and μ_2 such that

$$0 \leq (x, Wx)_1 \leq \omega \|x\|_1^2, \quad x \in H_1,$$

$$\mu_1 \|u\|_2^2 \leq (u, Uu)_2 \leq \mu_2 \|u\|_2^2, \quad u \in H_2.$$

Let x and u together satisfy (1.1), (1.2). We introduce a <u>quadratic performance criterion</u>:

$$(2.1) \quad C(x, u) = \int_0^\infty [(x(t), Wx(t))_1 + (u(t), Uu(t))_2] dt.$$

Thus if x and $\tilde{x}$ are solutions of (1.1), (1.2), corresponding to controls u and $\tilde{u}$ we say that $\tilde{u}$ gives better performance than u if $C(\tilde{x}, \tilde{u}) < C(x, u)$.

Suppose now that we restrict ourselves to <u>linear feedback control policies</u> (1.5). Then (1.1) becomes (1.6), all of whose solutions have the form

$$(2.2) \quad x(t) = e^{(A+BK)t} x_0.$$

If such a control policy is employed the cost $C(x, u)$ defined by (2.1) becomes

OPTIMAL CONTROL PROBLEMS

$$(2.3) \quad C(x, u) = \int_0^\infty [(e^{(A+BK)t} x_0, W e^{(A+BK)t} x_0)_1$$

$$+ (K e^{(A+BK)t} x_0, U K e^{(A+BK)t} x_0)_2] dt$$

$$= \int_0^\infty (x_0, e^{(A+BK)^* t} [W + K^* U K] e^{(A+BK)t} x_0)_1 dt$$

$$= (x_0, [\int_0^\infty e^{(A+BK)^* t} [W + K^* U K] e^{(A+BK)t} dt] x_0)_1$$

$$\equiv (x_0, Q_K x_0)_1 ,$$

a quadratic form in the initial state vector x_0. In (2.3) and hereafter an asterisk denotes the adjoint of the operator to which it is appended, defined by

$$(x, Ty)_1 = (T^* x, y)_1, \quad x, y \in H_1, \quad T: H_1 \to H_1 ,$$
$$(u, Tv)_2 = (T^* u, v)_2, \quad u, v \in H_2, \quad T: H_2 \to H_2 ,$$
$$(x, Tv)_1 = (T^* x, v)_2, \quad x \in H_1, v \in H_2, \quad T: H_2 \to H_1 .$$

Obviously (2.3) makes sense if and only if the self-adjoint operator

$$(2.4) \quad Q_K(t) = \int_0^t e^{(A+BK)^* s} [W + K^* U K] e^{(A+BK)s} ds$$

has the property that $\lim_{t \to \infty} (x_0, Q_K(t) x_0)$ exists for all $x_0 \in H_1$. In this connection we introduce the

<u>Controllability Assumption.</u> There exists a linear feedback operator $\tilde{K}: H_2 \to H_1$ and a bounded self-adjoint operator $M: H_1 \to H_1$ such that if $Q_{\tilde{K}}(t)$ is given by (2.4) with K replaced by $\tilde{K}$, then $M - Q_{\tilde{K}}(t)$ is positive semidefinite for $t_0 \leq t < \infty$.

Now $Q_{\tilde{K}}(t_2) - Q_{\tilde{K}}(t_1)$ is obviously positive semi-definite when $t_2 > t_1$. Thus we may apply the following well known result from functional analysis:

<u>Lemma 1.</u> <u>Let</u> $R(\tau)$, $0 \leq \tau < \infty$, <u>be a function whose values are bounded self-adjoint operators on</u> H_1 <u>such that</u> $R(\tau_2) - R(\tau_1)$ <u>is positive semidefinite when</u> $\tau_2 > \tau_1$. <u>If there is a bounded self-adjoint operator</u> M <u>on</u> H_1 <u>such that</u> $M - R(\tau)$ <u>is positive semidefinite for all</u> τ, <u>then there is a bounded self-adjoint operator</u> R <u>on</u> H_1 <u>such that</u> $R(\tau)$ <u>converges to</u> R <u>in the strong sense, i.e.</u>,

$$\lim_{\tau \to \infty} R(\tau) x_0 = R x_0, \quad x_0 \in H_1 .$$

The proof may be found, e.g., in [B]. Applying this result to $Q_{\tilde{K}}(t)$ we see that there is a bounded self-adjoint operator $Q_{\tilde{K}}$ on H_1 such that

$$\lim_{t \to \infty} Q_{\tilde{K}}(t) x_0 = Q_{\tilde{K}} x_0, \quad x_0 \in H_1 ,$$

which immediately shows that

$$\lim_{t \to \infty} (x_0, Q_{\tilde{K}}(t) x_0)_1 = (x_0, Q_{\tilde{K}} x_0)_1, \quad x_0 \in H_1 .$$

Thus there is at least one linear feedback control policy for which (2.3) is meaningful. Any such feedback operator K will be called an <u>admissible feedback operator.</u> Each admissible feedback operator yields, via (2.3), a bounded quadratic form $(x_0, Q_K x_0)_1$. If there is an admissible feedback operator $\hat{K}$ with associated quadratic form $(x_0, Q_{\hat{K}} x_0)_1$ such that

$$(x_0, Q_K x_0)_1 - (x_0, Q_{\hat{K}} x_0)_1 \geq 0, \quad x_0 \in H_1,$$

(i.e., $Q_K - Q_{\hat{K}}$ is positive semidefinite) for all Q_K arising from other admissible feedback operators K, then we say that $\hat{K}$ is an optimal linear feedback control policy relative to the quadratic performance criterion (2.1).

3. The Optimal Feedback Operator

The construction of the optimal feedback operator $\hat{K}$ described in the preceding section was first carried out by R. E. Kalman [C] in 1960. Since then the problem has received much attention in the literature. The papers of Lukes [D], Wonham [E], Lukes and Russell [A], and Datko [F] generalize Kalman's results and provide a background for this article.

Rather than trying to minimize (2.1) directly one fixes $t > 0$ and considers a cost computed over a finite interval:

(3.1) $\quad C_t(x, u) = \int_0^t [(x(s), Wx(s))_1 + (u(s), U u(s))_2] ds$.

If we restrict attention to <u>time varying linear feedback control policies</u>

(3.2) $\quad u(s) \equiv K(s) x(s), \quad \int_0^t \|K(s)\|_{2,1}^2 \, ds < \infty,$

the solution of (1.1), (1.2) takes the form

$$x(s) = \Phi(s, 0) x_0$$

where $\Phi(s, \sigma)$ is the solution of

(3.3) $$\dot{\Phi} = (A + BK(s))\Phi$$

which satisfies $\Phi(\sigma, \sigma) = I$, the identity operator on H_1. Then (3.1) becomes

$$C_t(x, u) = \int_0^t [(x_0, \Phi(s, 0)^* W \Phi(s, 0) x_0)_1$$

$$+ (x_0, \Phi(s, 0)^* K(s)^* U K(s) \Phi(s, 0) x_0)_1] ds$$

$$= (x_0, [\int_0^t \Phi(s, 0)^* (W + K(s)^* U K(s)) \Phi(s, 0) ds] x_0)_1 .$$

Thus, defining a bounded self-adjoint operator

(3.4) $$P_K(t, \tau) = \int_{t-\tau}^t \Phi(s, t-\tau)^* (W + K(s)^* U K(s)) \Phi(s, t-\tau) ds$$

we have, when (3.2) holds,

$$C_t(x, u) = (x_0, P_K(t, t) x_0)_1 .$$

Let $\dot{P}_K(t, \tau)$ denote the derivative of (3.4) with respect to τ. Then we compute

(3.5) $$\dot{P}_K(t, \tau) = [A + BK(t-\tau)]^* P_K(t, \tau) + P_K(t, \tau)[A + BK(t-\tau)]$$

$$+ W + K(t-\tau)^* U K(t-\tau) .$$

OPTIMAL CONTROL PROBLEMS

To compare the operators $P_K(t,t)$ and $P_{K_0}(t,t)$ arising from two different feedback operator functions $K(s)$ and $K_0(s)$ we note that

$$(3.6) \quad \dot{P}_{K_0}(t,\tau) - \dot{P}_K(t,\tau) = [A + BK(t-\tau)]^*(P_{K_0}(t,\tau) - P_K(t,\tau))$$

$$+ (P_{K_0}(t,\tau) - P_K(t,\tau))[A + BK(t-\tau)] + (BK_0(t-\tau))^* P_{K_0}(t,\tau)$$

$$+ P_{K_0}(t,\tau) BK_0(t-\tau) + K_0(t-\tau)^* U K_0(t-\tau)$$

$$- (BK(t-\tau))^* P_{K_0}(t,\tau) - P_{K_0}(t,\tau) BK(t-\tau)$$

$$- K(t-\tau)^* U K(t-\tau) \ .$$

Since U and $P_K(t,\tau)$ are both self-adjoint and U is positive definite we can write

$$(BK_0(t-\tau))^* P_{K_0}(t,\tau) + P_{K_0}(t,\tau) BK_0(t-\tau) + K_0(t-\tau)^* U K_0(t-\tau)$$

$$= (UK_0(t-\tau) + B^* P_{K_0}(t,\tau))^* U^{-1}(UK_0(t-\tau) + B^* P_{K_0}(t,\tau))$$

$$- P_{K_0}(t,\tau) B U^{-1} B^* P_{K_0}(t,\tau) \ .$$

The last three terms in (3.6) can be treated similarly. Then, if we let $\psi(t,\tau,\sigma)$ be the solution of

$$(3.7) \quad \frac{d\psi}{d\tau} = (A \psi B K(t-\tau))\psi, \ \psi(t,\sigma,\sigma) = I \ ,$$

481

and note that $P_K(t, 0) = P_{K_0}(t, 0) = 0$, we can obtain

$$(3.8) \quad P_{K_0}(t,t) - P_K(t,t) = \int_0^t \psi(t,t,\sigma)^* [(UK_0(t-\sigma) + B^*P_{K_0}(t,\sigma))^*$$

$$U^{-1}(UK_0(t-\sigma) + B^*P_{K_0}(t,\sigma)) - (UK(t-\sigma) + B^*P_K(t,\sigma))^*$$

$$U^{-1}(UK(t-\sigma) + B^*P_{K_0}(t,\sigma))]\psi(t,t,\sigma)\,d\sigma \ .$$

The equation (3.9) immediately yields the following results. If

$$(3.9) \quad K_0(t-\tau) \equiv -U^{-1}B^*P_{K_0}(t,\tau)$$

then

$$P_{K_0}(t,t) - P_K(t,t) = -\int_0^t \psi(t,t,\sigma)^*(UK(t-\sigma) + B^*P_{K_0}(t,\sigma))^*$$

$$U^{-1}(UK(t-\sigma) + B^*P_{K_0}(t,\sigma))\psi(t,t,\sigma)\,d\sigma$$

is negative semidefinite. On the other hand if (3.9) does not hold identically for $0 \leq \tau \leq t$ then by setting

$$K(t-\tau) \equiv -U^{-1}B^*P_{K_0}(t,\tau)$$

we get

$$P_{K_0}(t, t) - P_K(t, t) = \int_0^t \psi(t, t, \sigma)^* (U K_0(t-\sigma) + B^* P_{K_0}(t, \sigma))$$

$$U^{-1}(U K_0(t-\sigma) + B^* P_{K_0}(t, \sigma)) \psi(t, t, \sigma) d\sigma \quad ,$$

which is positive semidefinite and not zero.

From (3.5) we see that if $K_0(t-\tau)$ and $P_{K_0}(t, \tau)$ are to satisfy (3.9) then $P_{K_0}(t, \tau)$ must be an operator valued solution of

$$(3.10) \quad \dot{P} = [A - BU^{-1}B^*P]^* P + P[A - BU^{-1}B^*P] + W$$

$$+ (-U^{-1} B^* P)^* U(-U^{-1} B^* P) \quad ,$$

which reduces to the <u>Riccati differential equation</u>

$$(3.11) \quad \dot{P} = A^* P + PA + W - PBU^{-1}B^*P \quad .$$

Moreover, $P_{K_0}(t, \tau)$ satisfies the initial condition

$$(3.12) \quad P_{K_0}(t, 0) = 0 \quad .$$

Now (3.11) is a nonlinear equation and it is not immediately clear that the solution satisfying (3.12) exists over the whole interval $[0, t]$. To prove this we have to do a little more work. Let $P_0(t, \tau)$ be given by (3.4) with $K(s) \equiv 0$. Then

$$P_0(t, \tau) = \int_{t-\tau}^t e^{A^*(s-t-\tau)} W e^{A(s-t+\tau)} ds$$

and thus

$$(x_0, P_0(t, \tau)x_0)_1 \le \tau e^{2\tau \|A\|_1} \|W\|_1 \|x_0\|_1^2, \quad x_0 \in H_1.$$

Replacing $K(t-\tau)$ by 0 and assuming (3.9) holds in (3.8) we see that $P_0(t, \tau) - P_{K_0}(t,\tau)$ is positive semidefinite. Therefore

$$(x_0, P_{K_0}(t, \tau)x_0)_1 \le \tau e^{2\tau \|A\|_1} \|W\|_1 \|x_0\|_1^2, \quad x_0 \in H_1.$$

Since $P_{K_0}(t, \tau)$ is self-adjoint, we have

(3.13) $$\|P_{K_0}(t, \tau)\|_1 \le \tau e^{2\tau \|A\|_1} \|W\|_1.$$

Now the differential equation (3.11) is autonomous and the solution which satisfies the zero initial condition for $\tau = 0$ is a function of τ alone. Thus, denoting this solution by $P(\tau)$, we have

(3.14) $$P(\tau) = P_{K_0}(t, \tau), \quad 0 \le \tau \le t < \infty,$$

$$\|P(\tau)\|_1 = \|P_{K_0}(t, \tau)\|_1 \le \tau e^{2\tau \|A_1\|} \|W\|_1.$$

Thus $P(\tau)$ is uniformly bounded in norm on any finite interval. This enables us to use the usual extension techniques [G] to show that the solution $P(\tau)$ continues to exist for $0 \le \tau < \infty$.

Thus we have proved

Theorem 1. **Let** $P(\tau)$ **be the solution of** (3.11) **satisfying** $P(0) = 0$ **and let**

(3.15) $\qquad K_0(t-\tau) \equiv -U^{-1} B^* P(\tau), \quad 0 \leq \tau \leq t$.

Then $P_K(t,t) - P_{K_0}(t,t) = P_K(t,t) - P(t)$ **is positive semi-definite for all feedback operator functions** $K(t)$. **Moreover, if** $K_0(t)$ **does not satisfy** (3.15) **then there exists** $K(t)$ **such that** $P_K(t,t) - P_{K_0}(t,t)$ **is not positive semidefinite.**

Thus a time varying feedback operator $K_0(t)$ is optimal relative to the quadratic performance criterion (3.1) defined over the finite interval $[0, t]$ if and only if (3.15) holds.

We are now ready to complete our study and obtain the optimal feedback operator $\hat{K}$ for the performance criterion (2.1) as described at the close of the preceding section.

Let $\tilde{K}$ be the feedback operator described in the Controllability Assumption. Using (2.4) and Theorem 1 we see that

$$Q_{\tilde{K}}(t) - P_{K_0}(t,t) = P_{\tilde{K}}(t,t) - P_{K_0}(t,t)$$

is positive semidefinite for any $t > 0$. The Controllability Assumption then implies that $M - P_{K_0}(t,t) = M - P(t)$ is positive semidefinite for $t > 0$.

We claim now that if $t_2 > t_1$, $P(t_2) - P(t_1)$ is positive semi-definite. To prove this, let $\psi(t, \tau)$ satisfy

$$\frac{d\psi}{dt} = [A - BU^{-1} B^* P(t)]\psi, \quad \psi(\tau, \tau) = I .$$

From (3.10) we see that

$$(3.16) \quad P(t_2) = \int_0^{t_2} \psi(t_2, \tau)^* [W + (-U^{-1}B^*P(\tau))^* U(-U^{-1}B^*P(\tau))]$$
$$\psi(t_2, \tau) d\tau .$$

Then let

$$(3.17) \quad \bar{P}(t_2, t_1) = \int_{t_2-t_1}^{t_2} \psi(t_2, \tau)^* [W + (-U^{-1}B^*P(\tau))^* U(-U^{-1}B^*P(\tau))]$$
$$\psi(t_2, \tau) d\tau .$$

Since W and U are positive semidefinite and positive definite, respectively, $P(t_2) - \bar{P}(t_2, t_1)$ is positive semidefinite.

Let

$$\bar{K}(t_1 - \tau) = -U^{-1} B^* P(\tau + t_2 - t_1), \quad 0 \leq \tau \leq t_1 ,$$

and let $\psi(t_1, \tau, \sigma)$ be given by (3.7) with $t = t_1$ and $K(t_1 - \tau)$ replaced by $\bar{K}(t_1 - \tau)$. Then

$$\psi(t_1, \tau, \sigma) = \psi(\tau + t_2 - t_1, \sigma + t_2 - t_1)$$

and (3.17) becomes

$$(3.18) \quad \bar{P}(t_2, t_1) = \int_0^{t_1} \psi(t_1, t_1, \sigma)^* [W + \bar{K}(t_1 - \sigma)^* U \bar{K}(t_1 - \sigma)] \psi(t_1, t_1, \sigma) d\sigma$$

$$= P_{\bar{K}}(t_1, t_1) .$$

But $P_{\bar{K}}(t_1, t_1) - P_{K_0}(t_1, t_1) = P_{\bar{K}}(t_1, t_1) - P(t_1)$ is positive semidefinite by Theorem 1. Combining this with the remark which follows (3.17) we see that $P(t_2) - P(t_1)$ is positive semi-definite if $t_2 > t_1$. Taking this along with the fact that $M - P(t)$ is positive semi-definite for all $t \geq 0$ we can use Lemma 1 to see that there is a bounded self-adjoint positive semidefinite operator P such that

(3.19) $$\lim_{t \to \infty} P(t) = P$$

in the strong sense. The positive semidefiniteness of P follows from the fact that $P(t)$ has this property for all $t > 0$, as shown by (3.16).

Theorem 2. <u>Let P be the operator defined by (3.19). Then</u>

(i) $A^*P + PA + W - PBU^{-1}B^*P = 0$;

(ii) <u>if</u>

(3.20) $$\hat{K} = -U^{-1}B^*P$$

<u>then</u>, $Q_{\hat{K}}$ <u>being given by</u> (2.3) <u>with</u> $K = \hat{K}$,

(3.21) $$Q_{\hat{K}} = P ;$$

(iii) $Q_K - Q_{\hat{K}}$ <u>is positive semidefinite for all admissible feedback operators</u> K, <u>so that the feedback control policy</u> (3.20) <u>is optimal relative to the quadratic performance criterion</u> (2.1).

Proof. The property (i) is an immediate consequence of (3.11) and (3.19). To prove (3.21) we note from (3.4) and (3.14) that

(3.22) $$P(t) = \int_0^t \Phi(s,0)^* [W + K_0(s)^* U K_0(s)] \Phi(s,0) ds$$

where $\Phi(s, 0)$ is defined by (3.3), $K(s)$ replaced by $K_0(s)$. Now $K_0(s)$ satisfies (3.15) so (3.22) becomes

$$(3.23) \quad P(t) = \int_0^t \Phi(t, s, 0)^* [W + (-U^{-1}B^*P(t-s))^* U(-U^{-1}B^*P(t-s))] \Phi(t, s, 0)\,ds$$

and $\Phi(t, s, 0)$ satisfies

$$\frac{d\Phi}{ds} = [A - BU^{-1}B^*P(t-s)]\Phi, \quad \Phi(t, 0, 0) = I.$$

Fix $T > 0$ and set

$$(3.24) \quad P(t, T) = \int_0^T \Phi(t, s, 0)^* [W + (-U^{-1}B^*P(t-s))^* U(-U^{-1}B^*P(t-s))] \Phi(t, s, 0)\,ds.$$

It is clear that $P(t, T_2) - P(t, T_1)$ is positive semidefinite for $T_2 > T_1$ and $P(t) - P(t, T) = P(t, t) - P(t, T)$ is positive semidefinite if $t > T$. Since $M - P(t)$ is positive semidefinite for each t, $M - P(t, T)$ is positive semidefinite, $t > T$. If we let $t \to \infty$ in (3.24) Lemma 1 and the results in [H] can be used to show that there is a bounded self-adjoint positive semidefinite $P(\infty, T)$ such that

$$\lim_{t \to \infty} P(t, T) = P(\infty, T)$$

$$= \int_0^T \Phi(\infty, s, 0)^* [W + (-U^{-1}B^*P)^* U(-U^{-1}B^*P)] \Phi(\infty, s, 0)\,ds$$

where $\Phi(\infty, s, 0)$ satisfies

$$\frac{d\Phi}{ds} = [A - BU^{-1}B^*P]\Phi, \quad \Phi(\infty, 0, 0) = I.$$

Moreover it is clear that $M - P(\infty, T)$ and $P(\infty, T_2) - P(\infty, T_1)$ are positive semidefinite for all $T \geq 0$ and $T_2 > T_1$. Applying Lemma 1 we see that

$$(3.25) \quad \lim_{T \to \infty} P(\infty, T) = \int_0^\infty \Phi(\infty, s, 0)^* [W + (-U^{-1}B^*P)^*$$

$$U(-U^{-1}B^*P)]\Phi(\infty, s, 0)\,ds \equiv P_1$$

exists in the strong sense. Since we also have (3.20), we want to show that $P = P_1$. Since $P(t) = P(t, t)$ and $P(t, t) - P(t, T)$ is positive semidefinite for $t \geq T$, we can use (3.19) with

$$\lim_{T \to \infty} (\lim_{\substack{t \to \infty \\ t \geq T}} P(t, T)) = P_1,$$

(a rewritten form of (3.25)) to see that $P - P_1$ is positive semidefinite. On the other hand, with $\hat{K}$ given by (3.20) it is clear that $P_1 - P_{\hat{K}}(t, t)$ is positive semidefinite, $P_{\hat{K}}(t, t)$ be given by (3.4) with $K(s)$ replaced by $\hat{K}$. But Theorem 1 shows that $P_{\hat{K}}(t, t) - P(t)$ is positive semidefinite. Thus $P_1 - P(t)$, and hence $P_1 - P$ is positive semi-definite. Since $P_1 - P$ and $P - P_1$ are both positive semidefinite and self-adjoint, we conclude that $P_1 = P$. Thus, in the strong sense,

$$P = \int_0^\infty \Phi(\infty, s, 0)^* [W + (-U^{-1}B^*P)^*$$

$$U(-U^{-1}B^*P)] \Phi(\infty, s, 0) ds = Q_{\hat{K}}$$

if $Q_{\hat{K}}$ is given by (2.3) and $\hat{K}$ by (3.20). This proves (3.21). Now if there were an admissible K and $x_0 \in H_1$ such that

$$(x_0, [Q_K - Q_{\hat{K}}]x_0)_1 < 0$$

we would also have

$$(x_0, [Q_K(t) - P(t)]x_0)_1 < 0$$

for t sufficiently large, as follows from (2.4) and (3.19). But this contradicts Theorem 1. Thus $Q_K - Q_{\hat{K}}$ must be positive semidefinite and we have proved (iii). The proof of Theorem 2 is therefore complete.

Remarks

Theorem 2 is proved in [A] and [F] by slightly different techniques and it is also shown in [A] that (3.20) is the only optimal feedback operator.

We have chosen in this paper to consider only linear feedback control policies and we have found an optimal control policy of this type. But it can be shown very easily (see [A], e.g.) that if $u(s)$ is any norm-square integrable control on $[0, t]$, and if $\hat{u}(s)$ is the control generated by the feedback law (3.15) then $C_t(\hat{x}, u) \le C_t(x, u)$, where $\hat{x}, \hat{u}$ and x, u both satisfy (1.1), (1.2). Similarly, if $u(s)$ is any norm-square integrable control on $[0, \infty)$ which yields a finite value for $C(x, u)$ and if $\hat{u}(s)$ is generated by (3.20) then $C(\hat{x}, \hat{u}) \le C(x, u)$. Moreover, the existence of $\hat{K}$ can be proved under a slightly weaker Controllability Assumption; namely, that for each $x_0 \in H_1$ there is a control u_0 such that if x, u_0 satisfy (1.1), (1.2) then $C(x, u_0) \leq M \|x_0\|_1^2$ for some fixed $M > 0$.

In order to prove that

(3.26) $$\lim_{t \to \infty} \|x(t)\|_1 = 0$$

for solution of the closed loop system

$$\dot{x} = (A + B\hat{K})x$$

some assumptions must be made on W. For example, if W is positive definite then the finiteness of $C(x, u)$ implies

$$\int_0^\infty \|x(t)\|_1^2 \, dt < \infty .$$

In [1] R. Datko has shown that this implies

$$\|e^{(A+B\hat{K})t}\| \leq e^{-ct}, \quad 0 \leq t < \infty$$

for some positive real number c. In this case

$$\|x(t)\|_1 \leq \|e^{(A+B\hat{K})t}\|_1 \|x_0\|_1 \leq e^{-ct} \|x_0\|_1$$

and it is clear that (3.26) holds.

4. Finite Dimensional Approximations

We have seen that the optimal feedback operator $\hat{K}$ is given by

$$\hat{K} = -U^{-1} B^* P$$

where P is a positive semidefinite solution of

(4.1) $\qquad A^* P + PA + W - PBU^{-1}B^* P = 0$.

The task of finding P is complicated by a number of factors: the equation (4.1) is nonlinear, P is an operator on a space of possibly infinite dimension, and (4.1) has, in general, infinitely many solutions.

We first address ourselves to the problem of dimension. We will introduce the

<u>Assumption on A</u>. <u>There is a sequence</u> $\{E_k\}$ <u>of projections on</u> H_1 <u>(not necessarily orthogonal projections) such that</u>

(a) $\lim\limits_{k \to \infty} E_k x = x$, $\lim\limits_{k \to \infty} E_k^* x = x$, $x \in H_1$;

(b) $A_k \equiv E_k A = A E_k$;

(c) $E_k(H_1) = R_k$, <u>a finite dimensional subspace of</u> H_1 .

(d) $\|E_k\|_1 \leq \beta$, $k = 1, 2, 3, \ldots$, <u>where</u> β <u>is a fixed positive constant</u>.

Property (b) shows that R_k is invariant under A_k. We may therefore treat A_k as an operator on R_k and, since (c) states that R_k is finite dimensional, we can consider A_k to be a matrix if we have in mind a fixed coordinate system for R_k.

We remark that one important case where the Assumption on A is satisfied occurs when A is a normal (or self-adjoint) operator on H_1 possessing a countable sequence $\{\lambda_k\}$ of eigenvalues corresponding to eigenvectors ϕ_k, such that $\{\phi_k\}$ is an orthonormal basis for H_1. The projection

E_k can then be defined by

(4.2) $$E_k x = E_k \left(\sum_{j=1}^{\infty} \alpha_j \phi_j \right) = \sum_{j=1}^{k} \alpha_j \phi_j$$

where $x \in H_1$ has the unique orthonormal expansion $x = \sum_{j=1}^{\infty} \alpha_j \phi_j$. In this case E_k is an orthogonal projection (i.e., $E_k^* = E_k$) whose range R_k is spanned by the set $\{\phi_1, \phi_2, \ldots, \phi_k\}$. In this case we can take $\beta = 1$. Other examples, involving non-orthogonal projections E_k with $\beta > 1$ arise when A is a <u>spectral operator</u> [J].

We define

$$x_k = E_k x, \quad x \in H_1,$$
$$B_k = E_k B,$$
$$W_k = E_k^* W E_k,$$

and consider the control system

(4.3) $$\dot{x}_k = A_k x_k + B_k u.$$

Given $x_k(0) = x_{0k} = E_k x_k \in R_k$ we seek to minimize

(4.4) $$C_k(x_k, u) = \int_0^{\infty} [(x_k(t), W_k x_k(t))_1 + (u(t), U u(t))_2] dt$$

as we did in Sections 2 and 3 for (2.1). Assuming that each of the finite dimensional systems (4.3) satisfies the Controllability Assumption, one shows, just as before that there are optimal feedback operator $\hat{K}_k$ given by

$$\hat{K}_k = -U^{-1} B_k^* P_k$$

493

where, in the strong sense,

$$(4.5) \quad \dot{P}_k(\tau) = A_k^* P_k(\tau) + P_k(\tau) A_k + W_k - P_k(\tau) B_k U^{-1} B_k^* P_k(\tau),$$

$$P_k(0) = 0.$$

Moreover, P_k satisfies the nonlinear equation

$$(4.6) \quad A_k^* P_k + P_k A_k + W_k - P_k B_k U B_k^* P = 0$$

and the optimal feedback operator for the problem (4.3), (4.4) is given by

$$(4.7) \quad \hat{K}_k = -U^{-1} B_k^* P_k, \quad k = 1, 2, 3, \ldots.$$

In the following theorem we establish both the validity of the Controllability Assumption for the systems (4.3) and the strong convergence of the operators P_k to P. In order to do this we have to introduce a fairly mild

__Assumption on__ W_k. There is a positive number d such that

$$(4.8) \quad (x, W_k x)_1 \leq d\,(x, Wx)_1, \quad x \in H_1.$$

The inequality (4.8) can readily be shown to hold if there are positive numbers ω_1 and ω_2 such that

$$(4.9) \quad \omega_1 \|x\|_1^2 \leq (x, Wx)_1 \leq \omega_2 \|x\|_1^2, \quad x \in H_1$$

or if $E_k^* W = W E_k$ for all k (which is true, e.g., if A is a normal operator with the E_k defined as in (4.2) and W commutes with A).

OPTIMAL CONTROL PROBLEMS

<u>Theorem 3.</u> <u>If the Assumption on W_k is valid then each of the systems (4.3) satisfies the Controllability Assumption so that the operator P_k of (4.5) exists. Moreover</u>

$$\lim_{k \to \infty} P_k x_0 = P x_0, \quad x_0 \in H_1$$

and

$$\lim_{k \to \infty} \hat{K}_k x_0 = \hat{K} x_0, \quad x_0 \in H_1 .$$

<u>Proof.</u> Let $\hat{K}$ be the optimal feedback operator for the problem (1.1), (2.1). Given $x_0 \in H_1$, let $\hat{x}(t)$ solve the closed loop equation

$$\dot{x} = (A + B\hat{K})x$$

with $\hat{x}(0) = x_0$. Let

$$\hat{u}(t) \equiv \hat{K} \hat{x}(t) .$$

Since the optimal $\hat{K}$ yields a cost less than or equal to the cost yielded by $\tilde{K}$ of the Controllability Assumption,

$$(4.10) \quad C(\hat{x}, \hat{u}) = \int_0^\infty [(\hat{x}(t), W\hat{x}(t))_1 + (\hat{u}(t), U\hat{u}(t))_2] dt$$

$$\leq M \|x_0\|^2 .$$

Let $x_{0k} = E_k x_0$. Then

$$\hat{x}_k(t) \equiv E_k \hat{x}(t)$$

495

is the solution of

$$\dot{\hat{x}}_k = A_k \hat{x}_k + B_k \hat{u}(t), \quad 0 \le t < \infty,$$

with $\hat{x}_k(0) = x_{0k}$. For this pair $\hat{x}_k$, $\hat{u}$ the cost (4.4) becomes

$$(4.11) \quad C_k(\hat{x}_k, \hat{u}) = \int_0^\infty [(\hat{x}_k(t), W_k \hat{x}_k(t))_1 + (\hat{u}(t), U\hat{u}(t))_2] dt$$

$$= \int_0^\infty [(\hat{x}(t), W_k \hat{x}(t))_1 + (\hat{u}(t), U\hat{u}(t))_2] dt$$

$$\le \max(d, 1) \int_0^\infty [(\hat{x}(t), W\hat{x}(t))_1 + (\hat{u}(t), U\hat{u}(t))_2] dt$$

$$= \max(d, 1) C(\hat{x}, \hat{u}).$$

The second equality in (4.11) is valid because E_k, being a projection, satisfies $E_k^2 = E_k$ and the inequality in (4.11) follows from the Assumption on W_k. Noting the remarks at the end of Section 3 we see that (4.3) satisfies the Controllability Assumption. Moreover, since the cost (4.11) is greater than or equal to the optimal cost for (4.3), (4.4) we have

$$(4.12) \quad (x_{0k}, P_k x_{0k})_1 = (x_0, P_k x_0)_1 \le C_k(\hat{x}_k, \hat{u}) =$$

$$= \max(d, 1) C(\hat{x}, \hat{u}) = \max(d, 1)(x_0, P x_0)_1, \quad x_0 \in H_1.$$

Let $T > 0$ be fixed and define

$$\hat{X}_T = \{x \in H_1 \mid x = \hat{x}(t), \ t \in [0, T]\}.$$

Then $\hat{X}_T$ is a compact set with respect to the topology induced on H_1 by $\|\ \|_1$. Now for each $x \in \hat{X}_T$ we have, since W_k converges strongly to W

(4.13)
$$\lim_{k \to \infty} (x, W_k x)_1 = (x, Wx)_1 ,$$

i.e. $(x, W_k x)_1$ converges <u>pointwise</u> to $(x, Wx)_1$ on $\hat{X}_T$. But, if $x_1, x_2 \in \hat{X}_T$

(4.14)
$$|(x_1, W_k x_1)_1 - (x_2, W_k x_2)_1|$$

$$= |\int_0^1 \frac{d}{d\lambda} (x(\lambda), W_k x(\lambda))_1 d\lambda|$$

where

$$x(\lambda) = \lambda x_2 + (1-\lambda) x_1 .$$

Now

$$\frac{d}{d\lambda} (x(\lambda), W_k x(\lambda))_1 = 2(W_k x(\lambda), \frac{dx(\lambda)}{d\lambda})_1$$

$$= 2(W_k x(\lambda), x_2 - x_1) \leq 2 M_T \|W_k\|_1 \|x_2 - x_1\|_1$$

$$\leq 2 M_0 \|E_k^* W E_k\|_1 \|x_2 - x_1\|_1 \leq 2 M_T \beta^2 \|W\| \|x_2 - x_1\|_1$$

where

$$M_T = \sup_{x \in \hat{X}_T} \{\|x\|_1\} .$$

Therefore, from (4.14),

(4.15) $|(x_1, W_k x_1)_1 - (x_2, W_k x_2)_1| \leq 2 M_T \beta^2 \|W\| \|x_2 - x_1\|_1 .$

Thus the functions $(x, W_k x)_1$ are equicontinuous on $\hat{X}_T$ and, setting $x_2 = 0$ in (4.15), they are uniformly bounded as well. Then we may combine (4.13) with the Arzelà-Ascoli theorem to see that

$$\lim_{k \to \infty} (\sup_{x \in \hat{X}_T} |(x, W_k x)_1 - (x, Wx)_1|) = 0 .$$

Using this result together with the Assumption on W_k gives

$$\limsup_{k \to \infty} |C(\hat{x}, \hat{u}) - C_k(\hat{x}_k, \hat{u})|$$

$$\leq \lim_{k \to \infty} |\int_0^T (\hat{x}(t), W \hat{x}(t))_1 dt - \int_0^T (\hat{x}(t), W_k \hat{x}(t))_1 dt|$$

$$+ \limsup_{k \to \infty} (\int_T^\infty (\hat{x}(t), W \hat{x}(t))_1 dt + \int_T^\infty (\hat{x}(t), W_k \hat{x}(t)) dt)$$

$$\leq 0 + (1+d) \int_T^\infty (\hat{x}(t), W \hat{x}(t))_1 dt .$$

But, since $C(\hat{x}, \hat{u})$ is finite, $\int_T^\infty (\hat{x}(t), W \hat{x}(t))_1 dt$ can be made arbitrarily small by taking T sufficiently large. Thus we conclude that

OPTIMAL CONTROL PROBLEMS

$$\lim_{k \to \infty} |C(\hat{x}, \hat{u}) - C_k(\hat{x}_k, \hat{u})| = 0 .$$

Since $C(\hat{x}, \hat{u}) = (x_0, Px_0)_1$ and $C_k(\hat{x}_k, \hat{u}) \geq (x_{0k}, P_k x_{0k})_1 = (x_0, P_k x_0)_1$ we see that

(4.16) $$\limsup_{k \to \infty} (x_0, P_k x_0)_1 \leq (x_0, Px_0)_1$$

for all $x_0 \in H_1$.

It follows that there is a positive number M_{x_0} for each $x_0 \in H_1$ such that

(4.17) $$0 \leq (x_0, P_k x_0)_1 \leq M_{x_0} .$$

Suppose there were an $x_0 \in H_1$ and a subsequence P_{k_ℓ} such that

(4.18) $$\lim_{\ell \to \infty} (x_0, P_{k_\ell} x_0)_1 = (x_0, Px_0)_1 - \delta, \quad \delta > 0 .$$

Since $P(\tau)$ converges strongly to P as $\tau \to \infty$ we can find τ_0 such that

(4.19) $$|(x_0, Px_0)_1 - (x_0, P(\tau_0)x_0)_1| < \delta/2 .$$

Using the results in [H] on strong continuity of solutions of operator differential equations we can show that

(4.20) $$\lim_{k \to \infty} P_k(\tau), \quad 0 < \tau < \infty ,$$

499

in the strong sense. Thus, taking ℓ_0 sufficiently large we see from (4.18) and (4.20) that we can ensure that

(4.21) $\quad |(x_0, P_{k_{\ell_0}}(\tau_0) x_0)_1 - (x_0, P(\tau_0) x_0)_1| < \delta/4$,

(4.22) $\quad |(x_0, P_{k_{\ell_0}} x_0)_1 - [(x_0, P x_0) - \delta]| < \delta/4$.

Combining (4.19), (4.21) and (4.22) we have

(4.23) $\quad (x_0, P_{k_{\ell_0}}(\tau_0) x_0)_1 > (x_0, P_{k_{\ell_0}} x_0)_1$.

But (4.5) holds and $(x_0, P_k(\tau) x_0)_1$ is monotone increasing with τ so

(4.24) $\quad (x_0, P_{k_{\ell_0}} x_0)_1 \geq (x_0, P_{k_{\ell_0}}(\tau) x_0)_1, \quad \tau \in [0, \infty)$.

Thus (4.23) and (4.24) provide a contradiction and (4.18) cannot hold for any subsequence P_{k_ℓ}. Combined with (4.16) and (4.17) this shows that

(4.25) $\quad \lim_{k \to \infty} (x_0, P_k x_0)_1 = (x_0, P x_0)_1, \quad x_0 \in H_1$.

Let $\Omega = \{1, 2, 3, \ldots, \infty\}$ denote the set consisting of the positive integers and the symbol ∞. We construct a topology on Ω by specifying a neighborhood system: if $k \in \Omega$ and $k \neq \infty$, then any subset of Ω which includes k is a neighborhood of k; for $k = \infty$ a subset of Ω is a neighborhood of ∞ if and only if that subset contains ∞ and all but finitely many of the integers. With this topology Ω is a compact topological space.

The results (4.20) and (4.25) show that, if we put $P = P_\infty$, $P(\tau) = P_\infty(\tau)$, then $(x_0, P_k x_0)_1$ is a continuous function of k for $k \in \Omega$ and $(x_0, P_k(\tau) x_0)_1$ is similarly continuous, $0 \leq \tau < \infty$. We have, for $\tau_2 > \tau_1$,

$$(x_0, P_k(\tau_2) x_0)_1 \geq (x_0, P_k(\tau_1) x_0)_1, \quad k \in \Omega ,$$

and

(4.26) $$\lim_{\tau \to \infty} (x_0, P_k(\tau) x_0)_1 = (x_0, P_k x_0)_1, \quad k \in \Omega .$$

Under these conditions Dini's theorem shows (see, e.g. [K]) that the convergence (4.26) is <u>uniform</u> with respect to k, $k \in \Omega$. In Lemma 1 we replace the $R(\tau)$ by $P_k(\tau)$, $k \in \Omega$. Examining the proof of Lemma 1 in [B] we see that the uniformity of the convergence (4.26) for $k \in \Omega$ implies that

$$\lim_{\tau \to \infty} P_k(\tau) x_0 = P_k x_0, \quad x_0 \in \Omega ,$$

<u>uniformly</u> for $x_0 \in H_1$. From this, using

$$\| P x_0 - P_k x_0 \|_1 \leq \| P x_0 - P(\tau) x_0 \|_1 + \| P(\tau) x_0 - P_k(\tau) x_0 \|_1$$

$$+ \| P_k(\tau) x_0 - P_k x_0 \|_1 .$$

One can show without difficulty that

(4.27) $$\lim_{k \to \infty} P_k x_0 = P x_0, \quad x_0 \in H_1 .$$

Since $\hat{K}_k = -U^{-1}B_k^*P_k$, $K = -U^{-1}B^*P$ and B_k^*, P_k converge strongly to B^*, P respectively, we have

(4.28) $$\lim_{k \to \infty} \hat{K}_k x_0 = \hat{K} x_0, \quad x_0 \in H_1 ,$$

and the proof of Theorem 3 is complete.

Remarks. Theorem 3 is somewhat easier to prove if $E_k^*W = WE_k$, $k = 1, 2, 3, \ldots$ and $E_k E_\ell = E_\ell E_k = E_\ell$, $k > \ell$. In this case $(x_0, P_k x_0)_1$ is monotone increasing with respect to k and (4.25) implies (4.27) directly via Lemma 1, replacing the continuous variable τ by the index k.

If the control space H_2 is finite dimensional, as it usually is in practice, the strong convergence in (4.28) becomes uniform convergence, i.e.,

$$\lim_{k \to \infty} \|P - P_k\|_1 = 0, \quad \lim_{k \to \infty} \|\hat{K} - \hat{K}_k\|_{2,1} = 0 .$$

See [A] for more details.

We see then that in order to construct approximations to P in the strong operator topology, and hence approximations to the optimal feedback operator $\hat{K}$ in the same topology, it is sufficient to be able to calculate the operators P_k. We have defined R_k as the range of E_k. If we similarly define R_k^* as the range of E_k^* then we have

(4.29)
$$A_k : R_k \to R_k ,$$
$$A_k^* : R_k^* \to R_k^* ,$$
$$B_k U^{-1} B_k^* : R_k^* \to R_k ,$$
$$W_k : R_k \to R_k^* ,$$
$$P_k : R_k \to R_k^* .$$

502

OPTIMAL CONTROL PROBLEMS

Let n_k be the common finite dimension of R_k and R_k^* and let $\xi_1, \xi_2, \ldots, \xi_{n_k}$ be a basis for R_k. For $i = 1, 2, \ldots, n_k$ there is exactly one vector ξ_i^* such that

$$(\xi_i^*, (I - E_k)x)_1 = 0, \quad x \in H_1$$

$$(\xi_i^*, \xi_j)_1 = \delta_{ij} = \begin{cases} 1 & \text{if } i = j, \\ 0 & \text{if } i \neq j \end{cases} \quad i,j = 1, 2, \ldots, n_k.$$

The vectors ξ_i^* constitute a basis for R_k^* dual to the basis $\xi_1, \xi_2, \ldots, \xi_{n_k}$ for R_k. With respect to these bases the operators (4.29) have unique $n_k \times n_k$ matrix representations, which we will denote by the same symbols, A_k, A_k^*, W_k, P_k, etc. Moreover, the last three operators in (4.29) will have symmetric (or Hermitian) matrix representations.

Thus P_k can be considered as a positive-semidefinite matrix solution of a matrix quadratic equation (4.6). Such equations have been studied in some detail by Potter [L], who shows that P_k can be calculated explicitly in terms of the eigenvectors of the $2n_k \times 2n_k$ matrix

$$\begin{bmatrix} A_k^* & W_k \\ B_k U^{-1} B_k^* & -A_k \end{bmatrix}$$

This method of Potter's, together with our strong convergence results (4.27), (4.28) provides an algorithm for the approximate computation of the optimal feedback operator $\hat{K}$.

REFERENCES

[A] D. L. Lukes and D. L. Russell: "The quadratic criterion for distributed systems", SIAM J. on Control, Vol. 7 (1969) No. 1, pp. 101-121.

[B] F. Riesz and B. Sz.-Nagy: "Functional Analysis", Frederick Ungar Pub. Co., New York, 1955. (See page 263).

[C] R. E. Kalman: "Contributions to the theory of optimal control", Bol. Soc. Mat. Mexicana, Vol. 5 (1960), pp. 102-119.

[D] D. L. Lukes: "Stabilizability and optimal control", Funk. Ekvac., Vol. 11 (1968), pp. 39-50.

[E] W. M. Wonham: "On a matrix Riccati equation of stochastic control", SIAM J. on Control, Vol. 6 (1968), pp. 681-697.

[F] R. Datko: "A linear control problem in an abstract Hilbert space", to appear in Jour. Diff. Eqns.

[G] E. A. Coddington and N. Levinson: "Theory of Ordinary Differential Equations", McGraw-Hill Book Co., Inc., New York, 1955.

[H] D. L. Russell: "Continuity in the strong topology of operator-valued solutions of nonlinear differential equations with an application to optimal control", SIAM J. on Control, Vol. 7 (1969), pp. 132-140.

[I] R. Datko: "Extending a theorem of A. M. Liapounov to Hilbert space", to appear in J. Math. Anal. Appl.

[J] N. Dunford: "Spectral operators", Pac. J. Math., Vol. 4 (1954), pp. 321-354.

[K] R. P. Boas: "<u>A Primer of Real Functions</u>", Carus Mathematical Monographs, published by the Mathematical Association of America, 1960. (See footnote on p. 89 and note 21 on p. 158.)

[L] J. E. Potter: "<u>Matrix quadratic solutions</u>", J. SIAM Appl. Math., Vol. 14 (1966), pp. 496-501.

This paper was supported in part by the Office of Naval Research under Contract NR-041-404.

Department of Mathematics
University of Wisconsin
Madison, Wisconsin 53706

Visiting 1970-71:

Department of Mathematics
University of California, Los Angeles
Los Angeles, California 90024

Complementary Variational Principles

PETER D. ROBINSON

CONTENTS

1. Introduction 509
2. Bounds for the Action in Dynamics 513
   - 2.1 The Euler-Lagrange approach to an upper bound. 513
   - 2.2 The Euler-Hamilton variational principle. . . . 516
   - 2.3 A lower bound for $S(p,q)$. 518
   - 2.4 Complementary upper and lower bounds for $S(p,q)$ 519
3. General Theory 520
   - 3.1 Introduction 520
   - 3.2 Abstract formulation 522
   - 3.3 Boundary conditions, with appropriate functionals $\Gamma(U,\Phi)$. 526
   - 3.4 Sufficient conditions for complementary upper and lower bounds. 529
   - 3.5 Uniqueness of (u,φ). 533
   - 3.6 Choices for the linear operators 534
4. The Equation $T^*T\Phi + f(\Phi) = 0$ 538
   - 4.1 Preliminaries. 538

4.2	Complementary bounds	539
4.3	Bounds involving $L = T^*T$	541
4.4	More specific conditions for bounds	543
4.5	Error bounds for an approximate solution	546
4.6	A related equation	547

5. <u>The Hammerstein Integral Equation</u> 548

5.1	Relationship to $T^*T\Phi + f(\Phi) = 0$	548
5.2	The complementary functionals	549
5.3	Pairs of complementary bounds	551

6. <u>Some Illustrative Applications</u> 552

6.1	The Poisson-Boltzmann equation	552
6.2	The Thomas-Fermi equation	554
6.3	The Föppl-Hencky equation	557
6.4	An integral equation in communication theory	558
6.5	A non-linear Kirkwood integral equation	560
6.6	A transportation network problem	562
6.7	Compressible fluid flow	567
6.8	Heat loss from a cell	568

Appendix: Convex functions and functionals . . . 570

References and Author Index 572

COMPLEMENTARY VARIATIONAL PRINCIPLES

1. Introduction

Problems in applied mathematics can often be framed in terms of variational principles. The equations which describe the problem to be solved turn out to be the conditions that a certain functional should be stationary. In many cases, a problem can be formulated variationally in two different ways which are related to each other. One way involves minimizing some functional, the other way involves maximizing a related functional, and the respective minimum and maximum values of the two functionals are the same. In these cases the two variational principles are said to be complementary (or dual or reciprocal).

The practical value of complementary variational principles is clear when this minimum-maximum value is a quantity of physical interest which needs to be calculated. Trial functions can be inserted in the complementary functionals, immediately yielding upper and lower bounds on the quantity of interest. This process will usually be far simpler than solving the equations which are the conditions that the functionals be stationary, and then calculating the quantity. Even in situations where the stationary functional-value is not of immediate physical interest, the closeness of the complementary bounds can often be used as a measure of the accuracy of approximate solutions of the equations which describe the problem.

Some isolated examples of variational principles that are now recognized as complementary were discovered in the nineteenth century. For example there are the Dirichlet and Thomson principles in electrostatics (see Courant and Hilbert [1]), and Lord Rayleigh [2] gave upper and lower limits for the acoustic end-correction for an open tube. That the minimum potential energy theorem for an elastic body has a dual related to "complementary" energy seems to have been first discovered by Castigliano in 1873 [1]. The first systematic account of complementary variational principles is due to Friedrichs (1929), and the treatment by Courant and Hilbert [1] is based on this.

Subsequent accounts have been given in various areas

of applied mathematics. In particular, the following techniques have been exploited:
(a) the Prager-Synge hypercircle method [3];
(b) use of the Schwarz inequality, especially by Diaz and Weinstein [4];
(c) the operator decomposition $L = T^*T$, by Kato and Fujita [5,6].

Applications of the principles have been made in elasticity [7], plasticity [8], fluid flow [9-11], nonlinear networks [12-14], linear and non-linear programming [15,16], as well as in branches of mathematical physics such as quantum mechanics, electromagnetic theory and diffusion [e.g. 17]. Many of the problems are linear, and so are not of direct interest in these proceedings of a Seminar on Non-Linear Functional Analysis, but some additional references to linear problems are mentioned for completeness [55-63].

The discussion of complementary variational principles by Courant and Hilbert takes familiar Euler-Lagrange type variational principles, and shows how complementary principles can then be developed by involutory (Legendre) transformations. This approach is clarified by Sewell in an excellent review article [18], where the important individual ideas are spotlighted and applications given principally in continuum mechanics.

The present account is based on the canonical Euler-Hamilton type equations

$$(1.1) \qquad T\Phi = \frac{\delta}{\delta U} W(U,\Phi) ,$$

$$(1.2) \qquad T^*U = \frac{\delta}{\delta \Phi} W(U,\Phi) ,$$

where T and T^* are adjoint <u>linear</u> operators. Problems in applied mathematics can often be described by such equations. The functional $W(U,\Phi)$ can be <u>non-linear</u>, and is a generalization of the Hamiltonian in classical mechanics. Complementary upper and lower bounds are obtained for the stationary value of the generalized "action". In many applications this quantity is of direct interest. This approach was suggested by Noble in 1964 [19, 20], and put into a functional-analytic framework by Rall [21]. It is perhaps more aesthetically pleasing than the approach based on Legendre transformations, since the pair (U, Φ) can play the roles of pairs of related variables which occur in applied mathematics. Some examples are (position, momentum), (stress, strain), (voltage, current), (potential, its gradient) and (profitability of selling in a town, flow of goods between towns). However the two approaches are very often equivalent, and indeed there are situations where Sewell's treatment emphasizing the Legendre transformation has distinct advantages [11, 18].

A new book by Arthurs [17] also adopts the Noble approach and serves as a good introduction to the theory with both linear and non-linear applications. The conditions for complementary upper and lower bounds are given in terms of the derivatives of the functional W, namely

(1.3) $$\frac{\delta^2 W}{\delta U^2} \geq 0, \quad \frac{\delta^2 W}{\delta \Phi^2} \leq 0,$$

these derivatives being evaluated at the solution-pair (u, φ) of equations (1.1) and (1.2). The emphasis by Arthurs is on local bounds, and terms of third order in δu and $\delta \varphi$ are neglected. However Sewell [18] and Noble [22] realized that the crucial condition is for $W(U, \Phi)$ to have saddle-point behavior, namely that

(1.4) $W(U, \Phi)$ be convex in U and concave in Φ.

In these circumstances the complementary bounds are global, not merely local. Moreover if the convex-concavity is strict, then the solution-pair (u, φ) is unique, favorable boundary conditions being assumed. It is these convex-concavity properties which are exploited in the present review.

In Section 2, complementary bounds are developed for the action in a simple situation in dynamics, where the Euler-Hamilton equations have been familiar since the last century.

This is by way of an introductory example. The theory is generalized in Section 3, using a formalism which can readily be applied to differential, integral or matrix equations. Four different types of boundary conditions are considered, and it is shown how each can be embroidered into the theory. In Section 4 equations of the type

(1.5) $$T^*T\Phi + f(\Phi) = 0$$

are discussed; Hammerstein integral equations are examples of this type, and are treated briefly in Section 5. Finally in Section 6 some illustrative examples are presented.

One important extension of the theory is not discussed here, but is mentioned in Sewell's review [18] and will also feature in a forthcoming article by Noble and Sewell. If equation (1.1) is replaced by the <u>inequality constraints</u>

(1.6) $$T\Phi \leq \frac{\delta W}{\delta U}, \quad U \geq 0$$

together with the scalar product condition

(1.7) $$\left\{U, \ T\Phi - \frac{\delta W}{\delta U}\right\} = 0 \ ,$$

then equations (1.2), (1.6) and (1.7) also lead to complementary variational principles which have applications in non-linear programming [15, 16].

Another topic of interest is the possibility of introducing equality constraints into the theory [cf 23].

Perhaps one should make a final remark that it is by no means always possible to find variational principles which are equivalent to the equations describing a given problem [24].

I should like to record my sincere gratitude to Professors Arthurs, Noble and Sewell for helpful discussions and for making available to me material in advance of publication. I should also like to thank the Mathematics Research Center, University of Wisconsin, for inviting me to present this paper.

2. Bounds for the Action in Dynamics

2.1. The Euler-Lagrange approach to an upper bound

Suppose that a particle of unit mass moves in a straight line under the influence of a conservative force $f(q)$, $q(t)$ being the actual distance of the particle along the line at time t. If the potential energy of the particle is $-F(q)$ where

$$(2.1) \qquad f(q) = \frac{d}{dq} F(q) ,$$

then the Lagrangian is

$$(2.2) \qquad L(\dot{q}, q) = \frac{1}{2} \dot{q}^2 + F(q) ,$$

$\dot{q}$ denoting dq/dt. If $t_0 \le t \le t_1$ is a time-interval of interest, the action[†] for the motion during that time is

$$(2.3) \qquad S(\dot{q}, q) = \int_{t_0}^{t_1} L(\dot{q}, q) dt .$$

Let

$$(2.4) \qquad q(t_0) = \alpha(t_0), \; q(t_1) = \alpha(t_1)$$

be the initial and final q-values, assumed known. Then Hamilton's principle states that the action functional $S(\dot{Q}, Q)$ is stationary for variations of $Q(t)$ around the actual distance function $q(t)$ provided that also

$$(2.5) \qquad Q(t_0) = \alpha(t_0), \; Q(t_1) = \alpha(t_1) .$$

[†] Sometimes called Hamilton's principal function.

The condition for this variational principle is that the Euler-Lagrange equation

(2.6) $$\frac{d}{dt}\left(\frac{\partial L}{\partial \dot{Q}}\right) - \frac{\partial L}{\partial Q} = 0 \quad (L = L(\dot{Q}, Q))$$

should hold for $Q = q$. In our simple example, (2.6) reduces to Newton's equation of motion for the particle, namely

(2.7) $$-\ddot{q} + f(q) = 0 \ .$$

Let us suppose that we want to know the value of $S(\dot{q}, q)$, but that we are unable to determine $q(t)$ exactly. We seek conditions under which the action functional

(2.8) $$S(\dot{Q}, Q) = \int_{t_0}^{t_1} [\tfrac{1}{2} \dot{Q}^2 + F(Q)] dt$$

provides a bound for $S(\dot{q}, q)$, $Q(t)$ being an arbitrary function of t satisfying (2.5). Let

(2.9) $$\xi = Q - q \ ,$$

then

(2.10) $$S(\dot{Q}, Q) - S(\dot{q}, q) = \int [\tfrac{1}{2} \dot{\xi}^2 + \ddot{q}\xi + F(q + \xi) - F(q)] dt \ ,$$

the limits of integration here and subsequently in this section being t_0 and t_1. Since from (2.4) and (2.5)

(2.11) $$\xi(t_0) = 0, \ \xi(t_1) = 0 \ ,$$

we may integrate by parts to give

(2.12) $$\int \dot{q}\dot{\xi}\, dt = -\int \ddot{q}\xi\, dt = -\int f(q)\xi\, dt ,$$

the last step following from (2.7). Substitution in equation (2.10) yields

(2.13) $$S(\dot{Q},Q) - S(\dot{q},q) = \int \frac{1}{2}\dot{\xi}^2\, dt + \int [F(q+\xi) - F(q) - \xi f(q)]dt .$$

The first term on the right of (2.13) is non-negative, and if $F(Q)$ is a convex function of Q the second term is also non-negative (see the Appendix). Thus

(2.14) $$S(\dot{Q},Q) \geq S(\dot{q},q) \text{ if } F(Q) \text{ is convex} .$$

If $f(Q)$ is differentiable, we have from the mean value theorem

(2.15) $$F(q+\xi) - F(q) - \xi f(q) = \frac{1}{2}\xi^2 f'(q + \epsilon\xi), \quad 0 < \epsilon < 1 ,$$

and so a more useful result is perhaps (using (2.13) and (2.15))

(2.16) $$S(\dot{Q},Q) \geq S(\dot{q},q) \text{ if } f' \text{ is non-negative.}$$

Notice that the conditions in (2.14) and (2.16) are <u>sufficient</u> for $S(\dot{Q},Q)$ to be an upper bound on $S(\dot{q},q)$, but they are not <u>necessary</u>. The second term on the right of (2.13) could be negative, but be dominated by the positive first term. It is also worth emphasizing that we have not neglected third-order terms in ξ or anything like that; the bound is not merely a local one.

2.2. The Euler-Hamilton variational principle

In the Euler-Lagrange treatment of the action functional, variations in $\dot{Q}$ are dependent on those in Q; indeed they are their time derivatives. In the Euler-Hamilton approach, functionals are expressed in terms of q and of p, the momentum conjugate to q, defined by

$$(2.17) \qquad p = \partial L / \partial \dot{q} \;.$$

The Hamiltonian is

$$(2.18) \qquad H(p,q) = p \frac{dq}{dt} - L \;,$$

and the action is written as

$$(2.19) \qquad S(p,q) = \int [p \frac{dq}{dt} - H(p,q)] dt$$

or equivalently as

$$(2.20) \qquad S(p,q) = \int [-\frac{dp}{dt} q - H(p,q)] dt + [p\alpha]_{t_0}^{t_1} \;.$$

Hamilton's principle now states that the functional $S(P,Q)$ is stationary for <u>independent</u> variations of $P(t)$ and $Q(t)$ around the actual functions $p(t)$ and $q(t)$, again provided that $Q(t)$ satisfies the boundary conditions (2.5) but with no restriction on $P(t)$. The conditions for the variational principle are that the Euler-Hamilton canonical equations

$$(2.21) \qquad \frac{dQ}{dt} = \frac{\partial H}{\partial P} \;,$$

$$(2.22) \qquad -\frac{dP}{dt} = \frac{\partial H}{\partial Q} \;, \qquad (H = H(P,Q))$$

should hold for $P = p$, $Q = q$. The admission of variations in P which are independent of those in Q gives an extra degree of freedom to the action functional, and suggests the possibility of a bound on $S(p,q)$ different from $S(\dot{Q},Q)$.

For our simple example we have from (2.2) and (2.17)

(2.23) $$p = \dot{q} .$$

Thus from (2.18)

(2.24) $$H(P,Q) = \frac{1}{2} P^2 - F(Q)$$

and the canonical equations (2.21) and (2.22) are

(2.25) $$\dot{Q} = P ,$$

(2.26) $$-\dot{P} = -f(Q) ,$$

which hold at $P = p$, $Q = q$. Elimination of p here brings us back to the Euler-Lagrange equation (2.6). The equivalent forms for the action functional are

(2.27) $$S(P,Q) = \int [P\dot{Q} - \frac{1}{2} P^2 + F(Q)] dt$$

and

(2.28) $$S(P,Q) = \int [-\dot{P}Q - \frac{1}{2} P^2 + F(Q)] dt + [P\alpha]_{t_0}^{t_1} .$$

We notice that if the first canonical equation (2.25) should hold for <u>arbitrary</u> P and Q, then expression (2.27) reduces to the Lagrangian form of the action functional $S(\dot{Q},Q)$. This functional is then stationary at $Q = q$ provided that the second canonical equation (2.26) also holds at (p,q). We have seen that $S(\dot{Q},Q)$ can provide a bound on $S(\dot{q},q)$, i.e. $S(p,q)$, and so it is natural also to investigate the behavior of $S(P,Q)$ if Q is determined in terms of P from the second canonical equation (2.26).

2.3. A lower bound for $S(p,q)$

Suppose that the second Euler-Hamilton equation (2.26) holds identically, so that

$$(2.29) \qquad Q(P) = f^{-1}(\dot{P})$$

for arbitrary P, where f^{-1} is the inverse of f. Together (2.28) and (2.29) specify a form of the action functional dependent on P only, which we call $G(P)$. Thus

$$(2.30) \quad G(P) = S[P, f^{-1}(\dot{P})] = \int [-\dot{P} f^{-1}(\dot{P}) - \frac{1}{2}P^2 + F(f^{-1}(\dot{P}))] dt$$
$$+ [P\alpha]_{t_0}^{t_1} .$$

This functional is <u>complementary</u> to $S(\dot{Q}, Q)$ in (2.8), in the sense that each can be obtained from $S(P,Q)$ by eliminating P or Q with the help of one or the other canonical equation assumed to hold identically. The stationary property of $G(P)$ is a variational principle complementary to that of $S(\dot{Q}, Q)$.
Let

$$(2.31) \qquad \eta = P - p ,$$

then from (2.28)

$$(2.32) \quad G(p) - G(P) = \int [\frac{1}{2} \eta^2 + p\eta - \dot{p}q + F(q) + \dot{P}Q - F(Q)] dt$$
$$- [\eta\alpha]_{t_0}^{t_1} ,$$

$Q(P)$ and $q(p)$ being specified by (2.29). Now since the first canonical equation (2.25) also holds at (p,q), we have

$$(2.33) \qquad \int \dot{p}\eta \, dt = \int \dot{q}\eta \, dt = [\alpha\eta]_{t_0}^{t_1} - \int q\dot{\eta} \, dt ,$$

which yields from (2.32)

$$G(p) - G(P) = \int [\frac{1}{2}\eta^2 - q(\dot{p}+\dot{\eta}) + Q\dot{P} + F(q) - F(Q)]dt$$

(2.34)
$$= \int \frac{1}{2}\eta^2 dt + \int [F(q) - F(Q) - (q-Q)f(Q)]dt .$$

The second term in (2.34) is non-negative if F is convex, in which case

(2.35) $\qquad S(p,q) = G(p) \geq G(P) .$

Again the bound is not merely local.

2.4. Complementary upper and lower bounds for $S(p,q)$

In Section 2.1, the functional $S(\dot{Q},Q)$ depends on Q only, and so for convenience let us define

(2.36) $\qquad\qquad J(Q) = S(\dot{Q},Q) ,$

so that

(2.37) $\qquad J(q) = S(\dot{q},q) = S(p,q) = G(p) .$

The stationary properties of $J(Q)$ and $G(P)$ are called complementary variational principles, and we have established the complementary upper and lower bounds

(2.38) $\qquad\qquad J(Q) \geq S(p,q) \geq G(P)$

given that F is convex (or that f' is non-negative).

To construct these complementary bounds for the physical problem posed in the form of Newton's equation (2.7) together with the boundary conditions (2.4), the procedure is as follows:

1. Decompose Newton's equation into a pair of Euler-

Hamilton canonical equations by introducing the variable $p = \dot{q}$.

2. Deduce the necessary form of Hamiltonian H.

3. Write down the action functional $S(P,Q)$.

4. Derive the complementary functionals $J(Q)$ and $G(P)$, by assuming in turn that each of the canonical equations holds identically.

Since the sum of the kinetic and potential energies is a constant of the motion, from (2.3) various alternative expressions are possible for $S(p,q)$.

Problems in dynamics are more often *initial* value problems than boundary value problems as indicated by (2.4). However, as we shall see, it is not difficult to adapt the action functional to cope with such requirements.

3. General Theory

3.1. Introduction

The functional analysis of Section 2 is couched in the language of analytical mechanics; in that context there has been discussion of action, Euler-Hamilton equations and so forth in the textbooks for the past century. But the analysis could arise in other contexts. Instead of Newton's equation of motion

(3.1) $$-\ddot{q} + f(q) = 0$$

we might well have started with the Poisson-Boltzmann equation

(3.2) $$-d^2\Phi/dx^2 + f(\Phi), \quad x_0 \le x \le x_1 ,$$

subject to

(3.3) $$\Phi(x_0) = \alpha(x_0), \quad \Phi(x_1) = \alpha(x_1)$$

of which the solution is presumed to be $\Phi = \varphi$. With the increasing function

(3.4) $$f(\Phi) = \exp\Phi - \exp(-\Phi)$$

this equation arises in the theories of both colloids [25] and plasmas [26] and $\varphi(x)$ is related to the electric potential. We can decompose (3.2) into the pair of equations

(3.5) $$d\Phi/dx = U = \partial H/\partial U ,$$

(3.6) $$-dU/dx = -f(\Phi) = \partial H/\partial \Phi ,$$

which imply a 'Hamiltonian'

(3.7) $$H(U,\Phi) = \frac{1}{2} U^2 - F(\Phi) ,$$

and proceed in a manner precisely analogous to that for the example in dynamics. The distance x replaces the time t, $\varphi(x)$ replaces $q(t)$, and u replaces p. The 'action' turns out to be a measure of the field energy, and bounds for it are of physical interest.

The key features are the roles of the two linear operators d/dx and $-d/dx$, and the decomposition of the equation (3.2) into the pair of canonical equations (3.5) and (3.6). The operators d/dx and $-d/dx$ are <u>adjoints</u> of each other in the sense that

(3.8) $$\int_{x_0}^{x_1} U \frac{d}{dx} \Phi \, dx = \int_{x_0}^{x_1} -\frac{dU}{dx} \Phi \, dx + [U\Phi]_{x_0}^{x_1}$$

for arbitrary U and Φ.

We shall find quite generally that if a problem can be described by a pair of generalized Euler-Hamilton equations

(3.9) $\quad T\Phi = \delta W/\delta U, \quad T^*U = \delta W/\delta\Phi$,

then complementary variational principles exist which can lead to upper and lower bounds for the stationary value of the generalized action. Here T and T^* are <u>any</u> adjoint linear operators, $W(U,\Phi)$ is an appropriate generalized Hamiltonian functional, and $\delta/\delta U$, $\delta/\delta\Phi$ are the appropriate functional derivatives.

We present the theory in a form which covers the cases where T and T^* are first-order differential operators in one or more dimensions, integral operators or matrix operators. Other choices for T and T^* are possible (e.g. second-order differential operators) but are best considered separately (cf [31,63]).

3.2. Abstract formulation

Let $\mathcal{H}_U$ and $\mathcal{H}_\Phi$ be real Hilbert spaces of vector-valued functions defined on a compact convex subset V of E^n with a smooth boundary ∂V (smooth except for $n = 1$, when ∂V merely consists of two endpoints). The scalar products on $\mathcal{H}_U$ and $\mathcal{H}_\Phi$ are denoted by $\{\,,\}$ and $<,>$ respectively. Linear operators T, T^*, σ and σ^* map according to the table

$$T: \quad \mathcal{H}_\Phi \to \mathcal{H}_U \quad \text{in } V ,$$

$$T^*: \quad \mathcal{H}_U \to \mathcal{H}_\Phi \quad \text{in } V ,$$

$$\sigma: \quad \mathcal{H}_\Phi \to \mathcal{H}_U \quad \text{on } \partial V ,$$

$$\sigma^*: \quad \mathcal{H}_U \to \mathcal{H}_\Phi \quad \text{on } \partial V .$$

Here T^* is the adjoint of T in the sense that

(3.10) $\qquad \{U, T\Phi\}_V = <T^*U, \Phi>_V + \{U, \sigma\Phi\}_{\partial V}$

and σ^* is the adjoint of σ in the sense that

(3.11) $\qquad \{U, \sigma\Phi\}_{\partial V} = <\sigma^*U, \Phi>_{\partial V}$.

The operator σ depends on T and should be written σ_T; we suppress the suffix T for convenience. A more general boundary term can be included in (3.10) but Arthurs [27] shows that it must reduce to the form $\{U, \sigma\Phi\}$ if the resulting generalized action is to have extremal properties.

Let $W(U, \Phi)$ be a Fréchet differentiable functional defined on $\mathcal{H}$, the cartesian product space of $\mathcal{H}_U$ and $\mathcal{H}_\Phi$. Then the class of boundary value problems which we consider is described by the canonical equations

(3.12) $\qquad T\Phi = \delta W/\delta U$ in V ,

(3.13) $\qquad T^*U = \delta W/\delta\Phi$ in V ,

subject to a boundary condition

(3.14) $\qquad Z(U, \Phi) = 0$ on ∂V .

The derivatives $\delta/\delta U$ and $\delta/\delta\Phi$ are partial Fréchet derivatives.

Suppose that the solution-pair of equations (3.12)–(3.14) is

(3.15) $\qquad U = u, \quad \Phi = \varphi$

and that this solution is unique. The basic problem is to find an 'action' functional $I(U, \Phi)$ whose stationary behavior occurs at (u, φ). The notation I rather than S is customary for this functional; Vainberg [24] calls it a 'potential', but we will retain the term 'action' as everything can be thought of as a generalization of the situation in dynamics.

Following Noble [19], Komkov [28] and Arthurs [17] we introduce an action functional of the form

(3.16) $\quad I(U,\Phi) = \{U, T\Phi\}_V - W(U,\Phi)_V + \Gamma(U,\Phi)_{\partial V}$,

or equivalently

(3.17) $\quad I(U,\Phi) = <T^*U, \Phi>_V - W(U,\Phi)_V + [\Gamma(U,\Phi) + <\sigma^*U, \Phi>]_{\partial V}$.

From (3.16) and (3.17) we obtain the Fréchet derivatives

(3.18) $\quad \dfrac{\delta I}{\delta U} = [T\Phi - \dfrac{\delta W}{\delta U}]_V + [\dfrac{\delta \Gamma}{\delta U}]_{\partial V}$,

(3.19) $\quad \dfrac{\delta I}{\delta \Phi} = [T^*U - \dfrac{\delta W}{\delta \Phi}]_V + [\dfrac{\delta \Gamma}{\delta \Phi} + \sigma^* U]_{\partial V}$.

A necessary and sufficient condition for stationary behavior at (u, φ) is that

(3.20) $\quad \dfrac{\delta I}{\delta U} = 0$ and $\dfrac{\delta I}{\delta \Phi} = 0$ at (u, φ) .

Thus $I(U,\Phi)$ is stationary at (u,φ) provided that the canonical equations (3.12) and (3.13) hold, and provided also that

(3.21) $\quad \delta\Gamma/\delta U = 0$ on ∂V

and

(3.22) $\quad \delta\Gamma/\delta\Phi + \sigma^* U = 0$ on ∂V

all when $U = u$ and $\Phi = \varphi$. Hence, in order that the solution-pair of the boundary-value problem should coincide with the functions making $I(U,\Phi)$ stationary, we require equations (3.21) and (3.22) together to be equivalent to (3.14). This leads to different functionals $\Gamma(U,\Phi)$ according to the precise nature of the relationship between U and Φ which is implied by (3.14), and we list the four principal types below in Section 3.3.

The variational principles for the complementary functionals $J(\Phi)$ and $G(U)$ derived from $I(U,\Phi)$ are as follows. Notation such as $\delta W/\delta u$ or $\delta W/\delta\varphi$ means $\delta W/\delta U$ or $\delta W/\delta\Phi$ evaluated at (u,φ).

First variational principle

Let Φ be arbitrary and define from (3.16)

$$(3.23) \quad J(\Phi) = I[U(\Phi),\Phi] = \{U(\Phi), T\Phi\}_V - W[U(\Phi),\Phi]_V + \Gamma[U(\Phi),\Phi]_{\partial V},$$

$U(\Phi)$ being determined from the equations

$$(3.24) \quad \frac{\delta I}{\delta U} = 0, \text{ i.e. } T\Phi = \frac{\delta W}{\delta U} \text{ in } V, \quad \frac{\delta \Gamma}{\delta U} = 0 \text{ on } \partial V,$$

assumed to hold identically. Then $J(\Phi)$ is stationary at $J(\varphi) = I(u,\varphi)$ provided that

$$(3.25) \frac{\delta I}{\delta \varphi} = 0, \text{ i.e. } T^* u = \frac{\delta W}{\delta \varphi} \text{ in } V, \quad \frac{\delta \Gamma}{\delta \varphi} + \sigma^* u = 0 \text{ on } \partial V.$$

Second variational principle

Let U be arbitrary and define from (3.17)

$$(3.26) \quad G(U) = I[U,\Phi(U)] = <T^* U, \Phi(U)>_V - W[U,\Phi(U)]_V$$
$$+ \Gamma[U,\Phi(U)]_{\partial V} + <\sigma^* U, \Phi(U)>_{\partial V},$$

$\Phi(U)$ being determined from the equations

(3.27) $\quad \dfrac{\delta I}{\delta \Phi} = 0$, i.e. $T^* U = \dfrac{\delta W}{\delta \Phi}$ in V, $\dfrac{\delta \Gamma}{\delta \Phi} + \sigma^* U = 0$ on ∂V,

assumed to hold identically. Then $G(U)$ is stationary at $G(u) = I(u, \varphi)$ provided that

(3.28) $\quad \dfrac{\delta I}{\delta u} = 0$, i.e. $T\varphi = \dfrac{\delta W}{\delta u}$ in V, $\dfrac{\delta \Gamma}{\delta u} = 0$ on ∂V.

3.3. Boundary conditions, and the appropriate functionals $\Gamma(U, \Phi)$

We will consider four basic types of boundary condition (3.14):

(3.29)　　I:　　　$\varphi = \alpha$ on ∂V ;

(3.30)　　II:　　　$\sigma \varphi = a(u)$ on ∂V ;

(3.31)　　III:　　　$\sigma^* u = \beta$ on ∂V ;

(3.32)　　IV:　　　$\sigma^* u = b(\varphi)$ on ∂V.

Types I and III are <u>unmixed</u>, and α and β are prescribed functions or merely constants. Types II and IV are <u>mixed</u>; $a(U)$ and $b(\Phi)$ are known functions with inverses a^{-1} and b^{-1} and integrals

(3.33) $\quad A(U) = \displaystyle\int_{U_0}^{U} a(U') dU, \quad B(\Phi) = \displaystyle\int_{\Phi_0}^{\Phi} b(\Phi') d\Phi'$.

The choices of U_0 and Φ_0 are to some extent arbitrary. We can often conveniently take them as zero, but not if this leads to singular integrals.

In each case we need to determine the functional $\Gamma(U, \Phi)$ so that the two equations (3.21) and (3.22) are together equivalent to the appropriate boundary condition. Sometimes the boundary conditions are <u>essential</u> (they must be satisfied by the trial functions) and at other times they are <u>natural</u> (need not be satisfied by the trial functions). The results are given in Table 1, together with the nature of the boundary conditions for the functionals $J(\Phi)$ and $G(U)$. We also list the boundary contributions to $J(\Phi)$ and $G(U)$, as specified by (3.23) and (3.26). We remember that $\delta\Gamma/\delta U$ is identically zero for $J(\Phi)$, and that $\delta\Gamma/\delta\Phi + \sigma^{*}U$ is identically zero for $G(U)$.

The scalar product $\{1, A(U)\}$ is used in Table 1 for the <u>direct integral</u> of $a(U)$. This notation is convenient, but not strictly correct. Rather one should use the more clumsy expression

(3.34) $$\{U - U_0, \int_{t=0}^{1} a[U_0 + t(U - U_0)]dt\}$$

(with the understanding that the direct integral vanishes when $U = U_0$), or, if U_0 can be taken as zero, then the simpler

(3.35) $$\{U, \int_0^1 a(tU)dt\}$$

(see Vainberg [24]. There is a similar reservation concerning the scalar product $<1, B(\Phi)>$ in Table 1. A possible practical advantage of these strictly correct direct integrals is that they do not demand explicit knowledge of $A(U)$ or $B(\Phi)$.

In certain problems the boundary may be split up into a number of parts $\partial V_1, \partial V_2, \ldots$, with a different boundary condition satisfied on each part. In a situation like this, the appropriate functionals $\Gamma_1, \Gamma_2, \ldots$ for each part are added together to give the composite Γ for the problem (see Section 6.8).

Table 1. Boundary conditions and functionals $\Gamma(U, \Phi)_{\partial V}$

Condition on ∂V	Type I $\varphi = \alpha$	Type II $\sigma\varphi = a(u)$	Type III $\sigma^* u = \beta$	Type IV $\sigma^* u = b(\varphi)$
Nature for $J(\Phi)$	essential	essential	natural	natural
Nature for $G(U)$	natural	natural	essential	essential
$\Gamma(U, \Phi)_{\partial V}$	$\langle \sigma^* U, \alpha - \Phi \rangle$	$-\langle \sigma^* U, \Phi \rangle + \{1, A(U)\}$	$-\langle \beta, \Phi \rangle$	$-\langle 1, B(\Phi) \rangle$
$\delta\Gamma/\delta U$	$\sigma(\alpha - \Phi)$	$-\sigma\Phi + a(U)$	0	0
$\delta\Gamma/\delta\Phi + \sigma^* U$	0	0	$-\beta + \sigma^* U$	$-b(\Phi) + \sigma^* U$
$\Gamma[U(\Phi), \Phi]_{\partial V}$	0	$-\{a^{-1}(\sigma\Phi), \sigma\Phi\} + \{1, A(a^{-1}(\sigma\Phi))\}$	$-\langle \beta, \Phi \rangle$	$-\langle 1, B(\Phi) \rangle$
$\Gamma[U, \Phi(U)]_{\partial V} +$ $\langle \sigma^* U, \Phi(U) \rangle_{\partial V}$	$\langle \sigma^* U, \alpha \rangle$	$\{1, A(U)\}$	0	$-\langle 1, B(b^{-1}(\sigma^* U)) \rangle$ $+ \langle \sigma^* U, b^{-1}(\sigma^* U) \rangle$

528

COMPLEMENTARY VARIATIONAL PRINCIPLES

3.4. Sufficient conditions for complementary upper and lower bounds

We first investigate the conditions under which $J(\Phi)$ is an upper bound to $J(\varphi) = I(u,\varphi)$, assuming that equations (3.25) hold so that $J(\Phi)$ is stationary. From (3.16) we have

$$J(\Phi) - J(\varphi) = [W(u,\varphi) - W(U,\Phi) + \{U,T\Phi\} - \{u,T\varphi\}]_V + [\Gamma(U,\Phi)$$
$$- \Gamma(u,\varphi)]_{\partial V}$$

(3.36)
$$= [W(u,\Phi) - W(U,\Phi) - \{u-U, T\Phi\}]_V$$
$$- [W(u,\Phi) - W(u,\varphi) - \{u, T(\Phi-\varphi)\}]_V + [\Gamma(U,\Phi)$$
$$- \Gamma(u,\varphi)]_{\partial V} .$$

Since from (3.24) $T\Phi = \delta W/\delta U$ in V, and also from (3.10)

$$\{u, T(\Phi-\varphi)\}_V = <T^*u, \Phi-\varphi>_V + <\sigma^*u, \Phi-\varphi>_{\partial V}$$

(3.37)
$$= <\frac{\delta W}{\delta \varphi}, \Phi-\varphi>_V + <\sigma^*u, \Phi-\varphi>_{\partial V} ,$$

we can eliminate T from (3.36) to give

(3.38) $J(\Phi) - J(\varphi) = [J(\Phi) - J(\varphi)]_V + [J(\Phi) - J(\varphi)]_{\partial V}$

where

(3.39)
$$[J(\Phi) - J(\varphi)]_V = [W(u,\Phi) - W(U,\Phi) - \{u-U, \frac{\delta W}{\delta U}\}]_V$$
$$- [W(u,\Phi) - W(u,\varphi) - <\frac{\delta W}{\delta \varphi}, \Phi-\varphi>]_V$$

and

$$(3.40) \quad [J(\Phi) - J(\varphi)]_{\partial V} = [\Gamma(U, \Phi) - \Gamma(u, \varphi) + <\sigma^* u, \Phi - \varphi>]_{\partial V} .$$

In expressions (3.39) and (3.40), $U(\Phi)$ is determined from equations (3.24). This separation into contributions from V and ∂V is <u>not</u> unique (because of (3.10)), but it is convenient here.

If the functional $W(U, \Phi)$ is both <u>convex</u> in U and <u>concave</u> in Φ, the contributions on the right of (3.39) are each non-negative (see the Appendix: a functional is concave if minus that functional is convex); thus $[J(\Phi) - J(\varphi)]_V$ is nonnegative.

Let us now look at the boundary term (3.40) for the separate types of boundary conditions I-IV discussed in Section 3.3, remembering that $\delta\Gamma/\delta U$ is identically zero for $J(\Phi)$. We obtain the following expressions for $[J(\Phi) - J(\varphi)]_{\partial V}$ as given by (3.40):

(3.41)   I: zero;

(3.42)   II: $-\{1, A(u) - A(U) - (u - U) a(U)\}_{\partial V}$ with $U = a^{-1}(\sigma\Phi)$;

(3.43)   III: zero;

(3.44)   IV: $-<1, B(\Phi) - B(\varphi) - (\Phi - \varphi) b(\varphi)>_{\partial V}$.

Thus if

$$(3.45) \quad \{1, A(U)\}_{\partial V} \text{ is concave in } U,$$

and

$$(3.46) \quad <1, B(\Phi)>_{\partial V} \text{ is concave in } \Phi,$$

none of the boundary contributions (3.41)-(3.44) is negative. With these extra restrictions on the functions $a(U)$ and $b(\Phi)$, we therefore have the result

(3.47) $J(\Phi) \geq J(\varphi)$ if W is convex in U and concave in Φ.

We can proceed in a similar manner with the complementary functional $G(U)$, assuming that equations (3.28) hold to make $G(U)$ stationary. We start with (3.17) and find that

$$G(u) - G(U) = [W(U,\varphi) - W(u,\varphi) - <T^*(U-u),\varphi>]_V$$

$$-[W(U,\varphi) - W(U,\Phi) - <T^*U, \varphi - \Phi>]_V + [\Gamma(u,\varphi) + <\sigma^*u,\varphi>$$

(3.48)
$$- \Gamma(U,\Phi) - <\sigma^*U,\Phi>]_{\partial V} .$$

Replacing T^*U by $\delta W/\delta \Phi$ from (3.27), and using the result that

$$<T^*(U-u),\varphi>_V = \{U-u, T\varphi\}_V - \{U-u, \sigma\varphi\}_{\partial V}$$

(3.49)
$$= \{U-u, \frac{\delta W}{\delta u}\}_V - <\sigma^*(U-u),\varphi>_{\partial V},$$

we can eliminate T^* from (3.48). We obtain

(3.50) $G(u) - G(U) = [G(u) - G(U)]_V + [G(u) - G(U)]_{\partial V}$

where

$$[G(u) - G(U)]_V = [W(U,\varphi) - W(u,\varphi) - \{U-u, \frac{\delta W}{\delta u}\}]_V$$

(3.51)
$$- [W(U,\varphi) - W(U,\Phi) - <\frac{\delta W}{\delta \Phi}, \varphi - \Phi>]_V$$

and

(3.52) $[G(u) - G(U)]_{\partial V} = [\Gamma(u,\varphi) - \Gamma(U,\Phi) + <\sigma^*U, \varphi - \Phi>]_{\partial V} .$

Again if W is convex in U and concave in Φ it is evident from (3.51) that $[G(u) - G(U)]_V$ is non-negative.

With $\delta\Gamma/\delta\Phi + \sigma^*U$ identically zero, the boundary contribution $[G(u) - G(U)]_{\partial V}$ as given by (3.52) takes the following forms for the various types I-IV of boundary condition:

(3.53) I: zero;

(3.54) II: $-\{1, A(U) - A(u) - (U-u)a(u)\}_{\partial V}$;

(3.55) III: zero;

(3.56) IV: $-<1, B(\varphi) - B(\Phi) - (\varphi - \Phi)b(\Phi)>_{\partial V}$ with $\Phi = b^{-1}(\sigma^*U)$.

If restrictions (3.45) and (3.46) again hold, none of these expressions is negative. Thus the sign of $G(u) - G(U)$ is not disturbed by these boundary conditions.

We have derived the complementary bounds, which are not merely local,

(3.57) $\qquad J(\Phi) \geq J(\varphi) = I(u,\varphi) = G(u) \geq G(U)$

if

(3.58) $\qquad W(U, \Phi)$ is convex in U and concave in Φ ,

and if, when relevant,

(3.45) $\qquad\qquad \{1, A(U)\}_{\partial V}$ is concave in U

and

(3.46) $\qquad\qquad <1, B(\Phi)>_{\partial V}$ is concave in Φ .

If in (3.58), (3.45) and (3.46) concave replaced convex, and convex replaced concave, then the inequality signs in (3.57) would be reversed.

3.5 Uniqueness of (u,φ)

The saddle-point behavior of the functional $W(U,\Phi)$, i.e. convexity in U, concavity in Φ, is a sufficient condition for complementary upper and lower bounds. Also in many cases this behavior is sufficient to establish the uniqueness of the solution-pair (u,φ) of the boundary-value problem described by equations (3.12)-(3.14)(cf. [18,22]).

Suppose that (u_1,φ_1) and (u_2,φ_2) are possible different solution-pairs. Then if W is strictly convex in U,

$$(3.59) \quad W(u_1,\varphi_2) - W(u_2,\varphi_2) - \{u_1 - u_2, \frac{\delta}{\delta u_2} W(u_2,\varphi_2)\}_V > 0 .$$

Also, if W is strictly concave in Φ.

$$(3.60) \quad -W(u_1,\varphi_2) + W(u_1,\varphi_1) + <\frac{\delta}{\delta\varphi_1} W(u_1,\varphi_1), \varphi_2 - \varphi_1>_V > 0 .$$

If we substitute $T\varphi_2$ for $\delta W/\delta u_2$ in (3.59) and T^*u_1 for $\delta W/\delta\varphi_1$ in (3.60) and add the two inequalities, we obtain

$$(3.61) \quad \begin{aligned} & W(u_1,\varphi_1) + <T^*u_1, \varphi_2-\varphi_1>_V \\ & > W(u_2,\varphi_2) + \{u_1-u_2, T\varphi_2\}_V . \end{aligned}$$

But from (3.10)

$$(3.62) \quad <T^*u_1, \varphi_2-\varphi_1> = \{u_1, T(\varphi_2-\varphi_1)\}_V - \{u_1, \sigma(\varphi_2-\varphi_1)\}_{\partial V} .$$

Substituting this in (3.61) we find that

$$(3.63) \quad \begin{aligned} W(u_1,\varphi_1) - \{u_1, T\varphi_1\}_V & > W(u_2,\varphi_2) - \{u_2, T\varphi_2\}_V \\ & + \{u_1, \sigma(\varphi_2-\varphi_1)\}_{\partial V} . \end{aligned}$$

But the labelling of the solution-pairs is arbitrary, and so we could interchange suffices 1 and 2 to give

$$W(u_2,\varphi_2) - \{u_2, T\varphi_2\}_V > W(u_1, \varphi_1) - \{u_1, T\varphi_1\}_V$$
(3.64)
$$+ \{u_2, \sigma(\varphi_1 - \varphi_2)\}_{\partial V} .$$

Addition of the inequalities (3.63) and (3.64) yields

(3.65) $\quad 0 > \{u_1 - u_2, \sigma(\varphi_2 - \varphi_1)\}_{\partial V} = <\sigma^*(u_1 - u_2), \varphi_2 - \varphi_1>_{\partial V} .$

Thus if in fact

(3.66) $\quad \{u_1 - u_2, \sigma(\varphi_2 - \varphi_1)\}_{\partial V} = <\sigma^*(u_1-u_2), \varphi_2-\varphi_1>_{\partial V} \geq 0 ,$

it follows that we have a contradiction and that (u,φ) is unique. Condition (3.66) is satisfied if $\varphi_1 = \varphi_2$ or $u_1 = u_2$ on ∂V, and thus it certainly holds with Type 1 and Type III boundary conditions. It holds with Type II conditions if

(3.67) $\qquad \{u_1 - u_2, a(u_2) - a(u_1)\}_{\partial V} \geq 0$

and with Type IV if

(3.68) $\qquad <b(\varphi_1) - b(\varphi_2), \varphi_2 - \varphi_1>_{\partial V} \geq 0 .$

It can also be shown that if $W(U,\Phi)$ is only convex in U but <u>strictly</u> concave in Φ then φ is unique (see [18, 22]), again with suitable reservations concerning the boundary conditions. Similarly if W is strictly convex in U but only concave in Φ then u is unique.

3.6. <u>Choices for the linear operators</u>

We now list various possibilities for the operators T and T^*, and give σ, σ^* and the appropriate scalar products in each case. We also give simple conditions for (3.45) and (3.46) to hold.

3.6.1. First-order differential operators in one dimension (x)

$$T = d/dx, \quad T^* = -d/dx, \quad \sigma = 1, \quad \sigma^* = 1 \ ;$$

U and Φ both scalar functions of x, $x_0 \leq x \leq x_1$;

$$\{U, T\Phi\}_V = \int_{x_0}^{x_1} U \frac{d\Phi}{dx} \, dx \ ;$$

$$<T^*U, \Phi>_V = \int_{x_0}^{x_1} -\frac{dU}{dx} \Phi \, dx \ ;$$

$$\{U, \sigma^*\Phi\}_{\partial V} = <\sigma^*U, \Phi>_{\partial V} = [U\Phi]_{x_0}^{x_1} \ ;$$

$\{1, A(U)\}_{\partial V}$ is concave if $a'(U) \leq 0$ at $x = x_1$, $a'(U) \geq 0$ at $x = x_0$;

$<1, B(\Phi)>_{\partial V}$ is concave if $b'(\Phi) \leq 0$ at $x = x_1$, $b'(\Phi) \geq 0$ at $x = x_0$.

(This is rather a special case since ∂V merely consists of the two end-points x_0 and x_1).

3.6.2. First-order differential operators in three dimensions (x,y,z)

<u>Case (i)</u>. $\quad T = \text{grad}, \quad T^* = -\text{div}, \quad \sigma = \underline{n}, \quad \sigma^* = \underline{n}. \ ;$

<u>U</u> a Euclidean vector function of (x,y,z) ;

Φ a scalar function of (x,y,z) ;

<u>dx</u> an element of V ;

dS   an element of ∂V ;

$\underline{n}$ the unit normal vector to ∂V, outwards from V ;

$$\{U, T\Phi\}_V = \int_V \underline{U} \cdot \text{grad } \Phi \, d\underline{x} \; ;$$

$$<T^*U, \Phi>_V = \int_V (-\text{div } \underline{U}) \Phi \, d\underline{x} \; ;$$

$$\{U, \sigma \Phi\}_{\partial V} = \int_{\partial V} \underline{U} \cdot (\underline{n} \, \Phi) \, dS \; ;$$

$$<\sigma^* U, \Phi>_{\partial V} = \int_{\partial V} (\underline{n} \cdot \underline{U}) \Phi \, dS \; ;$$

$$\{1, A(U)\}_{\partial V} \text{ is concave if } a' \leq 0 \; ;$$

$$<1, B(\Phi)>_{\partial V} \text{ is concave if } b' \leq 0 \; .$$

<u>Case (ii)</u>.   T = curl, T^* = curl, $\sigma = \underline{n}\wedge$, $\sigma^* = -\underline{n}\wedge$;
$\underline{U}$ and $\underline{\Phi}$ both vector functions of (x,y,z) ;

$$\{U, T\Phi\}_V = \int_V \underline{U} \cdot (\text{curl } \underline{\Phi}) d\underline{x} \; ;$$

$$<T^*U, \Phi>_V = \int_V (\text{curl } \underline{U}) \cdot \underline{\Phi} \, d\underline{x} \; ;$$

$$\{U, \sigma \Phi\}_{\partial V} = \int_{\partial V} \underline{U} \cdot (\underline{n} \wedge \underline{\Phi}) dS \; ;$$

$$<\sigma^* U, \Phi>_{\partial V} = \int_{\partial V} (-\underline{n} \wedge \underline{U}) \cdot \underline{\Phi} \, dS \; ;$$

COMPLEMENTARY VARIATIONAL PRINCIPLES

$\{1, A(U)\}_{\partial V}$ is concave if $a' \leq 0$;

$<1, B(\Phi)>_{\partial V}$ is concave if $b' \geq 0$.

3.6.3. Integral operators in n dimensions

$$\underline{x} = (x_1, x_2, \ldots, x_n) \ ;$$

T   an integral operator with kernel $\mathfrak{J}(\underline{x}, \underline{x}')$;

T^* an integral operator with kernel $\mathfrak{J}(\underline{x}', \underline{x})$;

$\sigma = 0$, $\sigma^* = 0$ (no boundary terms) ;

$$\{U, T\Phi\}_V = \int_V U(\underline{x})d\underline{x} \int_{V'} \mathfrak{J}(\underline{x}, \underline{x}') \Phi(\underline{x}')d\underline{x}' \ ;$$

$$<T^*U, \Phi>_V = \int_V d\underline{x} \int_{V'} \mathfrak{J}(\underline{x}', \underline{x}) U(\underline{x}')d\underline{x}' \Phi(\underline{x}) \ .$$

3.6.4. Matrix operators

U   a column vector with n elements (transpose $U^\dagger$) ;

Φ a column vector with m elements (transpose $\Phi^\dagger$) ;

T   a matrix operator represented by an $n \times m$ matrix A ;

T^* a matrix operator represented by the transposed $m \times n$ matrix $A^\dagger$;

$\sigma = 0$, $\sigma^* = 0$ (no boundary terms) ;

$$\{U, T\Phi\}_V = U^\dagger A\Phi \ ;$$

$$<T^*U, \Phi>_V = (A^\dagger U)^\dagger \Phi \ .$$

4. The Equation $T^*T\Phi + f(\Phi) = 0$

4.1. Preliminaries

In this section we consider the class of problems described by the equation

(4.1) $$T^*T\Phi + f(\Phi) = 0 \quad \text{in } V ,$$

where $f: \mathcal{H}_\Phi \to \mathcal{H}_\Phi$ is a given function or operator, and Φ is subject to suitable boundary conditions (from Section 4.3 onwards we assume Type I boundary conditions for simplicity). We suppose that a unique solution $\Phi = \varphi$ exists, which is impossible to determine exactly. Newton's equation (2.7) and the Poisson-Boltzmann equation (3.2) are special cases of (4.1), and many other non-linear examples of it are available [29-38]. Arthurs [39] discusses complementary variational principles arising from this equation, and derives local bounds when φ is given as some function of $T\varphi$ on ∂V. When U plays the role of $T\Phi$, the equation readily fits into the general theory of Section 3.

Besides obtaining complementary bounds $J(\Phi)$ and $G(U)$ for the stationary value of the generalized action, let us suppose that we are interested in finding good approximations to φ. If we consider trial functions of the form $U = T\Phi$, then the closeness of the bounds $J(\Phi)$ and $G(T\Phi)$ for the same Φ will give a measure of the 'goodness' of that Φ. This maneuver can also effect useful simplification in the expression for G; sometimes individual knowledge of T and T^* is not even required. The function U, having played its part on an equal footing with Φ in the development of the variational principles, takes its bow and leaves the stage.

The canonical equations associated with (4.1) are

(4.2) $$T\Phi = U = \frac{\delta W}{\delta U} \quad \text{in } V ,$$

and

(4.3) $$T^*U = -f(\Phi) = \frac{\delta W}{\delta \Phi} \quad \text{in } V.$$

Hence we take

(4.4) $$W(U,\Phi) = \frac{1}{2}\{U,U\}_V - <1,F(\Phi)>_V$$

where

(4.5) $$F(\Phi) = \int_{\Phi_0}^{\Phi} f(\Phi')d\Phi',$$

and the notation $<1,F(\Phi)>$ is used for the direct integral of $f(\Phi)$ (cf. Section 3.3).

4.2. Complementary bounds

The functional $W(U,\Phi)$ of (4.4) is strictly convex in U, and so from (3.58) we shall get complementary bounds $J(\Phi)$ and $G(U)$ if W is concave in Φ, i.e. if

(4.6) $$<1,F(\Phi)>_V \text{ is convex in } \Phi.$$

In all the cases listed in Section 3.6, condition (4.6) holds if for arbitrary Φ (see the Appendix)

(4.7) $$f'(\Phi) \geq 0.$$

We suppose that the boundary condition is one of the types I-IV, with $u = T\varphi$, and that if relevant the restrictions (3.45) and (3.46) hold good. Strict convexity of $<1,F(\Phi)>_V$ also guarantees the uniqueness of the solution φ if (3.66) is satisfied.

From equations (3.23), (3.24), (4.2) and (4.4) we find that

(4.8) $\quad J(\Phi) = \frac{1}{2}\{T\Phi, T\Phi\}_V + <1, F(\Phi)>_V + \Gamma[U(\Phi), \Phi]_{\partial V}$.

Likewise from equations (3.26), (3.27), (4.3) and (4.4) it follows that

$$G(U) = <T^*U, f^{-1}(-T^*U)>_V - \frac{1}{2}\{U, U\}_V + <1, F[f^{-1}(-T^*U)]>_V$$
(4.9)
$$+ \Gamma[U, \Phi(U)]_{\partial V} + <\sigma^*U, \Phi(U)>_{\partial V} .$$

The appropriate forms of the boundary terms in (4.8) and (4.9) are given in Table 1. These expressions for $J(\Phi)$ and $G(U)$ are generalizations of those in (2.8) and (2.30) for the dynamical example.

The inverse function f^{-1} appears in expression (4.9) for $G(U)$, and it is assumed that this inverse exists. One simple example where the inverse does not exist arises in the <u>linear</u> case

(4.10) $\quad\quad\quad\quad\quad f(\Phi) = f_0 \text{ in } V$,

f_0 being a known function of coordinates [40, 41] (the problem may still be non-linear via the boundary conditions). Now the second canonical equation becomes a <u>constraint</u> on U, and (4.9) is replaced by

$$G(U) = -\frac{1}{2}\{U, U\}_V + \Gamma[U, \Phi(U)]_{\partial V} + <\sigma^*U, \Phi(U)>_{\partial V} ,$$
(4.11)
$$(T^*U = -f_0 \text{ in } V) .$$

The Thomson lower bound in electrostatics is an elementary example of this type [1].

In the next section we mention one way of proceeding when f^{-1} exists but is difficult to cope with.

4.3. Bounds involving $L = T^*T$

Let us write

(4.12) $$L = T^*T \text{ in } V,$$

so that $\Phi = \varphi$ is now the solution of

(4.13) $$L\Phi + f(\Phi) = 0 \text{ in } V$$

(L is nothing to do with the Lagrangian $L(\dot{q},q)$ of Section 2). In the remainder of Section 4 we will particularize to the Type I boundary condition

(4.14) $$\varphi = \alpha \text{ on } \partial V$$

to avoid carrying the somewhat clumsy general boundary terms. Thus from Table 1

(4.15) $$\Gamma(U,\Phi)_{\partial V} = <\sigma^*U, \alpha - \Phi>_{\partial V},$$

and the expressions (4.8) and (4.9) simplify to

(4.16) $$J(\Phi) = \frac{1}{2}\{T\Phi, T\Phi\}_V + <1, F(\Phi)>_V, \quad (\Phi = \alpha \text{ on } \partial V),$$

and

(4.17) $$G(U) = <T^*U, f^{-1}(-T^*U)>_V - \frac{1}{2}\{U,U\}_V$$
$$+ <1, F[f^{-1}(-T^*U)]>_V + <\sigma^*U, \alpha>_{\partial V}.$$

Transforming $\{T\Phi, T\Phi\}_V$ with the help of (3.10), it follows from (4.16) that

(4.18) $$J(\Phi) = \frac{1}{2}<L\Phi, \Phi>_V + <1, F(\Phi)>_V + \frac{1}{2}<\sigma^*T\Phi, \alpha>_{\partial V}.$$

Thus in order to evaluate $J(\Phi)$ from (4.18), individual knowledge of T and T^* in V is not required. On ∂V an asymptotic form of T may suffice, or most favorably there may be no boundary terms at all (as is the case when $\alpha = 0$, or when $\sigma^* = 0$). In straightforward situations such as those listed in Section 3.6 there is no compelling reason to de-emphasize T in this way, but it may be that we are given an equation (4.13) and that the decomposition (4.12) is complicated and best avoided if possible. Mikhlin [42] has shown that positive-definite, self-adjoint operators L can always in principle be written as T^*T.

From (4.13) and (4.18) we obtain a convenient form for the stationary action value, namely

(4.19) $I(u, \varphi) = J(\varphi) = -\frac{1}{2} <f(\varphi), \varphi>_V + <1, F(\varphi)>_V + \frac{1}{2} <\sigma^* T\varphi, \alpha>_{\partial V}$.

Now we investigate what happens to $G(U)$ when we consider trial functions of the form

(4.20) $$U = T\tilde{\Phi}.$$

This is quite a sensible trial function, since the exact u is $T\varphi$. The term $\{T\tilde{\Phi}, T\tilde{\Phi}\}_V$ is transformed, and from (4.17) we find that

(4.21) $G(T\tilde{\Phi}) = <L\tilde{\Phi}, f^{-1}(-L\tilde{\Phi})>_V - \frac{1}{2} <L\tilde{\Phi}, \tilde{\Phi}>_V + <1, F[f^{-1}(-L\tilde{\Phi})]>_V$

$+ <\sigma^* T\tilde{\Phi}, \alpha - \frac{1}{2}\tilde{\Phi}>_{\partial V}$.

Again individual knowledge of T and T^* in V is not required. Notice that all the scalar products in expressions (4.18) and (4.21) for J and G are now on the space $\mathcal{H}_\Phi$.

From (4.18) and (4.21) with Φ for $\tilde{\Phi}$ there follows the result

(4.22) $J(\Phi) - G(T\Phi) = <L\Phi, \Phi - f^{-1}(-L\Phi)>_V + <1, F(\Phi) - F[f^{-1}(-L\Phi)]>_V$

$(\Phi = \alpha$ on $\partial V)$,

the boundary terms cancelling out.

The f^{-1} terms in (4.21) and (4.22) suggest a connection with the iterative scheme

$$(4.23) \qquad L\Phi_{n+1} + f(\Phi_n) = 0, \qquad n = 0, 1, 2, \ldots ,$$

for the solution of equation (4.13). Substitution of Φ_1 for $\tilde{\Phi}$ in (4.21) gives

$$(4.24) \qquad G(T\Phi_1) = <f(\Phi_0), \tfrac{1}{2}\Phi_1 - \Phi_0>_V + <1, F(\Phi_0)>_V$$

$$+ <\sigma^* T\Phi_1, \alpha - \tfrac{1}{2}\Phi_1>_{\partial V} ,$$

and so if the first iteration of (4.23) can be performed, the lower bound (4.24) is available without knowledge of the inverse function f^{-1}. If $\Phi_{n+1} = \alpha$ on ∂V, we have from (4.22)

$$(4.25) \qquad J(\Phi_{n+1}) - G(T\Phi_{n+1}) = <1, F(\Phi_{n+1}) - F(\Phi_n)>_V$$

$$- <f(\Phi_n), \Phi_{n+1} - \Phi_n>_V ,$$

indicating that J and G approach one another if the iterative scheme (4.23) converges.

4.4. More specific conditions for bounds

We have seen that J and G are complementary upper and lower bounds if f' is non-negative. This is the sufficient condition arising from the general theory. However one can give more specific conditions on J and G separately. Persisting with the Type I boundary condition (4.14), we find from (3.38)-(3.40), (4.4) and (4.15) that

$$J(\Phi) - J(\varphi) = \frac{1}{2} \{T(\Phi-\varphi), T(\Phi-\varphi)\}_V + <1, F(\Phi) - F(\varphi) - (\Phi-\varphi)f(\varphi)>_V \quad (4.26)$$

$$= \frac{1}{2} <(\Phi-\varphi)[L+f'(\varphi_\epsilon)](\Phi-\varphi)>_V, \quad \Phi = \alpha \text{ on } \partial V,$$

where

$$(4.27) \quad \varphi_\epsilon = \varphi + \epsilon(\Phi-\varphi), \quad 0 < \epsilon < 1.$$

Thus we have the result

(4.28) $J(\Phi) \geq J(\varphi) = I(u,\varphi)$ if the operator $(L+f')$ is non-negative.

Similarly from (3.50)-(3.52), (4.4) and (4.15) it follows that

$$G(u) - G(U) = \frac{1}{2} \{U-u, U-u\}_V + <1, F(\varphi) - F(\Phi) - (\varphi-\Phi)f(\Phi)>_V \quad (4.29)$$

$$\text{where } \Phi = f^{-1}(-T^*U).$$

Using Φ_0 and Φ_1 from (4.23), this gives

$$G(T\varphi) - G(T\Phi_1) = $$
$$(4.30) \quad \frac{1}{2} <(\Phi_1-\varphi), L(\Phi_1-\varphi)>_V + <1, F(\varphi) - F(\Phi_0) - (\varphi-\Phi_0)f(\Phi_0)>,$$
$$(\Phi_1 = \alpha \text{ on } \partial V).$$

To combine the two terms in (4.30) one must approximate and assume that Φ_0 is close to the exact solution φ. Then

$$(4.31) \quad L(\Phi_1 - \varphi) = f(\varphi) - f(\Phi_0) \doteq (\varphi - \Phi_0)f'(\Phi_0),$$

and also

(4.32) $\quad F(\varphi) - F(\Phi_0) - (\varphi - \Phi_0)f(\Phi_0) \doteq \frac{1}{2}(\varphi - \Phi_0)^2 f'(\Phi_0)$.

Thus, neglecting terms in $(\varphi - \Phi_0)^3$, we have from (4.30)-(4.32) the result

(4.33) $\quad G(T\varphi) - G(T\Phi_1) = \frac{1}{2} <(\Phi_1-\varphi),[L+L\{f'(\Phi_0)\}^{-1}L](\Phi_1-\varphi)>_V$,

indicating that <u>locally</u>

(4.34) $\quad I(u,\varphi) = G(T\varphi) \geq G(T\Phi)$ if the operator $[L+L(f')^{-1}L]$ is non-negative.

We see from (4.28) that $J(\Phi)$ can still be an upper bound if f' is negative, provided that the positive operator L outweighs f'. In this situation one would expect the operator $[L + L(f')^{-1}L]$ to be negative, and from (4.33) $G(T\Phi)$ also to be a (local) <u>upper</u> bound [23,43].

It is interesting to note that when

(4.35) $\quad \sigma^* = 0$

then from (4.18) and (4.24) we have the results

(4.36) $\quad J(\Phi_n) - G(T\Phi_{n+1}) = \frac{1}{2} <L(\Phi_{n+1} - \Phi_n), (\Phi_{n+1} - \Phi_n)>_V$

and

(4.37) $\quad G(T\Phi_{n+1}) - J(\Phi_{n+1}) =$

$\quad -<1, F(\Phi_{n+1}) - F(\Phi_n) - (\Phi_{n+1} - \Phi_n)f(\Phi_n)>_V$,

$n = 0,1,2,\ldots$ (see also (4.25) with $\sigma^* \neq 0$). Thus when the operator L is positive and f' is negative (strictly $<1, F(\Phi)>$ concave), we have the sequence

(4.38) $\quad J(\Phi_0) \geq G(T\Phi_1) \geq J(\Phi_1) \geq G(T\Phi_2) \geq J(\Phi_2) \geq \ldots$

of interspersed functionals, which are decreasing upper
bounds if $L + f'$ is non-negative [cf. 43]. This condition is
related to the rough-and-ready condition for the convergence
of the iteration process (4.23). If Φ_n and Φ_{n+1} are close
to φ, it follows that

(4.39) $$[f'(\varphi)]^{-1} L(\Phi_{n+1} - \Phi_n) \doteq (\Phi_n - \Phi_{n-1})$$

and so we should expect convergence if the largest eigenvalue of the operator $[f'(\varphi)]^{-1} L$ is less than -1.

4.5. Error bounds for an approximate solution

If a trial function Φ is regarded as an approximation
to the exact solution φ, then error bounds can be derived
for Φ in terms of the difference between the functionals $J(\Phi)$
and $G(T\Phi)$. Specifically we find an upper bound on $\|\Phi - \varphi\|$,
the L_2-norm of $(\Phi - \varphi)$, which is the positive quantity defined by

(4.40) $$\|\Phi - \varphi\|^2 = <\Phi-\varphi, \Phi-\varphi>_V .$$

Suppose that the operator $(L+f')$ is bounded below by
a positive number m, so that

(4.41) $$<(\Phi-\varphi), [L+f'(\Phi_\epsilon)](\Phi-\varphi)>_V \geq m<\Phi-\varphi, \Phi-\varphi>_V .$$

Assuming that

(4.42) $$f'(\Phi) \geq \delta \geq 0 \quad \text{for any} \quad \Phi \in \mathcal{H}_\Phi ,$$

we could conveniently take

(4.43) $$m = \lambda_0 + \delta ,$$

where $\lambda_0 > 0$ is the smallest eigenvalue of the positive operator L. Or if λ_0 is unknown, a positive lower bound $(\lambda_0)_-$ might be used instead. Then since f' is non-negative, $J(\Phi)$ and $G(T\Phi)$ are complementary upper and lower bounds which implies that

(4.44) $$J(\Phi) - G(T\Phi) \geq J(\Phi) - J(\varphi) \ .$$

From (4.26), (4.41) and (4.44) it follows that

(4.45) $$\|\Phi - \varphi\| \leq [\tfrac{2}{m} \{J(\Phi) - G(T\Phi)\}]^{1/2} \ .$$

This result is similar to one derived by Shampine [44], but his did not involve $G(T\Phi)$.

Provided that third-order terms in $(\Phi - \varphi)$ can be neglected, an analogous bound on $\|(\Phi - \varphi)\|$ can be derived from (4.33). Since

(4.46) $$J(\Phi) - G(T\Phi) \geq G(T\varphi) - G(T\Phi) \ ,$$

we find that

(4.47) $$\|\Phi - \varphi\| \leq [\tfrac{2}{\tilde{m}} \{J(\Phi) - G(T\Phi)\}]^{1/2} \ ,$$

where it is assumed that $[L + L(f')^{-1}L]$ is bounded below by a positive number $\tilde{m}$.

4.6. A related equation

An equation closely related to (4.1) is

(4.48) $$T^*(\omega T \Phi) + f(\Phi) = 0 \quad \text{in } V \ ,$$

where ω is a positive function of coordinates. The associated canonical equations are

547

$$(4.49) \qquad T\Phi = \omega^{-1} U = \frac{\delta W}{\delta U} \quad \text{in } V$$

and

$$(4.50) \qquad T^* U = -f(\Phi) = \frac{\delta W}{\delta \Phi} \quad \text{in } V \ .$$

These lead to the functional

$$(4.51) \qquad W(U,\Phi) = \frac{1}{2} \{U, \omega^{-1} U\}_V - <1, F(\Phi)>_V \ ,$$

which is convex in U. Thus again we get complementary bounds if $<1, F(\Phi)>_V$ is convex in Φ, and the analysis of Sections 4.2-4.5 can be generalized slightly to include this positive function ω.

The Föppl-Hencky equation discussed in Section 6 is of this type.

5. The Hammerstein Integral Equation

5.1. Relationship to $T^*T\Phi + f(\Phi) = 0$

The Hammerstein integral equation

$$(5.1) \qquad \Psi + K[e(\Psi)] = 0 \quad \text{in } V$$

(supposed solution $\Psi = \psi$) is also closely related to the equation discussed in Section 4. It is perhaps sufficiently important for us to give results for it separately. Let the integral operator K be self-adjoint and positive-definite so that we can write

$$(5.2) \qquad K = T^*T$$

where T^* and T are now integral operators with kernels $\mathcal{J}(\underline{x}, \underline{x}')$ and $\mathcal{J}(\underline{x}', \underline{x})$ which need not necessarily be known. There are no boundary terms at all to worry about here, since σ and σ^* are both zero and the functional Γ does not appear (see Section 3.6.3).

If we set

(5.3) $$\Phi = e(\Psi), \quad \Psi = f(\Phi)$$

so that the functions e and f are inverses of each other, then we can identify equation (5.1) with (4.1) and obtain complementary upper and lower bounds if $<1, F(\Phi)>_V$ is convex. This condition is related to $<1, E(\Psi)>_V$ being convex, where $E(\Psi)$ is the integral of $e(\Psi)$. We see this as follows. From (5.3) we have

(5.4) $$e(\Psi)d\Psi + f(\Phi)d\Phi = \Phi d\Psi + \Psi d\Phi$$

which is the differential of

(5.5) $$E(\Psi) + F(\Phi) = \Phi\Psi + \text{constant}.$$

Using (5.3) and (5.5) it is easy to show that, for arbitrary pairs (Φ_1, Ψ_1) and (Φ_2, Ψ_2),

(5.6)
$$<1, F(\Phi_1)>_V - <1, F(\Phi_2)>_V - <\Phi_1 - \Phi_2, f(\Phi_2)>_V$$
$$= <1, E(\Psi_2)>_V - <1, E(\Psi_1)>_V - <\Psi_2 - \Psi_1, e(\Psi_1)>_V.$$

The convexity of either $<1, F(\Phi)>_V$ or $<1, E(\Psi)>_V$ thus implies that the left-hand side of (5.6) is non-negative, which is the basic sufficient condition for complementary upper and lower bounds (cf. (3.39) and (3.51)). Also if $<1, E(\Psi)>_V$ is strictly convex it follows that the solution ψ of (5.1) is unique; and if a parameter λ is inserted multiplying K, we shall have uniqueness for positive λ (cf. Section 4.6).

5.2. The complementary functionals

Let us arrange the constant in (5.5) to be zero by appropriate choice of the lower integration limits for $E(\Psi)$ and $F(\Phi)$. Then the generalized action functional

(5.7) $$I(U,\Phi) = \{U,T\Phi\}_V - \frac{1}{2}\{U,U\}_V + <1,F(\Phi)>_V$$

becomes in terms of Ψ

(5.8) $$I[U,e(\Psi)] = \{U,Te(\Psi)\}_V - \frac{1}{2}\{U,U\}_V - <1,E(\Psi)>_V$$
$$+ <\Psi, e(\Psi)>_V$$

with

(5.9) $$\{U,Te(\Psi)\}_V = <T^*U, e(\Psi)>_V .$$

The canonical equations are

(5.10) $$Te(\Psi) = U ,$$

(5.11) $$T^*U = -\Psi ,$$

leading to complementary bounds

(5.12) $$J[e(\Psi)] \geq I[u,e(\psi)] \geq G(U)$$

where

(5.13) $$J[e(\Psi)] = \frac{1}{2}<e(\Psi), Ke(\Psi)>_V - <1, E(\Psi)>_V + <\Psi, e(\Psi)>_V ,$$

(5.14) $$I[u,e(\psi)] = \frac{1}{2}<\psi, e(\psi)>_V - <1, E(\psi)>_V .$$

and

(5.15) $$G(U) = -\frac{1}{2}\{U,U\}_V - <1, E(-T^*U)>_V .$$

If we set

(5.16) $$U = Te(\tilde{\Psi}) ,$$

then (5.15) becomes

550

(5.17) $\quad G[Te(\widetilde{\Psi})] = -\frac{1}{2}<e(\widetilde{\Psi}), Ke(\widetilde{\Psi})>_V - <1, E[-Ke(\widetilde{\Psi})]>_V$,

which does not require knowledge of the decomposition (5.2). Error bounds on $\|e(\Psi) - e(\psi)\|$ are available from (4.46) and (4.47).

5.3. Pairs of complementary bounds

If the equation (4.13) is a differential equation

(5.18) $\quad\quad\quad\quad L\Phi + f(\Phi) = 0 \text{ in } V$

with the homogeneous Type I boundary condition

(5.19) $\quad\quad\quad\quad \varphi = 0 \text{ on } \partial V$,

then (5.18) and (5.19) could be equivalent to a Hammerstein integral equation

(5.20) $\quad\quad\quad\quad \Phi + L^{-1}[f(\Phi)] = 0 \text{ in } V$

where L^{-1} is self-adjoint and positive-definite. Thus different pairs of complementary bounds might arise for the same boundary-value problem. But in fact they are intimately connected.

Let us regard (5.20) as a direct example of (5.1) with $K = L^{-1}$ and denote the functionals G, I, J of Section 5.2 by $\mathcal{G}, \mathcal{I}, \mathcal{J}$ when referring to (5.20). Thus

(5.21) $\quad \mathcal{I}[f(\Phi)] = \frac{1}{2}<f(\Phi), L^{-1}f(\Phi)>_V - <1, F(\Phi)>_V + <\Phi, f(\Phi)>_V$,

(5.22) $\quad\quad \mathcal{J}[u, f(\varphi)] = \frac{1}{2}<\varphi, f(\varphi)>_V - <1, F(\varphi)>_V$

and

(5.23) $\quad \mathcal{G}[Tf(\widetilde{\Phi})] = -\frac{1}{2}<f(\widetilde{\Phi}), Lf(\widetilde{\Phi})>_V - <1, F[-L^{-1}f(\widetilde{\Phi})]>_V$.

If we write the first member of (4.23) as

(5.24) $$\Phi_1 = -L^{-1}[f(\Phi_0)]$$

(which will now imply that $\Phi_1 = 0$ on ∂V), we see from (4.24), (4.19) and (4.18), all now with no boundary terms, that

(5.25) $$\mathcal{J}[f(\Phi_0)] = -G(T\Phi_1) ,$$

(5.26) $$\mathcal{J}[u,f(\varphi)] = -I[u,\varphi] ,$$

and

(5.27) $$\mathcal{G}[Tf(\Phi_0)] = -J(\Phi_1) .$$

It had previously been thought that the pairs of bounds were essentially different [45], but as we see the question reduces to one of different trial functions.

6. Some Illustrative Applications

There are many applications of complementary variational principles to linear problems, arising for example in potential theory, electromagnetic theory, diffusion and quantum mechanics. Some details are given in Arthurs' book [17]; see also [55-63]. Here our interest is in non-linear problems, and we present some illustrative examples. In Sections 6.1-6.5 we are concerned with some particular non-linear differential and integral equations, whereas in Sections 6.6-6.8 the examples are of a more general kind with no specific calculations.

6.1. The Poisson-Boltzmann equation

This boundary-value problem

(6.1) $$\begin{cases} -\dfrac{d^2\varphi}{dx^2} + (e^\varphi - e^{-\varphi}) = 0, & x_0 \leq x \leq x_1 , \\ \varphi(x_0) = \alpha(x_0), \; \varphi(x_1) = \alpha(x_1) , \end{cases}$$

COMPLEMENTARY VARIATIONAL PRINCIPLES

arises both in colloid and in plasma theory [25,26]. The solution $\varphi(x)$ is related to the electric potential, and the stationary value of the action functional is related to the field energy. The relevant operators and scalar products are given in Section 3.6.1.

We have

(6.2) $\quad f(\Phi) = 2\sinh\Phi, \ f'(\Phi) = 2\cosh\Phi \geq 2, \ F(\Phi) = 2\cosh\Phi - 2$.

The functionals satisfying $J(\Phi) \geq I \geq G(T\tilde{\Phi})$ are

(6.3)
$$J(\Phi) = \int_{x_0}^{x_1} \left[\frac{1}{2}\left(\frac{d\Phi}{dx}\right)^2 + 2\cosh\Phi - 2\right] dx,$$

$$\Phi(x_0) = \alpha(x_0), \ \Phi(x_1) = \alpha(x_1);$$

(6.4) $\quad I(u,\varphi) = \int_{x_0}^{x_1} \left[\frac{1}{2}\left(\frac{d\varphi}{dx}\right)^2 + 2\cosh\varphi - 2\right] dx;$

and

(6.5)
$$G(T\tilde{\Phi}) = \int_{x_0}^{x_1} \left\{-\frac{1}{2}\left(\frac{d\tilde{\Phi}}{dx}\right)^2 - \left(\frac{d^2\tilde{\Phi}}{dx}\right)\sinh^{-1}\left(\frac{1}{2}\frac{d^2\tilde{\Phi}}{dx^2}\right)\right.$$
$$\left. + 2\cosh[\sinh^{-1}(\frac{1}{2}\frac{d^2\tilde{\Phi}}{dx^2})] - 2\right\} dx$$
$$+ \left[\frac{d\tilde{\Phi}}{dx}\alpha\right]_{x_0}^{x_1}.$$

When $x_0 = 0, \ x_1 = 1, \ \alpha(0) = 0, \ \alpha(1) = 1$ the trial functions

$$\Phi = \frac{\sinh ax}{\sinh a}, \ \tilde{\Phi} = \frac{\sinh \tilde{a} x}{\sinh \tilde{a}}$$

are suggested by the linearized form of (6.1). If a and $\tilde{a}$ are chosen to optimize J and G, the results

553

$$a = 1.46, \quad I = 0.30830$$
$$\tilde{a} = 1.48, \quad G = 0.30785$$

are obtained [46]. The error bound from equation (4.45) is

$$\|\Phi - \varphi\| \leq 0.0087;$$

this uses the values $\lambda_0 = \pi^2$, $\delta = 2$ (see Section 4.6). The form of trial function is thus a good one, and yields a very close estimate of $I(u, \varphi)$.

6.2. The Thomas-Fermi equation

The Thomas-Fermi equation for a system containing an arbitrary number of electrons and nuclei is, in atomic units [47]

(6.6) $$-\nabla^2 \varphi + f(\varphi) = 0 \quad \text{in} \quad V = E^3,$$

where

(6.7) $$(4\pi)^{-1} f(\varphi) = \lambda^{-3/2} \left[\varphi - c + \sum_i \frac{Z_i}{|\underline{r} - \underline{r}_i|} \right]^{3/2}$$

is the electron number density, $\underline{r}_i$ is the position of the ith nucleus with charge Z_i, $\lambda = \frac{1}{2}(3\pi^2)^{\frac{2}{3}}$, and

$$\varphi + \sum_i \frac{Z_i}{|\underline{r} - \underline{r}_i|}$$

is the electrostatic potential. The constant c depends on the neutrality of the system, and the boundary condition is

(6.8) $$\varphi \sim c - \frac{1}{r} \sum_i Z_i \quad \text{on} \quad \partial V \quad (\text{i.e. as } r \to \infty).$$

We see that f' is non-negative, and so complementary bounds are available using the operators and scalar products from Section 3.6.2. The results are [29]

(6.9) $$J(\Phi) \geq -4\pi E \geq G(U)$$

where

(6.10) $$E = \int_V \left[\frac{\lambda}{10} \left\{ \frac{f(\varphi)}{4\pi} \right\}^{5/3} - \frac{1}{2} \left\{ \frac{f(\varphi)}{4\pi} \right\} \sum_i \frac{Z_i}{|\underline{r} - \underline{r}_i|} \right] d\underline{r}$$

is the total electron energy in the Thomas-Fermi approximation,

(6.11) $$J(\Phi) = \int_V \left[\frac{1}{2} (\text{grad } \Phi)^2 + \frac{8}{5} \pi\lambda^{-3/2} (\Phi - c + \sum_i \frac{Z_i}{|\underline{r}-\underline{r}_i|})^{5/2} \right] d\underline{r}$$

and

(6.12) $$G(U) = -\int_V \left[\frac{1}{2} \underline{U}^2 + \frac{3}{5} \lambda (4\pi)^{-2/3} (\text{div } \underline{U})^{5/3} \right.$$
$$\left. + (c - \sum_i \frac{Z_i}{|\underline{r}-\underline{r}_i|}) \text{div } \underline{U} \right] d\underline{r}$$
$$+ \int_{\partial V} (c - \frac{1}{r} \sum_i Z_i) \underline{U} \cdot \underline{n} \, dS \ .$$

If attention is restricted to a neutral atom in its ground state, with a single nucleus of charge Z at the origin, the Thomas-Fermi equation reduces to [47]

(6.13) $$-\frac{d^2 \psi}{dx^2} + x^{-1/2} \psi^{3/2} = 0, \qquad 0 \leq x < \infty$$

where

(6.14) $\quad \psi(0) = 1; \quad \psi \to 0, \quad x\psi' \to 0 \quad \text{as} \quad x \to \infty$.

Here

(6.15) $\quad \varphi(\underline{r}) = c + \dfrac{Z}{r} \psi(r)$

and

(6.16) $\quad r = bx \quad \text{where} \quad b = \dfrac{1}{2}(\dfrac{3}{4}\pi)^{2/3} Z^{-1/3}$.

The complementary bounds (6.11) and (6.12) reduce to [37]

(6.17) $\quad \dfrac{b}{4\pi Z^2} J(\Psi) = \int_0^\infty \left[\dfrac{1}{2}(\dfrac{d\Psi}{dx})^2 + \dfrac{2}{5}\Psi^{5/2} x^{-1/2}\right] dx$,

Ψ satisfying (6.14),

and

(6.18) $\quad \dfrac{b}{4\pi Z^2} G(T\tilde{\Psi}) = -\int_0^\infty \left[\dfrac{1}{2}(\dfrac{d\tilde{\Psi}}{dx})^2 + \dfrac{3}{5} x^{1/3}(\dfrac{d^2\tilde{\Psi}}{dx^2})^{5/3}\right] dx$

$\quad - \left[\dfrac{d\tilde{\Psi}}{dx}\right]_{x=0}$.

Roberts [48] suggests the simple trial function

$$\Psi(x) = (1 + \gamma x^{1/2})\exp(-\gamma x^{1/2}) \; ;$$

optimum values for the bounds are [37]

$$J = \dfrac{4\pi Z^2}{b}(0.6810) \quad \text{at} \quad \gamma = 1.905$$

and

$$G = \frac{4\pi Z^2}{b}(0.6699) \quad \text{at} \quad \gamma = 1.750 \; .$$

The numerical solution of equation (6.13) actually yields

$$-4\pi E = \frac{4\pi Z^2}{b}(0.6806) \; ,$$

sandwiched between J and G.

6.3. The Föppl-Hencky equation

The Föppl-Hencky equation can be written in the form

(6.19) $\quad -\dfrac{d}{dx}(x^3 \dfrac{d\varphi}{dx}) - 2x^3 \varphi^{-2} = 0, \quad 0 \leq x \leq 1 \; ,$

subject to the boundary conditions

(6.20) $\quad \varphi'(0) = 0, \quad \varphi(1) = \lambda > 0 \; .$

It arises in elastic membrane theory, and φ is essentially the dimensionless positive stress which develops in a circular membrane when subjected to a constant normal force. The solution is unique if λ is large enough [49].

In the notation of Section 4.6, we have

(6.21) $\quad \omega = x^3 \geq 0 \; ,$

and

(6.22) $\quad f(\Phi) = -2x^3 \Phi^{-2}, \; f'(\Phi) = 4x^3 \Phi^{-3} \geq 0, \; F(\Phi) = 2x^3 \Phi^{-1} \; ,$

so that complementary bounds

(6.23) $\quad J(\Phi) \geq I(u, \varphi) \geq G(x^3 \tilde{\Phi})$

are available. The boundary conditions (6.20) are a combination of Type I (essential for J) and Type III (essential for G).

The basic functionals are [35]

$$(6.24) \quad J(\Phi) = \int_0^1 [\frac{1}{2}x^3(\Phi')^2 + 2x^3\Phi^{-1}]dx, \quad \Phi = \lambda \text{ at } x = 1,$$

$$(6.25) \quad I(u,\varphi) = \int_0^1 [\frac{1}{2}x^3(\varphi')^2 + 2x^3\varphi^{-1}]dx,$$

and

$$(6.26)\ G(x^3\tilde{\Phi}) = \int_0^1 [2(2x^3)^{1/2}\{-\frac{d}{dx}(x^3\tilde{\Phi}')\}^{1/2} - \frac{1}{2}x^3(\tilde{\Phi}')^2]dx$$

$$+ \lambda\tilde{\Phi}'(1),$$

$$\Phi'(0) = 0,$$

dashes denoting x-derivatives. Trial functions of the form

$$\Phi = a + (\lambda-a)x^b, \quad \tilde{\Phi} = \tilde{a} + (\lambda-\tilde{a})x^{\tilde{b}}$$

yield optimum values

$$J = 0.48311 \quad \text{at} \quad a = 1.238, \ b = 1.70;$$

$$G = 0.48125 \quad \text{at} \quad \tilde{a} = 1.236, \ b = 1.93.$$

The closeness of the optimum parameters $(a,\tilde{a})$ and $(b,\tilde{b})$ indicates the soundness of this form of trial function.

Generalizations of equation (6.19) are considered by Dickey [50] and also furnish possible applications of the complementary variational principles.

6.4. An integral equation in communication theory

The nonlinear integral equation

$$L\varphi - \varphi^{-1} = 0$$

where

$$(6.28) \quad L\Phi(x) = \int_0^{\pi/2} \frac{\sin(x-x')}{\pi(x-x')} \Phi(x')dx'$$

arises in communication theory [51]. Nowosad [52] has shown that this integral operator L is self-adjoint and positive-definite, and that the solution φ is real. We treat this as an example of equation (4.1), with no boundary terms (cf. Section 3.6.3), although we could equally well regard (6.27) as a Hammerstein equation in φ^{-1}.

We have

$$(6.29) \quad f(\Phi) = -\Phi^{-1}, \quad f'(\Phi) = \Phi^{-2} \geq 0, \quad F(\Phi) = -\ln \Phi .$$

The complementary bounds are

$$(6.30) \quad J(\Phi) \geq I(u,\varphi) \geq G(T\tilde{\Phi})$$

where

$$(6.31) \quad J(\Phi) = \int_0^{\pi/2} [\frac{1}{2}\Phi L\Phi - \ln \Phi]dx ,$$

$$(6.32) \quad I(u,\varphi) = \int_0^{\pi/2} (\frac{1}{2} - \ln\varphi)dx ,$$

and

$$(6.33) \quad G(T\tilde{\Phi}) = \int_0^{\pi/2} [1 - \frac{1}{2}\tilde{\Phi}L\tilde{\Phi} + \ln(L\tilde{\Phi})]dx .$$

Calculations with trial functions

$$\Phi = a + bx^2, \quad \tilde{\Phi} = \tilde{a} + \tilde{b}x^2$$

yield optimum values [34]

$$J = 0.22364 \text{ at } a = 1.36, \quad b = 0.06 ;$$
$$G = 0.22193 \text{ at } \tilde{a} = 1.36, \quad b = 0.08 .$$

Using the iterated value $\tilde{\lambda}_0 = 2.1377$ and taking $\delta = 0$ we obtain from (4.45)

$$\|\Phi - \varphi\| \le 0.040 .$$

6.5. A nonlinear Kirkwood integral equation

The nonlinear equation

$$(6.34) \quad \psi(x) = \frac{\lambda}{4} \int_{-\infty}^{\infty} \mathcal{K}(x-x') \, x'[g(x') - 1]dx' \qquad (\lambda > 0)$$

where

$$(6.35) \quad \begin{cases} g(x) = \exp[\psi(x)/x] & |x| \ge 1 \\ \quad\quad = 0 & |x| < 1 \end{cases}$$

and

$$(6.36) \quad \begin{cases} \mathcal{K}(t) = t^2 - 1 & |t| \le 1 \\ \quad\quad = 0 & |t| > 1 \end{cases}$$

arises in the statistical mechanics of a fluid composed of rigid spherical molecules [53]. The radial distribution function $g(r)$ tends to unity as r tends to infinity, and is an even, non-negative function. It is possible to rewrite (6.34) in the form [38]

$$(6.37) \quad L\varphi + \frac{4x^2}{\lambda} \ln(\varphi + 1) - q(x) = 0, \quad 1 \le x < \infty ,$$

where

(6.38) $$\varphi(x) = g(x) - 1 \ ,$$

(6.39) $$L\varphi(x) = \int_{\max(1,x-1)}^{x+1} xx'[1-(x-x')^2]\varphi(x')dx' \ ,$$

and

(6.40) $$\begin{cases} q(x) = \frac{1}{12} x^5 - x^3 + \frac{4}{3} x^2, & 1 \leq x \leq 2 \ , \\ = 0, & 2 \leq x < \infty \ . \end{cases}$$

Equation (6.37) is another example of (4.1) with no boundary terms, and

(6.41) $$q + f(\Phi) = \frac{4x^2}{\lambda} \ln(\Phi+1), \quad f'(\Phi) = \frac{4x^2}{\lambda} \frac{1}{\Phi+1} \geq 0 \ .$$

The basic functionals are

(6.42) $$J(\Phi) = \int_0^\infty \{\frac{1}{2} \Phi L \Phi + \frac{4x^2}{\lambda} [(\Phi+1)\log(\Phi+1) - \Phi] + q\Phi\} dx \ ,$$

(6.43) $$I(u,\varphi) = \int_1^\infty \{\frac{2x^2}{\lambda} (\varphi+2)\ln(\varphi+1) - \frac{4x^2}{\lambda} \varphi - \frac{1}{2} q\varphi\} dx \ ,$$

and

(6.44) $$G(T\tilde{\Phi}) =$$
$$\int_1^\infty \{-\frac{1}{2} \tilde{\Phi} L \tilde{\Phi} + q - L\tilde{\Phi} + \frac{4x^2}{\lambda}[1 - \exp\{\lambda (q-L\tilde{\Phi})/4x^2\}]\} dx \ .$$

Calculations for $\lambda = 5$ with trial functions (suggested by [53])

$$\Phi = a \exp[-b(x-1)]\cos[c(x-1)], \quad \tilde{\Phi} = \tilde{a}\exp[-\tilde{b}(x-1)]\cos[\tilde{c}(x-1)]$$

yield optimum values [38]

$$J = -0.020833 \text{ at } a = 0.48, \ b = 3.3, \ c = 2.2 ;$$

$$G = -0.021164 \text{ at } \tilde{a} = 0.48, \ \tilde{b} = 3.3, \ \tilde{c} = 2.1 .$$

A weak error bound (using $\tilde{\lambda}_0 = 0$ and $\delta = 0.541$) is

$$\|\Phi - \varphi\| \le 0.035 .$$

6.6. A transportation network problem

This example is taken from a paper on non-linear networks by Noble [20]; the theory is adapted to fit into the framework of the present review. Suppose that goods are being produced and consumed in four towns numbered $r = 1$ to 4, connected by five roads $k = 1$ to 5, as illustrated in Figure 1. Directions are assigned arbitrarily along each of

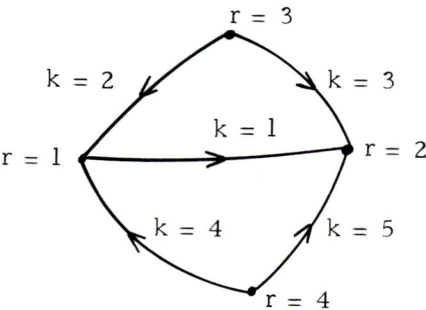

Figure 1

the roads, as shown by the arrows.

Assume that goods flow along branch k at a rate u_k per unit time, and that the excess of goods produced in town r over goods consumed in town r is e_r per unit time. If no goods are destroyed, e_r is equal to the flow of goods leaving the town r along the roads. Thus

(6.45) $$\begin{cases} u_1 & -u_2 & & -u_4 & & = e_1 \\ -u_1 & & -u_3 & & -u_5 & = e_2 \\ & u_2 & +u_3 & & & = e_3 \end{cases}$$

We regard e_4 as redundant, since

(6.46) $$e_1 + e_2 + e_3 + e_4 = 0 \ .$$

Let φ_r be the profitability of selling a unit amount of goods in town r. Since only profitability differentials are important in determining the movement of goods between towns, we can arbitrarily choose the profitability at one of the towns (say $r = 4$) to be zero and measure the other profitabilities relative to this zero level. Denote the profitability difference between the two towns at the end of road k by z_k. Then

(6.47) $$\begin{cases} \varphi_1 & -\varphi_2 & & = z_1 \\ -\varphi_1 & & +\varphi_3 & = z_2 \\ & -\varphi_2 & +\varphi_3 & = z_3 \\ -\varphi_1 & & & = z_4 \\ & -\varphi_2 & & = z_5 \ . \end{cases}$$

The sets of equations (6.45) and (6.47) can be written in matrix form

(6.48) $$A\varphi = z \ ,$$

(6.49) $$A^\dagger u = e \ ,$$

where A is the matrix

$$\begin{pmatrix} 1 & -1 & 0 \\ -1 & 0 & 1 \\ 0 & -1 & 1 \\ -1 & 0 & 0 \\ 0 & -1 & 0 \end{pmatrix}$$

and $A^\dagger$ is its transpose. In (6.48) and (6.49) φ, z, u and e denote the appropriate column vectors.

There will be a relationship between the profitability difference z_k and the flow of goods u_k along each road, and so we write

(6.50) $$z_k = z_k(u_k) \ .$$

In order to make the problem determinate, we need also to know either

    (i) the relationship between the excess goods e_r and profitability φ_r, say

(6.51) $$e_r = e_r(\varphi_r) \ ,$$

or alternatively

    (ii) suppose that

(6.52) $$e_r = \text{a known constant.}$$

Case (i)

Equations (6.48) and (6.49) are now of the form

COMPLEMENTARY VARIATIONAL PRINCIPLES

(6.53) $$A\varphi = z(u),$$

(6.54) $$A^\dagger u = e(\varphi),$$

which are canonical Euler equations associated with the action functional

(6.55) $$\begin{cases} I(U,\Phi) = \{U,T\Phi\}_V - W(U,\Phi)_V \\ \qquad\quad = <T^*U,\Phi>_V - W(U,\Phi)_V \end{cases}$$

where

(6.56) $$W(U,\Phi) = \{1, Z(U)\}_V + <1, E(\Phi)>_V .$$

Here $Z(U)$ and $E(\Phi)$ are the integrals of $z(U)$ and $e(\Phi)$, and the operators and scalar products are given in Section 3.6.4. There are no boundary terms. Thus

(6.57) $$\{U,T\Phi\}_V = U^\dagger A\Phi, \qquad <T^*U,\Phi>_V = (A^\dagger U)^\dagger \Phi,$$

etc.

We get complementary bounds

(6.58) $$J(\Phi) \geq I(u,\varphi) \geq G(U)$$

if W is convex in U and concave in Φ, i.e. certainly if

(6.59) $$\frac{dz_k}{dU_k} \geq 0, \quad \frac{de_r}{d\Phi_r} \leq 0, \qquad \text{for arbitrary } U_k, \Phi_r .$$

The bounding functionals are

(6.60) $$J(\Phi) = \{z^{-1}(T\Phi), T\Phi\}_V - \{1, Z[z^{-1}(T\Phi)]\}_V - <1, E(\Phi)>_V$$

and

(6.61) $G(U) = <T^*U, e^{-1}(T^*U)>_V - \{1, Z(U)\}_V - <1, E[e^{-1}(T^*U)]>_V$.

Case (ii)

If e is a known vector independent of the φ_r, we take

(6.62) $\qquad W(u, \Phi) = \{1, Z(U)\}_V + <\Phi, e>_V$.

The functionals are now

(6.63) $J(\Phi) = \{z^{-1}(T\Phi), T\Phi\}_V - \{1, Z[z^{-1}(T\Phi)]\}_V - <\Phi, e>_V$

and

(6.64) $\qquad G(U) = -\{1, Z(U)\}_V$ with $A^\dagger U = e$.

If the relationships (6.51) are linear, so that

(6.65) $\qquad z_k = R_k u_k$,

then

(6.66) $\qquad G(u) = I(u, \varphi) = -\frac{1}{2}\{u, Ru\}_V = -\frac{1}{2}\sum_k R_k u_k^2$.

If the network were an electrical one, this would be the elementary formula for minus the heat generated [cf. 20,17]. Maybe this name is also appropriate in economics.

Complementary variational principles for linear electrical systems have been known since the last century, but only comparatively recently has it been realized that there are similar principles for nonlinear networks [12-14].

6.7. Compressible fluid flow

The equations of hydrodynamics are usually highly non-linear and difficult to solve, and so associated variational principles can be valuable. Bateman [9] has considered variational methods for homentropic, irrotational, compressible fluid flow in two dimensions. The basic equations reduce to

$$\text{grad } \varphi = \frac{1}{\rho}\underline{u} = \delta W/\delta u, \tag{6.67}$$

$$-\text{div } \underline{u} = 0 = \delta W/\delta \varphi \tag{6.68}$$

and

$$\rho^2 f'(\rho) - \frac{1}{2} u^2 = 0. \tag{6.69}$$

Here $\underline{q} = \text{grad } \varphi$ is the velocity, φ is the velocity potential, ρ is the density and

$$p = f(\rho) - \frac{1}{2}\rho q^2 \tag{6.70}$$

is the pressure.

From equations (6.67)-(6.69) it is clear that $W(u,\varphi)$ is a function of u only, or equivalently of ρ only, and it can be shown that [9,11]

$$W(u,\varphi) = \int_V [f(\rho) + \frac{1}{2}\rho q^2] dV. \tag{6.71}$$

Thus, assuming homogeneous boundary conditions,

$$\begin{cases} I(u,\varphi) = \int_V [\underline{u} \cdot \text{grad } \varphi - f(\rho) - \frac{1}{2}\rho q^2] dV. \\ = \int_V [(-\text{div } \underline{u})\varphi - f(\rho) - \frac{1}{2}\rho q^2] dV. \end{cases} \tag{6.72}$$

The complementary functionals actually reduce to

(6.73) $$J(\Phi) = -\int_V p(\rho)\,dV, \quad \rho = \rho(\Phi),$$

and

(6.74) $$G(U) = -\int_V [p(\rho) + \rho q^2]\,dV, \quad \rho = \rho(U), \; q = q(U),$$

subject to div $\underline{U} = 0$ in V. Sewell [11] discusses these variational principles in a more general context, and from his treatment it can be shown that

(6.75) $$J(\Phi) \geq I(u,\varphi) \geq G(U)$$

for subsonic flow. The criterion is that p should be a concave function of the components of $\underline{q}$. See also the review article by Sewell [18].

6.8. Heat loss from a cell

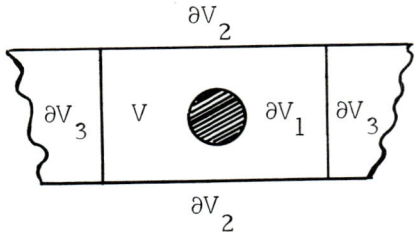

Figure 2

(cross-section)

Our final example illustrates the complementary functionals associated with a linear equation (Laplace's) subject to a <u>nonlinear</u> boundary condition.

This two-dimensional boundary-value problem is

(6.76) $$\begin{cases} \nabla^2 \varphi = 0 & \text{in } V(=E^2) \\ \varphi = \alpha & \text{on } \partial V_1 \\ \underline{n} \cdot \text{grad } \varphi = -\lambda \varphi^4 & \text{on } \partial V_2 \quad (\lambda > 0) \\ \underline{n} \cdot \text{grad } \varphi = 0 & \text{on } \partial V_3 \end{cases}$$

In practice φ might be the steady-state temperature distribution in a heat cell which is one of a linear chain of similar cells. The boundary ∂V_1 could then represent the perimeter of a cylindrical heat source at a known temperature α. There would be radiation loss according to Stefan's law from the external boundaries ∂V_2, and no heat flow across the internal boundaries ∂V_3. See Figure 2.

This example is interesting theoretically. The canonical equations are (cf. Section 3.6.2).

(6.77) $\qquad \text{grad } \Phi = \underline{U}, \; -\text{div } \underline{U} = 0 \text{ in } V,$

and so $G(\underline{U})$ is subject to a constraint in V (cf. equation (4.11)). The boundary conditions in (6.76) are a combination of Types I, III and IV, and the boundary terms in $J(\Phi)$ and $G(U)$ are given by adding appropriate Γ-functionals from Table 1. There are constraints for $J(\Phi)$ on ∂V_1 and $G(U)$ on ∂V_3. The radiation loss condition meets the concavity requirement (3.46).

The complementary upper and lower bounds reduce to

(6.78) $J(\Phi) = \frac{1}{2} \int_V (\text{grad } \Phi)^2 dv + \frac{1}{5}\lambda \int_{\partial V_2} \Phi^5 d(\partial V_2)$ with $\Phi = \alpha$ on ∂V_1,

and

(6.79) $G(U) = -\frac{1}{2} \int_V U^2 dV - \frac{4}{5} \lambda^{-1/4} \int_{\partial V_2} (-\underline{n} \cdot \underline{U})^{5/4} d(\partial V_2)$
$+ \int_{\partial V_1} (\underline{n} \cdot \underline{U}) \alpha \, d(\partial V_1)$, with div $\underline{U} = 0$ in V and $\underline{n} \cdot \underline{U} = 0$ on ∂V_3.

Calculations are in progress for a rectangular cell [54].

Appendix: Convex functions and functionals

The important property of the convex function $F(Q)$ which we used in equations (2.13) and (2.34) of Section 2 is that it never lies below its tangent, i.e.

(A.1) $$F(q + \xi) \geq F(q) + \xi F'(q) \qquad (F' = f)$$

for all q, $q+\xi$ belonging to the set $\{Q\}$ on which $F(Q)$ is defined (see Figure 3). This property does not actually define the convex function, but it follows readily from the definition. A function $F(Q)$ is defined to be convex if

(A.2) $$F(\lambda Q_1 + [1-\lambda]Q_2) \leq \lambda F(Q_1) + (1-\lambda) F(Q_2) ,$$

for all Q_1 and Q_2 belonging to $\{Q\}$ and for all λ such that $0 < \lambda < 1$. It is understood that the set $\{Q\}$ is a "convex" set, i.e. that $\lambda Q_1 + (1-\lambda)Q_2$ belongs to $\{Q\}$. This latter is a somewhat different use of the word convex.

If we rewrite (A.2) in the form

(A.3) $$F(Q_1) \geq F(Q_2) + \lambda^{-1} [F(Q_2 + \lambda[Q_1 - Q_2]) - F(Q_2)]$$

and use the fact that

(A.4) $$F(Q_2 + \lambda[Q_1 - Q_2]) - F(Q_2) = \lambda(Q_1 - Q_2) F'(Q_2) + O(\lambda^2) ,$$

then the result

(A.5) $$F(Q_1) \geq F(Q_2) + (Q_1 - Q_2) F'(Q_2)$$

is obtained in the limit as λ tends to zero. This is equivalent to (A.1).

If instead of functions we are dealing with functionals and $F(\Phi)$ is a functional of Φ, we merely replace Q by Φ in (A.2) and (A.3). But instead of (A.4) we have

(A.6) $$F(\Phi_2 + \lambda[\Phi_1 - \Phi_2]) - F(\Phi_2) = \lambda \{\Phi_1 - \Phi_2, \delta F/\delta \Phi_2\} + O(\lambda^2) ,$$

involving the scalar product { , } and the Fréchet derivative (see Vainberg [24]). This leads to the result

(A.7) $$F(\Phi_1) \geq F(\Phi_2) + \{\Phi_1 - \Phi_2, \delta F/\delta \Phi_2\},$$

generalizing (A.5).

We remark that

(i) the convexity is said to be <u>strict</u> if strict inequalities hold (assuming $Q \neq Q_2$, $\Phi_1 \neq \Phi_2$);

(ii) F is said to be concave if $-F$ is convex;

(iii) a sufficient condition for the convexity of the function $F(Q)$ is the positiveness of $F''(Q)$ - and of the functional $F(\Phi)$ the positive-definiteness of the operator $\delta^2 F/\delta \Phi^2$.

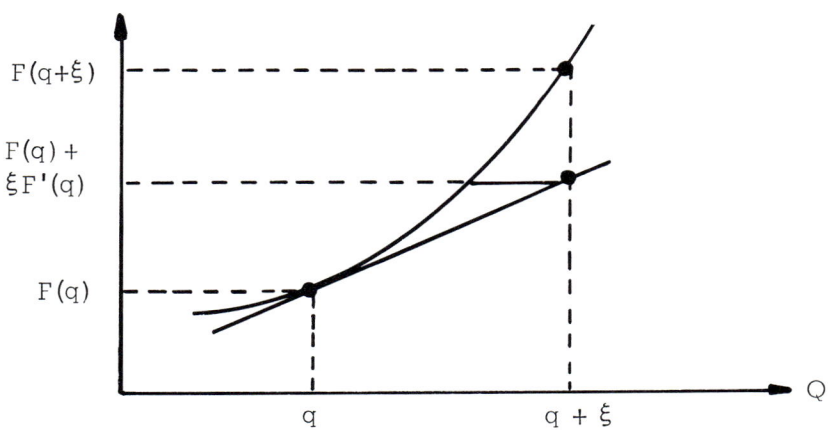

Figure 3. A convex function never lies below the tangent (the result generalizes for functionals).

REFERENCES AND AUTHOR INDEX

1. R. Courant and D. Hilbert, Methods of Mathematical Physics, Vol. I (Interscience, New York, 1953).

2. Lord Rayleigh, Phil. Mag. $\underline{47}$, 566 (1899).

3. J. L. Synge, The hypercircle in Mathematical Physics (Cambridge University Press, 1957).

4. J. B. Diaz, Upper and lower bounds for quadratic integrals, and at a point, for solutions of linear boundary value problems. (Proceedings of a symposium on boundary value problems in differential equations, Mathematics Research Center, University of Wisconsin, University of Wisconsin Press, 1959).

5. T. Kato, Math. Ann. $\underline{126}$, 253 (1953).

6. H. Fujita, J. Phys. Soc. Japan $\underline{10}$, 1 (1955).

7. E. Reissner, (article in) Problems of Continuum Mechanics, Soc. Industr. Appl. Math., 370 (Philadelphia 1961).

8. R. Hill, J. Mech. Phys. Solids $\underline{5}$, 66 (1956), and subsequent articles. See Sewell's review [18].

9. H. Bateman, Proc. Roy. Soc. $\underline{A125}$, 598 (1929).

10. P. E. Lush and T. Cherry, Quart. J. Mech. Appl. Math. $\underline{9}$, 6 (1953).

11. M. J. Sewell, J. Math. Mech. $\underline{12}$, 495 (1963).

12. G. Birkhoff, Quart. Appl. Math. 21, 160 (1963).

13. J. B. Diaz, Quart. Appl. Math., 13, 431 (1956).

14. W. Prager, Zeits. f. Angew Math. Physik 16, 185 (1965).

15. J. Abadie (ed.), Nonlinear programming. (Amsterdam, North Holland 1967).

16. J. J. Moreau, J. S.I.A.M. Control 4, 153 (1966).

17. A. M. Arthurs, Complementary Variational Principles (Oxford University Press, 1970).

18. M. J. Sewell, Phil. Trans. Roy. Soc. (London) 265, 319 (1969).

19. B. Noble, Complementary variational principles for boundary-value problems I. Basic principles, Report #473, Math. Research Center, University of Wisconsin (1964).

20. B. Noble, Complementary variational principles II. Nonlinear networks. Report #643, Math. Research Center, University of Wisconsin (1966).

21. L. B. Rall, J. Math. Anal. Applic. 14, 174 (1966).

22. B. Noble, Unpublished lecture notes, University of Wisconsin (1970).

23. P. D. Robinson, J. Math. Phys. 12, to appear (1971).

24. M. M. Vainberg, Variational methods for the study of non-linear operator equations (English edition: Holden-Day, San Francisco 1963).

25. R. P. Feynman, The Feynman Lectures on Physics, Vol. II (Addison-Wesley, Reading, Mas.. 1964).

26. F. Llewellyn-Jones, The Glow-Discharge (Methuen, London, 1966).

27. A. M. Arthurs, J. Math. Anal. Appl. (to appear) (1971).

28. V. Komkov, J. Math. Anal. Appl. 14, 511 (1966).

29. A. M. Arthurs and P. D. Robinson, Proc. Cambridge Phil. Soc. 65, 535 (1969).

30. N. Anderson and A. M. Arthurs, Nuovo Cimento 55B, 566 (1968).

31. N. Anderson and A. M. Arthurs, Nuovo Cimento 56B, 198 (1968).

32. N. Anderson and A. M. Arthurs, J. Math. Phys 9, 2037 (1968).

33. N. Anderson and A. M. Arthurs, Proc. Cambridge Phil. Soc. 68, 173 (1970).

34. N. Anderson and A. M. Arthurs, Nuovo Cimento Letters, 2, 631 (1969).

35. N. Anderson and A. M. Arthurs, J. Math. Phys. 11, 1048 (1970).

36. N. Anderson and A. M. Arthurs, Int. J. Electronics (to appear) (1971).

37. N. Anderson, A. M. Arthurs and P. D. Robinson, Nuovo Cimento 57B, 523 (1968).

38. N. Anderson, A. M. Arthurs and P. D. Robinson, Nuovo Cimento Letters 4, 628 (1970).

39. A. M. Arthurs, Proc. Cambridge Phil. Soc. <u>65</u>, 803 (1969).

40. A. M. Arthurs and P. D. Robinson, Proc. Roy. Soc. (London) <u>A303</u>, 497 (1968).

41. A. M. Arthurs and P. D. Robinson, Proc. Roy. Soc. (London) <u>A303</u>, 503 (1968).

42. P. G. Mikhlin, Variational Methods in Mathematical Physics (Pergamon Press, Oxford, 1964).

43. P. D. Robinson and A. M. Arthurs, J. Math. Phys. <u>9</u>, 1364 (1968).

44. L. F. Shampine, Num. Math. <u>12</u>, 410 (1968).

45. N. Anderson, A. M. Arthurs and P. D. Robinson, J. Inst. Math. Appl. <u>5</u>, 422 (1969).

46. A. M. Arthurs and C. W. Coles, J. Inst. Math. Appl. (to appear) (1971).

47. N. H. March, Advances in Physics <u>6</u>, 1 (1957).

48. R. E. Roberts, Phys. Rev. <u>170</u>, 8 (1968).

49. R. W. Dickey, Arch. Ratl. Mech. Anal. <u>26</u>, 219 (1967).

50. L. W. Dickey, J. Differential Equations, <u>4</u>, 399 (1968).

51. B. R. Salzberg and L. Kunz, Bell Systems Technical Journal <u>44</u>, 235 (1965).

52. P. Nowosad, J. Math. Anal. Appl. <u>14</u>, 484 (1966).

53. J. G. Kirkwood, E. K. Maun and B. J. Alder, J. Chem. Phys. <u>18</u>, 1040 (1950).

54. N. Anderson, A. M. Arthurs and P. D. Robinson, (unpublished).

Some additional recent references for <u>linear</u> problems:

Diffusion:

55. A. M. Arthurs, Proc. Roy. Soc. (London) $\underline{A298}$, 97 (1967).
56. G. C. Pomraning, J. Math. Phys. $\underline{8}$, 2096 (1967).

Quantum mechanics:

57. P. D. Robinson, J. Phys. A(Gen. Phys.) 2, 193 (1969).

58. P. D. Robinson, J. Math. Phys. $\underline{10}$, 472 (1968)
(In 57 and 58 it is shown how complementary bounds for eigenvalues can be derived).

59. P. D. Robinson, J. Phys. A(Gen. Phys.) $\underline{2}$, 295 (1969).

60. A. M. Arthurs, Phys. Rev. $\underline{176}$, 1730 (1968).

61. N. Anderson, A. M. Arthurs and P. D. Robinson, J. Phys. $\underline{A}$ (Gen. Phys.) $\underline{3}$, 1 (1970).

Magnetostatics

62. A. M. Arthurs and P. D. Robinson, Proc. Camb. Phil. Soc. $\underline{66}$, 433 (1969).

Biharmonic equation

63. A. M. Arthurs, Quart. Appl. Math. $\underline{28}$, 135 (1970).

Bradford University
Bradford, England

Index

A

Action functional, 524
Adjoints, 510, 521, 522, 523
  of a differentiable nonlinear mapping, 243
Akilov, 45
α-monotone operator, 257
Antitone operators, 17
Approximate minimizing sequence, 409
Approximation, 35
$G\Re$-derivative, 143
$G\Re$-differential, 143
Asymptotic derivatives, 265
  Fréchet, 266
  (H, σ), 266

B

Banach space with property (P), 224
Behavior, 511
Best approximate solution of minimal norm, 318
β-differentiability, 141
$(\beta_1, \ldots, \beta_{n-1})$-hypocontinuous, 166
Biargument equation, 10, 11, 21
Bidimensionally
  continuous, 180
  differentiable, 180
Bifurcation, 36
Boundary
  conditions, 526
  contributions, 530
  value problem, 11, 523, 520, 568
Bounded
  deteriation, 432, 442
  differentials, 133, 135
  operator, 366
  polynomial, 367
Brouwer's theorem, 16
Broyden's method, 458

C

Calculus of variations, 400, 408

INDEX

Canonical
 equations, 523, 550, 565
 Euler-Hamilton equations, 510
Carrier, 348
Cartesian product space, 523
Chain rule, 117, 125
 for higher order differentials, 164
Chandrasekhar equation, 31
Characterizations of higher order differentiability, 197
 at a point, 193
 over a region, 197
Closed
 linear operators with closed range on Banach spaces, 325
 operator, 327
 range operators, 322
 range theorem, 315
Coffman, 46
Collectively compact, 252
Colloid plasma, 553
Colloids, 521
Communication theory, 558
Compact map, 123
Compactness, 18
 of gradient, 225
Compact set, 215
Complementary
 bounds, 532, 539, 550, 551, 559, 565
 functionals, 520, 549, 568
 (or dual or reciprocal), 509
 upper and lower bounds, 519
 variational principles, 519
Completely
 compact mapping, 123
 continuous operators, 20

Compressible fluid flow, 567
Computational algorithms for minimization, 417
Concave, 511, 532
Conditions for weak lower semicontinuity, 227
Cones, 22
 of weak tangents, 236
Conjugate gradient method, 336
Connected sets of squares, 6
Consistent
 derivative approximation, 442
 discretization, 405
Constrained best approximation property, 323
Constrained best approximate solution property, 324
Constraint, 540
Continued fraction, 385
Contraction mappings, 13
Controllability assumption, 478
Control system, 473
Converses of Taylor's theorems, 193, 198
Convex, 511, 532
 functionals, 254
Convex functions, 404, 410, 515, 570
 hull, 181
 quasi, 404
 strictly, 404
 strongly quasi, 404
 uniformly quasi, 410
Courant, 77, 81
Cudia, D., 82, 87, 91

INDEX

D

Davidon methods, 438
Decomposable domain, 348
Decreasing upper bounds, 546
Defect, 26
Degenerate kernals, 34
de Lamadrid's differential, 141
Demicontinuous, 213
Derivative in the sense of
  Sendov, 4
Dieudonne, 45
Difference differentials, 202
Differential operators, 522, 536
Differentials
  axiomatic approach, 143
  in complex spaces, 205, 210
  in linear topological spaces, 140
Differentiation along a subspace, 136
Direct
  (H,β)-differentials, 202
  higher order differentials, 199
  integral, 527
Directional continuity, 110
Directionally demicontinuous, 213
Direct second order differentials
  Fréchet, 203
  in Gâteaux's sense, 203, 209
Direct second order variation, 203
Dirichlet, 509
Discretization, 404, 405
  optimal control, 402
Discretized constraints, 407

Distance, 1
  of functions, 4
Divided differences, 33
Duality map, 83

E

Eigenfunction expansions, 378
Elasticity, 31
Endbifurcation, 36
Equality constraints, 512
Equidifferentiability, 135
Equivalent norms, 117
Error analysis, 452
Error bounds, 546
Essential boundary conditions, 527
Euler, 48
  -Hamilton canonical equations, 516
  -Hamilton equations, 518, 522
  -Hamilton variation principle, 516
  -Lagrange equation, 48, 514
  -Lagrange variational principle, 510
Existence of solutions, 27

F

Feasible direction, 418
Field energy, 521
First kind, 369
Fixed point, 383
Fluids in a porous medium, 13
Föppl-Hencky, 548
  equation, 557

Fréchet
  asymptotic derivative, 266
  derivatives, 29, 55, 524, 571
  differentiable, 523
  differentials, 115, 117, 125, 148
  direct second order differential, 203
  higher order differentials,
    partial, 159, pointwise, 158,
    strong, 158, total, 159
  locally bounded derivative, 134
  locally uniform differential, 134
  partial differential, 137
  strong differential, 133
  total differential, 138
Fredholm operators, 328
Functional
  complementary, 518
  strongly quasiconvex, 232
Function of bounded variation, 18
Fundamental theorem of calculus, 173

G

Galerkin method, 29
Gâteaux
  derivative, 51
  differential, 109, 114
  direct second order differentials, 203, 209
  -Lévy differential, 114, 147
  linear variation - 129
  m^{th} order partial differential, 162

m^{th} order total differential 162
one-sided variations, 268
partial differential, 138
pointwise differential of order m, 161
strong differential, 162
total differential, 139
variations, 52, 109, 211
  sufficient conditions for linearity and continuity of, 113, 121, 122, 123, 128
Generalized action, 510
Generalized inverses, 339, 341, 344
  integral representations, 341
  limit representation, 341
  Neumann expansion, 340
  of a bounded operator, 316
  of bounded linear operators on Hilbert spaces, 315
  of linear topological homomorphisms, 325
  of nonlinear operator, 350
  of T relative to the projectors P and Q, 326
  of unbounded operator, 350
Generalized secant method, 444
Geometric mean value theorems, 174, 182
Globalizing the algorithm, 464
Goldstein, A. A., 92, 93
Golomb, M., 49, 77, 82, 84

INDEX

Gradients, 77, 130
  dependence on inner product, 131
  -like minimization algorithms, 417
  mapping, 130, 237, 239, 240, 241
  method, 80
Graves, L. M., 69
Green's function, 11

H

Hadamard
  differentiability, 149
  differentials, 123, 124, 125, 129, 147, 149
Hamiltonian, the, 510
  functional, 522
Hamilton's principle, 513
Hammerstein
  integral equations, 10, 31, 35, 512, 548
  operator, 18
Heat loss, 568
Hessian, 78
Higher order differentials, 151
Hilbert, 48, 77, 81
  spaces, 522
(H_n, σ_n, β_n)-derivative, 167

(H, σ)-
  continuity, 150
  derivative, 147
  differentiability, 146
(H, σ, β)-derivative, 147
Hyperbolic
  differential, 244
  integrability, 244

Hypercircle, 510

I

Inequality constraints, 512
Integral operators, 322
Infinite series of second kine, 388
Integrability conditions, 239, 242
Integral
  equations, 1
  operators, 537, 548
Integrofunctional-equations, 10, 12
Intervals, 1
Inverse function, 540
Inverses, 526
Involutory (Legendre) transformations, 510
Iteration procedure, 13
Iterative scheme, 543

J

Jacobian, 58
Joint continuity, 155
Jointly
  directionally continuous, 155
  strongly continuous, 122

K

Kantorovich, 45, 75
Kirkwood integral equation, 560
K-linear operator, 362
Krasnoselskij theorem, 20

Kuhn-Tucker conditions, 237

L

Lagrange, 47, 48
  Euler-equation, 48, 514
  Euler-variational principles, 510
  polynomial, 375
Lagrangian, 513
Lattice, 7
  of functions, 8
Least squares solution, 313, 318
  of minimal norm, 318
Legendre operator, 230
Leibniz's rule, 165
Lennard-Jones potential, 12
Leray-Schauder approximation, 222
Liénard equation, 26
Linear feedback control policy, 474
Linear operators, 510, 522, 534
Lipschitz condition, 13
Ljapunov-theorem, 20
Local approximation, 110
  property, 126
Logarithmic
  derivative, 270
  norm of bounded linear operators, 269
Lower semicontinuous, 215
  functional, 215
L_2-norm, 546
Majorant theory, 438
Map of a city, 6
Mathematical programming, 401

Matrix operators, 522, 537
Matrix representation, 389
Maximal generalized inverse, 330
Mazur, 91
McLeod, R., 92
Mean value theorems, 92, 171, 187
  for second order direct variations, 203
  in Banach spaces, 171, 174, 180, 182, 211, 212
  in linear topological spaces, 171, 181, 187
  in majorant form, 175
  using weak differentials, 211
Metric
  gradient, 83
  spaces, 13
Minimizing sequences, 409
Minty-Zarantonello monotonicity, 28
Monotone, 254
  operators, 254
Monotonically decomposible operators, 16, 24, 28
Monotonic type, 24
Moore-Penrose generalized inverse, 313, 314, 317
m^{th} order
  partial Gâteaux differential, 162
  total Gâteaux differential, 162
Multilinear operator, 152
Murav'ev variation, 114

INDEX

N

Natural, 527
Neumann expansions, 384
Newton-like methods, 30
Newton's
  equation, 514
  method, 29, 31, 33, 63, 75
Non compact operators, 9
None-expansive mappings, 14
Nonlinear
  integral equations, 9
  networks, 566
  vibrations, 11, 16, 20, 25
Normal
  cone, 23
  equation, 313, 316
  form, 369
Normally solvability, 345
Normally solvable, 322
n^{th}
  difference, 169
  variation, 168
Numerical range, 259
  of a bounded linear transformation, 259, 263
  of a nonlinear operator on a Banach space, 263

O

Oblique generalized inverse, 329
One-sided Gâteaux variations, 268
Operators of monotonic type, 24
Optimal control, 399, 406
  discretization of, 402
  problems, 473

Ordering of
  functions of several variables, 8
  knots of graphs, 5
  real vectors, 7
Orderings, 5
Ordering unit, 23

P

Pairs of related variables, 511
Parabolic distance, 3
Partial inverse, 321
Partially ordered Banach space, 17
Particularizing the problem, 465
Partitions, 7
Part metric, 3
Peano
  differentials, 200
  variation, 200
Pedrick, 46
Periodic solution, 11, 26
Perturbation, 35
Piecewise polynomial subspaces, 413
Plasmas, 521
Point of attraction, 264
Pointwise
  continuous, 154
  Gâteaux differential of order m, 161
Poisson-Boltzmann equation, 520, 552
Polar cone of weak tangents, 237
Polynomial, 363
  algebra, 374

Polynomial distance, 2
  operator, 363
Positive definiteness, 29
Potential operator, 130, 241
Primitive of a differential, 171
Projector on a linear topological space, 325
Projections P, Q, 316
Property M, 373
Pseudocondition number, 343
Pseudo
  derivative, 210
  inverse (see generalized inverse), 331
Pseudometric spaces, 14, 22
PY-equation, 12

Q

Quadratic performance criterion, 475
Quasibounded operators, 265, 266
Quasiconvex functional, 231
Quasi-Newton, 441
Quasinorm, 266
Quasiregular differention, 144

R

Rall, 46
Rayleigh-Ritz method, 402, 408, 412
Regula falsi, 32
Regular, 394
  method of differentiation, 144, 145
  value, 260
Resolvent, 35

Restricted generalized inverse, 323
Riccati differential equation, 483
Riemann
  differentials, 199
  integral, 69
  variation, 199
Riesz, F., 76

S

Saddle-point, 511
  behavior, 532
Scalar products, 522, 534, 571
Scattered sets, 8
Schauder
  -Leray theorem, 20
  theorem, 16, 22
Schwarz
  derivative, 115
  inequality, 510
  second derivative, 204
Second kind, 369, 383
Self-adjoint, 379
  operators, 542
Semi-inverse, 331, 350
  of linear operator, 331
  of nonlinear operator, 350
Semilocal analyses, 466
Sequentially
  lower semicontinuous, 403
  weakly continuous, 220
$(\sigma_1, \ldots, \sigma_n)$-continuous, 166
Sindalovskii variation, 114
Smulian, 91
Spectral radius formula, 264
Spectrum, 260

Stabilization problem, 473
Steepest descent, 334
  for least squares solutions, 335
Stone-Weierstrass Theorem, 375
Strictly
  convex, 533
  monotone, 254
  normally solvable, 347
Strongly convex functional, 256
Strong Gâteaux differential, 162
Strongly monotone, 254
Strong
  differentials, 133
  Fréchet differential, 133
  topology, 216
Subgradient, 258
Successive over-relaxation, 30
Symmetric operator, 369
Syntone, 31
  operators, 17

T

Tangent
  cone, 233
  ray, 233
Taylor
  differentials, 201
  theorem, 72
  theorem with integral remainder, 188
  theorem in Lagrange's form, 190
  theorem in Lagrange's form for functionals, 189
  theorem in Young's form in local convex spaces, 192

theorems and formulas, 187
  Banach space 187, 192
  locally convex space, 192
  variation, 201
Thomas-Fermi equation, 554, 555
Thomson
  lower bound, 540
  principles, 509
Topological
  direct sum, 325
  fixed point theorems, 16
  homomorphism, 326
  supplement, 325
Total
  derivative, 50
  differential, 76
  Fréchet differential, 138
  variation, 157
Totally bounded, 252
Transportation network, 562
Tschebyscheff-approximation, 35

U

Uniform differential, 135
Uniformly convex space, 346
Uniqueness, 533
Urysohn-equation, 10, 14, 35

V

Vainberg, 46
Variational principles, 509
Virtual solution (see least squares solutions), 330
Viscoelasticity, 11

Volterra-integral equation, 11

W

Weak
  derivative, 109, 210
  Gâteaux variation, 211
  neighborhood, 216
  topology, 217, 403
  closed set, 218
  compact set, 217
  continuous, 220
  lower semicontinuous, 218
  semicontinuous quadratic
  functionals, 230
  sequentially compact, 403
  (strongly) directly second
  order differentiable, 209
Weierstrass theorem, 373
Weight function, 15

Y

Young's, W. H., form of
  Taylor's theorem, 187

Z

0-linear operator, 363

OHIO UNIVERSITY LIBRARY

Please return this book as soon as you have finished with it. In order to avoid a fine it must be returned by the latest date stamped below.

DEC 17 1996

SEP 18 1996

CF